OKANAGAN UNIV/COLLEGE LIBRARY
02880920
P9-CKR-815

Results and Problems in Cell Differentiation

Series Editors:
W. Hennig, L. Nover, U. Scheer

28

Springer

Berlin
Heidelberg
New York
Barcelona
Hong Kong
London
Milan
Paris
Singapore
Tokyo

Ken McElreavey (Ed.)

The Genetic Basis of Male Infertility

With 27 Figures and 9 Tables

Springer

Dr. Ken McElreavey
Institut Pasteur
Immunogénétique Humaine
25 rue du Dr. Roux
75724 Paris Cedex 15
France

ISSN 0080-1844
ISBN 3-540-66264-2 Springer-Verlag Berlin Heidelberg New York

Library of Congress Cataloging-in-Publication Data
The genetic basis of male infertility / Ken McElreavey (ed.). p. cm. – (Results and problems in cell differentiation, ISSN 0080-1844 ; 28)Includes bibliographical references and index. ISBN 3-540-66264-2 (hardcover : alk. paper) 1. Infertility, Male – Genetic aspects. 2. Spermatogenesis – Genetic aspects. I. McElreavey, Ken, 1963– . II. Series. [DNLM: 1. Infertility, Male – genetics. 2. Spermatogenesis – genetics. W1 RE248X v. 28 1999]
RC889.G46 1999
616.6'92042 – dc21
DNLM/DLC
for Library of Congress

Printed in Germany

Production: PRO EDIT GmbH, D-69126 Heidelberg
Cover Concept: Meta Design, Berlin
Cover Production: design & production, D-69121 Heidelberg
Typesetting: Zechner Datenservice und Druck, D-67346 Speyer
SPIN: 10639152 39/3136 – 5 4 3 2 1 0 – Printed on acid-free paper

Preface

Mammalian spermatogenesis is a key component in the development of the testis. It continuously generates mature germ cells in large numbers from a small stem cell population. The success of this biological process depends on a precisely controlled cascade of developmental genes orchestrating spermatogonial cell proliferation, chromosomal reduction divisions to produce a haploid genome in each daughter cell and finally morphological differentiation of these latter cells into mature sperm. Mutations in any of the genes in this pathway can have a profound impact on the development of the testis as a whole and lead to male infertility. Progress in medical therapies for male infertility and the development of male contraceptives have been slow. One of the primary reasons for this is a lack of understanding of the aetiology of male infertility. Approximately 2% of human males are infertile due to severe defects in sperm production. In the United States alone approximately 2.3 million couples seek treatment for fertility problems each year. Data available over the past 20 years reveal that in approximately 30% of cases pathology is found in the man alone, and in another 20% the man and woman are affected. The male factor, therefore, is at least partly responsible in about 50% of infertile couples. Numerous diagnostic categories have been used in an attempt to subclassify the infertile male population yet, at present, around half are classified as being idiopathic. Except in a few well-defined categories the underlying cause is not understood, making a definitive diagnosis impossible and treatment empirical at best. Rare causes of infertility include infection, immunological factors or exposure to chemicals. Although environmental factors certainly play a role in male infertility, there is accumulating scientific evidence for a substantial genetic contribution.

The aim of this volume is to bring together recent information on genetic aspects of spermatogenesis and infertility. To achieve this, there are contributions exploring genetic aspects of spermatogenesis in man, mouse and *Drosophila*. If we are to continue to make advances in this field, it can only come from a fundamental understanding of the complex processes controlling human male germ cell development and maintenance. This need for such knowledge has been sharply focused recently with the introduction of sophisticated, assisted reproductive technologies such as intracytoplasmic sperm injection (ICSI). This procedure directly injects sperm from men with severely compromised spermatogenesis into ova, resulting in fertilisation and subsequent pregnancies in about 20% of cases. Without knowledge of the underlying male pathology we have no idea what undesirable traits those children, in particular the males, will have.

Paris, June 1999

Ken McElreavey

Contents

The Cell Biology and Molecular Genetics of Testis Determination
Craig A. Smith and Andrew H. Sinclair

The Sertoli Cell-Germ Cell Interactions and the Seminiferous Tubule Interleukin-1 and Interleukin-6 System
B. Jégou, J. P. Stéphan, C. Cudicini, E. Gomez, F. Bauché,
C. Piquet-Pellorce, and A. M. Touzalin

Leydig Cell Function and Its Regulation
M. P. Hedger and D. M. de Kretser

Mutations of the Cystic Fibrosis Gene
and Congenital Absence of the Vas Deferens
Pasquale Patrizio and Debra G. B. Leonard

Mitochondrial Function and Male Infertility
Thomas Bourgeron

The Human Y Chromosome and Male Infertility

Ken McElreavey, Csilla Krausz, and Colin E. Bishop

Spermatogenesis and the Mouse Y Chromosome: Specialisation Out of Decay

Michael J. Mitchell

The Comparative Genetics of Human Spermatogenesis: Clues from Flies and Other Model Organisms
Ron Hochstenbach and Johannes H. P. Hackstein

Clinical Aspects of Male Infertility

Csilla Krausz and Gianni Forti

1 Introduction

1.1 Definition of Couple and Male Infertility

The prevalence of infertile couples differs according to the definition of couple infertility. Since 90% of fertile couples successfully conceive within one year (Tietze 1956,1968), the most commonly used definition for couple infertility is the lack of pregnancy after one year of unprotected regular intercourse. Based on this definition, the prevalence of infertile couples is 10–15% in Western countries. A multicenter study (1982–1985) carried out by the World Health Organisation (WHO 1987) found that in 20% of cases the problem was predominantly male (pure male factor), in 38% female (pure female factor), in 27% both partners presented abnormalities and in the remaining 15% no clear cause of infertility was identified (unexplained infertility). Male factor infertility may be associated with a wide range of semen anomalies, such as sperm number, motility and morphology. The normal values for these parameters are defined by the WHO (Table 1). Azoospermia (absence of spermatozoids in the ejaculate), oligozoospermia (sperm number $<20\times10^6$/ml) asthenozoospermia

Table 1. Normal values of semen parameters according to WHO (1992)

Volume	>2 ml
pH	7.2–8.0
Sperm concentration	$>20\times10^6$ spermatozoa/ml
Total sperm count	$>40\times10^6$ spermatozoa/ejaculate
Motility	>50% with forward progression (categories a+b) or >25% with rapid progression (category a) at 60 min after ejaculation
Morphology	>30% with normal forms

Department of Clinical Physiopathology, Andrology Unit, University of Florence,
Viale Pieraccini, 6 50134 Florence, Italy

Results and Problems in Cell Differentiation, Vol. 28
McElreavey (Ed.): The Genetic Basis of Male Infertility

Table 2. Nomenclature for normal and pathological findings in semen analysis according to WHO (1992)

Normozoospermia	Normal ejaculate (as defined in Table1)
Oligozoospermia	Sperm concentration $<20\times10^6$/ml
Asthenozoospermia	<50% spermatozoa with forward motility (categories a+b) or <25% with motility (category a)
Teratozoospermia	<30% with normal forms
Oligo-astheno-teratozoospermia	Disturbance of all three variables
Azoospermia	No spermatozoa in the ejaculate
Aspermia	No ejaculate

Table 3. Classification of male infertility and type of treatment

Aetiology	**Treatment**
Pre-testicular	
Endocrine (gonadotrophic deficiency)	
– congenital (Kallman's syndrome)	Medical
– acquired (tumor, post-trauma, empty sella, iatrogenic)	Medical/surgical
Coital disorders	
– erectile dysfunction	Medical
– ejaculatory failure	ART (IUI; IVF)
Post-testicular	
Obstructive	
– epididymal (congenital or post-infective)	MESA/testicular biopsy+ICSI
– vasal (genetic or post-vasectomy)	MESA/testicular biopsy+ICSI
Accessory gland infection	Antibiotic treatment
Immunological (idiopathic or secondary)	Medical or ART (IUI or IPI)
Testicular	
Congenital	
– Anorchia	No therapy
– Cryptorchidism	Medical and/or surgical before age 2, ART
Genetic	
– Klinefelter sdr. and its variants	ICSI (testicular biopsy)
– Y chromosome deletions	ICSI (testicular biopsy)
– Monogenic anomalies	
– varicocele	Surgery in selected cases; ART
– antispermatogenic agents (environmental, drugs)	Identification and elimination of the noxa.
– chemotherapy, irradiation	Cryoconservation of sperm+ART
– vascular torsion, trauma, orchitis	Depending on the grade of spermatogenic failure (ART)
Idiopathic	ART

(sperm motility a+b<50%), teratozoospermia (normal morphology <30%; Table 2) are present in half of the cases of couple infertility and in 90% of infertile males. In about 10% of cases no abnormal routine semen parameters are found and in these cases specific metabolic (Huszar et al. 1992) and membrane defects (Calvo et al. 1989; Tesarik and Mendoza 1992) of spermatozoids could lead to reduced fertilizing ability and therefore to infertility. The aetiology of impaired sperm production and function could be due to different factors acting at pretesticular, post-testicular and directly at the testicular level (Table 3).

With the aid of an extensive medical workup (medical history, physical examination, semen analysis, hormone measurement, ultrasound examination of the genital tract, genetic assessement and sperm function tests) a definite diagnosis of the cause of male infertility can be obtained in over 70% of cases. In the remaining 30% of cases no cause for infertility is found (idiopathic infertility). There is no clear definition for idiopathic infertility for different reasons: (1) it is largely dependent on the intensity of the medical workup, some patients can be defined as suffering from idiopathic infertility in one clinic but could have a precise diagnosis in another clinic, and (2) for some abnormal andrological findings it is difficult to establish a direct cause-effect relationship since there is considerable variability of spermatogenic impairment between subjects. Therefore, varicocele, recurrent urogenital infections or monolateral cryptorchidism are considered as idiopathic infertility in some centers. In the majority of cases of idiopathic infertility an underlying genetic abnormality is the more plausible explanation.

1.2 Assisted Reproduction and Clinical Andrology

Given the success of assisted reproductive techniques (ART) available for any kind of male factor, there is an unfortunate tendency to ignore andrological evaluation. Intrauterine insemination (IUI) is used for mild forms of oligozoospermia, in vitro fertilization with embryo transfer (IVF-ET) for moderate oligozoospermia and intracytoplasmic sperm injection (ICSI) for the treatment of severe oligozoospermic or azoospermic patients. The latter technique, ICSI, consists of the selection of a single spermatozoid with the aid of a micromanipulator, and the subsequent injection of the single sperm into the cytoplasm of a metaphase II oocyte. The spermatozoids can be recovered from ejaculate or from epididymal fluid or directly from a testis following testis biopsy. The percentage of clinical pregnancy rate with this technique is 35% in experienced laboratories (Van Steirteghem et al. 1993). However, since some forms of male infertility are reversible with medical or surgical therapy, and others are due to genetic defects which can be transmitted to the probands, a complete medical work-up can avoid needless involvement of the female partner in pre-IVF or ICSI ovarian hyperstimulation protocols. In cases where genetic anomalies are found, appropriate genetic counselling can be offered to the couple.

2 The Elements of Standard Medical Workup

2.1 Medical and Familial History

The patient should be asked about chronic systemic disease, drug intake, lifestyle factors (smoke, alcohol), testicular injury, mumps orchitis, surgery for varicocele, inguinal hernia or cryptorchidism. The familial history should contain information concerning the familial fertility status and infertility related diseases, congenital abnormalities, mental retardation and miscarriages.

2.2 Physical Examination

Apart from a standard examination, scrotal palpation with the assessment of testicular volume and comparison with ellipsoids of known volume (Prader's orchidometer) and digital rectal examination must be included. The presence of penile abnormalities such as hypospadia, surgical or traumatic penile scars and induration plaques should be assessed in every subject.

2.3 Diagnostic Tests

Semen Analysis. This has to be performed according to World Health Organization recommended procedures. If the first semen analysis is normal, there is generally no need for repeat analysis. In all other circumstances in which an abnormal semen sample is obtained, an additional semen sample should be examined after a 6–12 week interval (since spermatogenesis takes approximately 3 months to be completed). The reason for a second sample is the large variability of semen parameters in the same individual over time (Manual WHO 1992). In azoospermia, after centrifugation of the semen sample, a careful analysis of the pellet is also necessary.

FSH, LH and Testosterone Measurement (Serum Inhibin B Levels in Expert Laboratories). The most important hormonal measurement in the infertile male, both from the diagnostic and from the prognostic point of view, is serum FSH. High FSH levels and normal LH and testosterone (T) levels are present in the majority of normally virilized infertile men with sperm concentrations lower than 5×10^6/ml and are usually correlated with spermatogenetic damage

(Bergmann et al. 1994). Serum inhibin B levels appear to be inversely related to FSH in infertile men, suggesting that inhibin B levels reflect Sertoli cell function and may be a good marker of testicular dysfunction (Anawalt et al. 1996).

Scrotal Ultrasound. Testicular ultrasonography can be useful in infertile males because such patients show a high rate of abnormalities (hydrocele, epididymal pathology, spermatocele, tumors) which can be overlooked during a physical examination. In particular, testicular tumors can be diagnosed at an early stage, before they become clinically evident (Nashan et al. 1990).

Genetic Assessement. Karyotyping must be performed in azoospermia and oligozoospermia since chromosome abnormalities are much more frequent in infertile males (5.3%) than in the general population (0.6%; Egozcue 1989). Of the various genes considered critical for the regulation of male fertility, genes on the long arm of the Y chromosome (Yq) seem to be the most promising. Macroscopic deletions in Yq11 are associated with impaired spermatogenesis (Tiepolo and Zuffardi 1976), but recently microdeletions of three regions of the Y chromosome were found in a large sample of oligozoospermic and azoospermic men and the regions were named azoospermia factor (AZF) AZFa, AZFb, and AZFc, respectively (Vogt et al. 1996). Screening for Y chromosome microdeletions should be done in all patients affected by severe oligo/azoospermia.

CFTR, androgen receptor and FSH receptor gene mutation screening should be done in selected cases (see below).

Bacteriological Examination and Transrectal Ultrasound. These are applied to the prostate if infection is suspected from the case history, from the physical examination, from semen analysis (i.e. high viscosity, pH >8, leucocytes $>2\times10^6$/ml), or from high free radical production after N-formyl chemotactic peptide (FMLP) stimulation test (Krausz et al. 1992).

Transrectal Ultrasound. This is used for the evaluation of the accessory glands with scrotal ultrasound in case of suspected agenesis of seminal vesicules and vas deferens.

Sperm Function Tests. Since routine semen analysis provides little information on sperm fertilizing ability, new methods of assessment of sperm function have been introduced. Several sperm function tests have been reported as good predictors of the ability of sperm to fertilize human oocytes in vitro (Calvo et al. 1989; Burkman et al. 1988; Huszar et al. 1992; Krausz et al. 1996). The hemizona assay, which measures the ability of spermatozoa to bind to the zona pellucida of a human oocyte, is the best predictor, but the scarcity of human oocytes makes this test unlikely to be used for routine purposes. A good alternative option is sperm responsiveness to progesterone. Progesterone is able to induce, through a non-genomic mechanism, both intracellular calcium increase and acrosome reaction in human spermatozoa. We have recently reported, in a large group of unselected couples, that both the acrosome reaction and intracellular free calcium concentration increase in response to progesterone were good predictors of in vitro fertilization rate (Krausz et al. 1996).

Testicular Biopsy. Until recently testicular biopsy was a procedure designed to confirm normal spermatogenesis in patients with suspected obstructive azoospermia (with normal FSH and normal testicular volume) in whom reconstructive microsurgery was planned. Today testicular biopsy can be considered more as a therapeutic than a diagnostic procedure. Testicular sperm can be obtained from testicular biopsies of men with azoospermia caused by obstruction, maturation arrest or Sertoli cell only syndrome and can be successfully used for ICSI treatment (Silber et al. 1996).

3 Aetiology of Male Infertility

3.1 Infertility Due to Antispermatogenic Agents

Drugs. Sulfasalazine, which is used to treat inflammatory bowel diseases can lead to oligozoospermia. Cytotoxic drugs – especially the alkalating agents – that are used to treat cancer and autoimmune disease cause gonadal failure. Anabolic steroid abuse has severe side effects on the testes, leading to oligo/azoospermia.

Enviromental Factors. Several chemical agents have been reported to impair reproductive function. There is strong evidence of spermatotoxic effects of pesticides, of some glycol esters (used in paintings, printing inks, and adhesives), and of the metals lead, cadmium, and mercury. Recently, the oestrogenic effects of several environmental toxins has been reported as a possible cause of defects in the male reproductive organs (Sharpe and Skakkebaek 1993).

Lyfestyle Factors. Alcohol and tobacco abuse, and use of recreational drugs may affect testicular function.

Diagnosis. A careful collection of medical, occupational and life style histories are fundamental.

3.2 Infertility Due to Endocrine Disorders (Hypogonadotrophic Hypogonadism)

The aetiology is either congenital or acquired. The deficit of LH and FSH in the periphery could be due to a primary lesion in the pituitary gland or could be secondary to insufficient hypothalamic GnRh production. Acquired causes include tumor (prolactinoma, craniopharyngioma), infection, infiltrative diseases, empty sella, radiation treatment and autoimmune hypophisitis. The congenital form of hypogonadotropic hypogonadism (Kallmann's syndrome) is due to an abnormality of the secretion of GnRH. The syndrome affects 1 in 10000 to 1 in 60000 people. Segregation analysis in familial cases has demonstrated diverse inheritance patterns (autosomal-dominant, autosomal-recessive, X-linked). There is a large excess of male over female patients, suggesting a preponderance of X-linked inheritance. A deletion of a single gene, KALIG - 1, has been found in the X-linked form of the disease (Bick et al. 1992). This gene encodes an adhesion protein involved in neuronal migration during fetal life and thus gives a link between deletions of KALIG-1 and defective neuronal migration observed in patients with Kallmann syndrome (Schwanzel-Fukuda et al. 1989). The lack of gonadotrophic hormones before and after birth seems to explain the frequently observed testicular maldescent as well as the absence of pubertal development.

Diagnosis. Testicular dysfunction due to an acquired gonadotropin defect is associated with such symptoms as the reduction of the volume of the ejaculate, reduction of beard growth and libido and asthenia. On the other hand, a eunuchoid habitus with infantile genitalia, sparse or nearly absent body hair, gynaecomastia and low testicular volume (5–10 ml), is typical of congenital gonadotropin deficiency which can be associated with hypo- or anosmia in Kallmann's syndrome. FSH, LH and testosterone plasma levels are low. Brain tomography of sella turcica completes the diagnostic workup.

3.3 Infertility Due to Impairment of Sperm Transport and/or Accessory Gland Infections

A normal epididymis (where maturation and storage of sperm occurs), vas deferens and accessory gland are a prerequisite for normal sperm transport. An obstruction or absence of any part of the seminal pathway leads to azoospermia in the presence of normal testicular function. Impaired or absent sperm transport may be caused by congenital (congenital bilateral absence of vas deferens,CBAVD) or acquired diseases (post-infective, inflammatory or post-vasectomy). CBAVD is now considered a mild from of cystic fibrosis. Mutations of the CF gene have been reported in 70–80% of such patients, the most frequent being the ΔF508 mutation (Chillon et al. 1995). Screening for CF mutations should be performed in all patients with absence of vas deferens. Due to the high carrier frequency in Europe and North America (1:25) genetic screening should also be done in the female partner, especially if an assisted reproductive technique (ART) attempt (usually epididymal sperm aspiration coupled with ICSI) is planned.

Diagnosis. Epididymal obstruction can be suggested by a history of sexually transmitted disease (gonorrhoea and chlamydia), epididymitis, prostatitis, urinary infection, or inguinal surgery. These men usually have a normal testicular volume. Signs of epididymal inflammation such as thickening, nodules and/or pain can be also found at scrotal palpation. Examination of the prostate gland by digital rectal examination may reveal an inflammed prostate gland. Absence of the vas deferens can be diagnosed by skilled scrotal palpation. Semen analysis with biochemical markers is of great diagnostic value, azoospermia with abnormally low semen volume, acidic pH, very low or indetectable fructose level (which is produced by seminal vesicles) is a characteristic finding in patients affected by CBAVD. Low levels of seminal α-glucosidase, the most common epididymal marker, are present in CBAVD, in men with epididymal obstruction but also in some men with azoospermia due to tubular damage (Casano et al. 1987). FSH and LH are normal. Scrotal ultrasound can detect epididymal anomalies, the absence of vas deferens and signs of obstruction.

3.4 Autoimmune Infertility

The testis is an immunologically privileged site and thanks to the hemato-testicular barrier, spermatozoal antigens are shielded from being recognised by the cells of the immune system. An autoimmune reaction against the spermatozoa as an isolated abnormality is seen in about 5% of infertile males. Circulating sperm antibodies are also present in association with epididymitis, varicocele, unilateral or bilateral obstruction of the genital tract, and after reversal of vasectomy.

Diagnosis. The patients present astheno/oligoasthenozoospermia. If the percentage of motile spermatozoa coated by antisperm antibodies is >50%, a pure immunological factor is likely and titration of sperm antibodies in serum will add weight and confirm the diagnosis. If semen has high viscosity, pH >8 and more than 10^6 leukocytes/ml and/or reduced seminal levels of prostatic markers (acidic phosphatase, citric acid, zinc), accessory gland infection (prostatitis or prostatovesiculitis) is likely. Although the clinical significance of isolated leukocytospermia as well as the role of subclinical genital infections in male infertility is controversial (Aitken and Baker 1995), the presence in the semen of activated leukocytes, which are able to produce large amounts of reactive oxygen radicals is associated with a lower fertilization rate in an IVF context (Krausz et al. 1994).

3.5 Infertility and Varicocele

Varicocele is the most frequent finding among infertile males (25–40%). However, its role in infertility, especially with respect to its management, is controversial (Hargreave 1997; Nieschlag et al. 1998). Although there is a strong association between severe and long-standing forms of varicocele and spermatogenic failure, the pathogenesis by which varicocele determines this condition is largely unknown. Among the potential pathogenetic factors are increased temperature and accumulation of CO_2 and other toxic substances brought on by the venous reflux.

Diagnosis. Scrotal palpation and scrotal ultrasound.

3.6 Infertility Due to Coital Disorders

Erectile dysfunction and ejaculatory disorders are a rare cause of infertility. Anejaculation or retrograde ejaculation may occur in diabetic patients as well as after retroperitoneal lymph-node dissection, spinal cord injury, bladder-neck surgery and in patients affected by multiple sclerosis.

Diagnosis. Medical history and the absence of spermatozoa in the seminal fluid with the presence of sperm in the urine are of diagnostical value for retrograde ejaculation.

3.7 Infertility and Cryptorchidism

Undescended testes is a frequent condition in newborns (2–3%) and its prevalence after the first year of life is about 0.8% to 1.6% (Jackson 1988). Cryptorchidism, in the majority of cases, is associated with impaired spermatogenesis although the extent of impairment may vary from a Sertoli cell only pattern to a slight form of hypospermatogenesis (Giwercman et al. 1989). Cryptorchidism can be mono- or bilateral (in 10–20% of cases it is bilateral), the testes may be located intra-abdominally or in the inguinal canal. During the first year of life, 66% of the undescended testes will spontaneously descend into the scrotum, in the remaining cases hormonal or surgical treatment is necessary. Anatomical obstacles (abnormalities of the head of the epididymis, of the vas deferens or of spermatic vessels etc.) represent known causes of cryptorchidism. In the majority of cases the aetiology is unknown, but the finding that spermatogenesis is also impaired in the contralateral testes (in patients with unilateral maldescent) indicates a possible congenital defect of the germ cell population. Cryptorchidism may be associated with numerous syndromes with congenital malformations such as Aarskog-Scott syndrome, Fanconi anemia, Noonan syndrome, persistent mullerian duct syndrome and Prader-Willi-Labhart syndrome. A careful follow up of patients affected by cryptorchidism is necessary since a significantly higher risk of testicular cancer has been observed (Giwercman et al. 1989).

Diagnosis. Scrotal palpation (with warm hands on a sitting and relaxed patient in order to distinguish between maldescent and retractile testes). Pelvic ultrasound and tomography for the detection of ectopic testes (located outside the normal path of descent).

4 Infertility Due to Genetic Disorders

4.1 Chromosomal Abnormalities

From the first large karyotype surveys on subfertile males (6982 individuals), it became evident that compared with newborns, infertile males have a higher prevalence of chromosomal abnormalities (Kjessler 1974; Koulischer and Schoysman 1975; Chandley 1979; Tiepolo et al. 1981). With the introduction of IVF and ICSI, concern has been voiced about the fact that patients undergoing these therapeutic procedures are more likely to be carriers of chromosomal abnormalities, and therefore may pose a genetic risk to their offspring. Several chromosome screenings were undertaken in the late 1980s and 1990s in order to establish the frequency and type of chromosomal abnormalities in patients affected either by severe oligozoospermia or those undergoing IVF/ICSI procedures (Retief et al. 1984; Bourrouillou et al. 1985; Hens et al. 1988; Baschat et al. 1996; Peschka et al. 1996; Mau et al. 1997).

The 47,XXY chromosome complement makes up the bulk of sex chromosome abnormalities and occurs eight times more frequently than among newborns. Micro- and macrodeletions of the long arm of the Y chromosome is a frequent finding in patients affected by azoospermia and severe oligozoospermia (see Chapter by McElrevey et al. this Vol). Among the structural abnormalities, reciprocal translocations, Robertsonian translocations, paracentric inversions and marker chromosomes are the most common. Heterozygotes for autosomal translocations are seven times more frequent in infertile male populations than in newborns. Carriers of marker chromosomes are eight times more frequent among infertile males (De Braekeleer and Dao 1991). In 60% of all cases of Robertsonian translocations, a (13;14)-translocation is found. This abnormality is rarely observed in azoospermic men but is often found in oligozoospermic patients. Although there is a significantly higher frequency of this translocation in the infertile population (about nine times higher than in newborns) the contribution of this chromosomal defect in oligozoospermia is still unclear. For example, robertsonian translocations have been found in normospermic fertile males in the same pedigree (Chandley 1975). The frequency of chromosomal inversions is 13 times higher in infertile males. Pericentric inversions in chromosomes 1, 3, 5, 6 and 10 probably interfere with meiosis, leading to a reduced rate of postmeiotic sperm development and thus to infertility.

4.2 Klinefelter Syndrome

The syndrome occurs in approximately 1 in 500 newborn males (Neilson and Wohlert 1991) and in the majority of cases, the syndrome is characterized by a 47,XXY chromosome complement. Variants such as 47,XXY/46XY mosaicism and other higher order X chromosome hyperploids have also been described. The 47,XXY karyotype is a frequent finding among infertile men (11% of azoospermic and 0.7% of oligozoospermic; de Braekeleer and Dao 1991; Yoshida et al. 1996). The classical signs of this syndrome are extremely small (<5 ml), firm testes, with eunuchoid habitus, gynaecomastia, high FSH and azoospermia. However, occasionally oligozoospermia has been found in patients affected by this syndrome, and cases of fertility (Kaplan et al. 1963) and proven paternity (Laron et al. 1982; Terzoli et al. 1992) have been reported. In the literature there are different opinions regarding the capacity of germ cells from a 47,XXY male to proceed through mitosis and meiosis to produce XX and XY hyperhaploid gametes. It is generally accepted that the activity of more than one X chromosome is fatal to the male germ cell (Steinberger et al. 1965; Skakkebaek et al. 1969, Lyon 1974). However, studies by fluorescent in situ hybridization (FISH) of spermatozoids from mosaic 47,XXY/46XY patients have challenged this hypothesis (Cozzi et al. 1994; Chevret et al. 1996; Martini et al. 1996). An increased incidence of hyperhaploid 24,XY in such patients suggests that these cells have a meiotic capacity. However, it is still uknown whether the hyperhaploid spermatozoa arises from the meiotic division of XXY germ cells or from an increase in the rate of meiotic non-dysjunction. With the introduction of ICSI, a new era has started for Klinefelter patients, although for the moment only a few cases of fertilization and pregnancy by ICSI procedure have been reported (Harari et al. 1995; Tournaye et al. 1996; Bourne et al. 1997).

4.3
Other Chromosomal Abnormalities

47,XYY Male. The frequency of males with this karyotype is 1:750. Carriers of this abnormality show a great diversity in the degree of spermatogenic impairment, ranging from severe oligozoospermia to apparent normality (Skakkebaek et al. 1973). Distortion of sex vescicule formation is probably the major cause of infertility in this patient group (Berthelsen et al. 1981).

XX-Male. This is a disorder of sex determination and occurs in about 1:20000 newborns. In about 80% of cases, XX-maleness can be explained by the translocation of the SRY gene (encoding the testis determining factor) to the X chromosome. The cause of SRY-negative XX-maleness remains to be elucidated. The phenotypic features of the syndrome are gynecomastia, female hair pattern and small testes with azoospermia. Genital malformations such as hypospadias are rare. In SRY-negative patients, ambiguous genitalia is a frequent finding (Abbas et al. 1990).

Aneuploidies of Autosomes. Most numerical aneuploidies of autosomes are lethal. Patients affected by Down's syndrome may be fertile or infertile (Zuhlke et al.1994).

4.4
Male Infertility from Defect in Meiosis

Blockage at any stage of spermatogenesis is found in an increasing number of infertile men. Abnormalities of the meiotic chromosomes in males with a normal karyotype can lead to infertility (Koulischer et al. 1982; Egozcue et al. 1983). It has been estimated that in about 7.9% of the patients, meiotic chromosome anomalies can explain their infertility (De Braekeleer and Dao 1991) Therefore, meiotic chromosome analysis is advisable in patients affected by idiopathic infertility.

5
Monogenic Diseases

5.1
The Kartagener Syndrome or Immotile Cilia Syndrome

This is an autosomal recessive disorder characterized by brochiectasis, sinusitis, dextrocardia, and infertility. Situs inversus may be absent in certain cases (Eliasson et al. 1977). Men with typical Kartagener syndrome but with normal spermatozoa (normal motility, ultrastructure, and fertilizing capacity) have been described (Samuel 1987). Infertility is due to sperm immobility. Various ultrastructural defects of ciliary structures have been described including absence of dynein arms (Afzelius 1976), defective radial spokes (Sturgess et al. 1979) and random ciliary orientation (Rutland and de Iongh 1990). The spermatozoa from these patients are functionally competent and can fertilize the egg in vitro (Aitken et al. 1983; Aitken 1991). The mutation responsible for this syndrome has been suggested to be on chromosome 14, since the human cytoplasmic dynein heavy chain is encoded by a gene mapped to 14qter by fluorescence in situ hybridization (Narayan et al. 1994).

Other monomorphic human sperm defects are: the "9+0" axonema defect (Neugebauer et al. 1990) and globozoospermia (Schill 1991). The "9+0" axonema defect is responsible for immotility of spermatozoa, while globozoospermia is characterized by the absence of acrosome which is indispensable for fertilization of the egg. Evidence from family studies suggest that these are genetically determined conditions, with an autosomal recessive or X-linked inheritance. The genes responsible for these defects have not been identified.

5.2 Androgen Insensitivity Syndromes

Androgen insensitivity syndromes are due to end-organ resistance to androgenic steroids. The androgen receptor (AR) is encoded by a single copy gene in the X-chromosome and consists of three main functional domains: the transactivation domain, the DNA binding domain, and the ligand-binding domain. Mutations in the gene leading to abolition of AR function lead to a female phenotype in an otherwise healthy, 46,XY individual (male pseudohermaphroditism or testicular feminization) while defects that do not totally disrupt AR action cause the partial androgen insensitivity syndrome (Refenstein syndrome). The most common mutations are single base mutations acting as missense, nonsense or splice mutations (for review see Gottlieb et al. 1996). The spectrum of defective virilization in the latter disorder ranges from gynecomastia, cryptorchidism and azoospermia to more severe defects such as hypospadias and presence of a pseudovagina.

5.3 The Infertile Male Syndrome

In a family study of the Refenstein syndrome, some men were noted to be infertile but otherwise phenotypically normal. They had the same degree of androgen resistance as the more severely affected first degree relatives (Wilson et al. 1974). Androgen insensitivity as a cause of infertility in men with uninformative family histories has also been described (Aiman et al. 1979). This syndrome has been called the "infertile male" syndrome and defines a new aetiological cause for infertility. It has been suggested that mutations of the AR gene may be a cause of infertility in men with unexplained infertility and elevated serum hormone parameters suggestive of mild androgen resistance (Aiman et al. 1979; Aiman and Griffin 1982). Mutations which induce defective trans-activation of the AR receptor have been reported in azoospermic men (Wang et al. 1992; Tut et al. 1997). The AR gene contains two polymorphic trinucleotide repeat loci CAG which code for a polyglutamine tract, and GGC, which codes for a polyglycine tract. Both repeats are in the first exon, which encodes the trans-activation domain of the receptor protein. The length of the CAG repeat is highly polymorphic between races with the average CAG repeat length, ranging from 17–26 (Edwards et al. 1992). Both lower (13–14 repeat; Komori et al. 1999) and higher >26 repeat numbers (Tut et al. 1997) can induce malfunction of AR.

5.4 X-Linked Spinal and Bulbar Muscular Atrophy (Kennedy Disease)

This disease belongs to a group of human disorders caused by the expansion of a trinucleotide repeat sequence. (Willems 1994). The repeat sequence responsible for Kennedy disease is the abovementioned AR receptor CAG repeat with affected individuals having at least 40 repeats (La Spada et al. 1992). Onset of clinical signs is usually after the age of 20. The characteristic neurological signs are muscular weakness, cramps and fasciculations. If laryngeal and pharyngeal muscles are affected, both dysarthria and dysfagia are also present. The reproductive system is affected in the majority of patients. Infertility is due to testicular atrophy (Harding et al. 1982).

5.5 Persistent Mullerian Duct Syndrome

Persistent Mullerian duct syndrome is an autosomal recessive syndrome due to the mutation of either the gene for anti-Mullerian hormone (AMH) or the gene for the AMH type II receptor. The absence of this AMH function is responsible for the presence of Fallopian tubes and a uterus in an otherwise normally virilised man. These patients present an inguinal hernia which contain Mullerian structures with occasionally an attached testes to the Fallopian tubes. Testicular function can be normal but inguinal or abdominal testes can lead to infertility if not treated at an early age. In some cases, malformations of epididymis and the proximal vasa deferentia has been reported (Imbeaud et al. 1996).

5.6 Inactivating FSH Receptor Mutation

FSH is considered to be essential for spermatogenesis; the hormone exerts its action on the Sertoli cells through the FSH receptor, a member of the G protein coupled receptor family (Minegishi et al. 1991). FSH receptors are localized to testicular Sertoli cells and ovarian granulosa cells and are coupled to activation of the adenylyl cyclase and other signaling pathways. Activation of FSH-Rs is considered essential for folliculogenesis in the female and spermatogenesis in the male. Recently, an inactivating point mutation of the FSH receptor gene causing male and female infertility was detected in six finnish families (Aittomaki et al. 1995). The mutation was a 566C→T transition in exon 7 of the FSH receptor, predicted to cause an Ala to Val substitution at residue 189 in the extracellular ligand-binding domain. Functional testing showed a clear-cut reduction in ligand binding and signal transduction by the mutated receptor. Whilst all women with the inactivating mutation are infertile, in males the mutation suppresses spermatogenesis to varying degrees, from cryptozoospermia to normozoospermia. In successive studies (Leifke et al. 1997; Tuerlings et al. 1998) 49 selected patients affected by severe testicular failure with high FSH values were screened for mutations in all the 10 exons of the gene and no mutations of the receptor gene were found. One hundred and fifty- one unselected infertile patients have been screened for the specific inactivating 566C→T mutation (Tapainen et al. 1997) and only two heterozygotes and no homozygotes were detected. This suggests that this genetic defect is a rare cause of infertility. Moreover, the fact that spermatogenesis is not completely abolished by the absence of FSH function (also in an animal model of FSHβ knockout mice) demonstrates that spermatogenesis can intiate and be maintained without FSH (Kumar et al. 1997). However, in selected patients (severe oligozoospermia or azoospermia with high FSH) with a positive family history for infertility in a brother, or of ovarian failure in a sister, a mutation screening of the FSH receptor gene should still be considered.

6 Clinical Considerations of Genetic Abnormalities

The rate of sex chromosome anomalies in ICSI fetuses has been reported to be approximately 1% (study using 585 prenatal diagnoses, Van Steirteghem et al. 1996), four times that found in naturally conceived live-born babies. Moreover, a considerable number of trisomy 18 and 21 cases were reported (Palermo et al. 1996; Van Opstal et al. 1997).

Although preliminary data on babies born from fathers affected by Klinefelter syndrome seems to be encouraging, the risk of producing an XXY or XXX pregnancy is still to be determined on a larger scale. Preimplantation or prenatal diagnosis is warranted for these cases.

Most autosomal aneuploidis in germ cells, if transferred by ART in embryos, are expected to be counterselected during early development. They may induce recurrent abortions (Hassold et al. 1980). Robertsonian translocations have a low to moderate risk (1–2% or less) for viable offspring with unbalanced karyotypes (Gardner and Sutherland 1996). This type of translocation may also predispose to uniparental disomy in the offspring (Ledbetter and Engel 1995), especially patients with a 13;14, 13;15 or 14;15 translocation (Meshede et al. 1998). Reciprocal translocations, inversions, and marker chromosomes need to be assessed on an individual basis. Among them, reciprocal translocations carry the highest risk for the birth of a handicapped aneuploid child (Meshede et al. 1998).

Screening for Y chromosome microdeletions is mandatory in patients affected by azoospermia and severe oligozoospermia, since in these two groups of patients there is a high risk (15% and 10%, respectively) for this genetic defect. For all that is known about the genetic makeup of the Y chromosome, no other risks than likely infertility are to be expected upon transmission of the microdeletion.

In the case of positive family history for infertility one should look specifically for the presence of the monogenic forms. A chromosome analysis at a high level of banding resolution is essential to exclude as far as possible a small familial balanced translocation.

7 Treatment of the Infertile Male

In contrast to infertile females, only a small percentage of infertile or subfertile males can undergo rational, effective treatment. For detailed consideration of the the therapy the reader is referred elsewhere. (Forti and Krausz 1998). ICSI has become the treatment of choice for severe male factor couple infertility. The directly procedure-related genetic risks of micromanipulative assisted reproduction have been discussed elsewhere (Meshede et al. 1995; Tesarik and Mendoza 1996). While an increased frequency of sex chromosome abnormalities has been reported in babies conceived through ICSI (Van Steirteghem et al. 1996), so far there is no evidence for an increased malformation rate (Bonduelle et al. 1996; Palermo et al. 1996; Wennerholm et al. 1996). However, this conclusion is based on a relatively small number of cases and more studies are needed to determine possible risks. Genetic disorders in human spermatogenesis have been commonly estimated at 30% and they are associated mostly with the phenotype of severe male factor infertility. It is possible that fertility problems in individulas conceived through ICSI will recur more frequently. To answer this question long-term follow-up studies on ICSI babies are necessary.

Andrological examination and genetic counselling (with comprehensive clinical and laboratory work-up) are fundamental for the diagnosis of male infertility. Ongoing and future genetic studies, especially on idiopathic infertile males and in familial cases of infertility, will probably give new aetiological causes for infertility. More information is needed about the pathogenetic mechanisms by which known factors determine testicular dysfunction in order to obtain a knowledge-based therapy in the future.

References

Abbas N, Toublanc J, Boucekkin C, Toublanc M, Affara NA, Job JC, Fellous M (1990) A possible common origin of Y negative human XX males and XX true hermaphrodites. Hum Genet 84:356–360

Afzelius BA (1976) A human syndrome caused by immotile cilia. Science 193:317–319

Aiman J, Griffin JE (1982) The frequency of androgen receptor deficiency in infertile men. J Clin Endocrinol Metab 54:725–732

Aiman J, Griffin JE, Gazak JM, Parker CR, Wilson JD, MacDonald PC (1979) The frequency of androgen insensitivity in infertile but otherwise normal men. Presented at the 26th annual meeting of the Society for Gynecologic Investigation, San Diego, March 21–24, 1979

Aitken RJ (1991) A clue to Kartagener's. Nature 353:306

Aitken RJ, Baker HWG (1995) Seminal leukocytes: passengers, terrorists or good Samaritans? Hum Reprod 10:1736–1739

Aitken RJ, Ross A, Lees MM (1983) Analysis of sperm function in Kartagener's syndrome. Fertil Steril 40:696–698

Aittomaki K, Lucena JL, Pakarinen P, Sistonen P, Tapanainen J, Gromoll J, Kaskikari R, Sankila EM, Lehvaslaiho H, Engel AR (1995) Mutation in the follicle-stimulating hormone receptor gene causes hereditary hypergonadotropic ovarian failure. Cell 82:959–968

Anawalt BD, Bebb RA, Matsumoto AM, Groome NP, Illingworth PJ, McNeilly AS, Bremner WJ (1996) Serum inhibin B levels reflect Sertoli cell function in normal men and men with testicular dysfunction. J Clin Endocrinol Metab 81:3341–3345

Baschat AA, Kupker W, Al Hasani S, Diedrich K, Schwinger E (1996) Results of cytogenetic analysis in men with severe subfertility prior to intracytoplasmic sperm injection. Hum Reprod 11:330–333

Bergmann M, Behre HM, Nieschlag E (1994) Serum FSH and testicular morphology in male infertility. Clin Endocrinol 40:133–136

Berthelsen JG, Skakkebaek N, Perboll O, et al. (1981) Electron microscopic demonstration of the extra Y chromosome in spermatocytes from human XYY males. In: Byskov AG, Peters H (eds) Development and function of reproductive organs. Excerpta Medica, Amsterdam, pp 328–337

Bick D, Franco B, Sherins RJ, Heye B, Pike L, Crawford J, Maddalena A, Incerti B, Pragliola A, Meitinger T, Ballabio A (1992) Brief report: intragenic deletion of the KALIG-1 gene in Kallmann's syndrome. N Engl J Med 326:1752–1755

Bonduelle M, Legein J, Buysse A, Van Assche E, Wisanto A, Devroey P, Van Steirteghem AC, Liebaers I (1996) Prospective follow-up study of 423 children born after intracytoplasmic sperm injection. Hum Reprod 7:1558–1564

Bourne H, Stern K, Clarke G, Pertile M, Speirs A, Baker G (1997) Delivery of normal twins following the intracytoplasmic injection of spermatozoa from a patient with 47, XXY Klinefelter's syndrome. Hum Reprod 12:2447–2450

Bourrouillou G, Dastugue N, Colombies P (1985) Chromosome studies in 952 males with a sperm count below 10 million/ml. Hum Genet 71:366–367

Burkman LJ, Coddington CC, Fraken DR, Kruger TF, Rosenwalks Z, Hodgen GD (1988) The hemizona assay: development of a diagnostic test for the binding of human spermatozoa to the human hemizona pellucida to predict ferilization potential. Fertil Steril 49:688–697
Calvo L, Vantman D, Banks SM, Tezon J, Koukoulis GN, Dennison L, Sherins RJ (1989) Follicular fluid-induced acrosome reaction distinguishes a subgroup of men with unexplained infertility not identified by semen analysis. Fertil Steril 52:1048–1054
Casano R, Orlando C, Caldini AL, Barbi T, Natali A, Serio M (1987) Simultaneous measurement of seminal L-carnitine, α,1-4-glucosidase, and glycerylphosphorylcholine in azoospermic and oligozoospermic patients. Fertil Steril 47:324–328
Chandley AC (1975) Human meiotic studies. In: Emery AEH (ed) Modern trends in human genetics. Butterworths, London, pp 31–82
Chandley AC (1979) The chromosomal basis of human infertility. Br Med Bull 35:181–186
Chevret E, Rousseaux S, Monteil M, Usson Y, Cozzi J, Pellettier R, Sele B (1996) Increased incidence of hyperhaploid 24, XY spermatozoa detected by three colour FISH in a 46, XY/47, XXY male. Hum Genet 97:171–175
Chillon M, Casals T, Mercier B, Bassas L, Lissens W, Silber S, Romey MC, Ruiz-Romero J, Verlingue C, Claustres M, Nunes V, Ferec C, Estivill X (1995) Mutations in the cystic fibrosis gene in congenital absence of the vas deferens. N Engl J Med 332:1475–1480
Cozzi J, Chevret E, Rousseaux S, Pelletier R, Benitz V, Jalbert H, Sele B
(1994) Achievement of meiosis in XXY germ cells: study of 543 sperm karyotypes from an XY/XXY mosaic patient. Hum Genet 93:32–34
De Braekeleer M, Dao TN (1991) Cytogenetic studies in male infertility: a review. Hum Reprod 6:245–250
Edwards A, Hammond HA, Jin L, Caskey CT, Chakraborty R (1992) Genetic variation at five trimeric and tetrameric tandem repeat loci in four human population groups. Genomics 12: 241–253
Egozcue J (1989) Chromosomal aspects of male infertility. In: Serio M (ed) Perspectives in andrology, Serono Symposia Publications, vol 54. Raven Press, New York, pp 341–346
Egozcue J, Templado C, Vidal F, Navarro J, Morer-Fargas F, Marina S (1983) Meiotic studies in a series of 1100 infertile and sterile males. Hum Genet 65:185–188
Eliasson R, Mossberg B, Cammer P, Azfelius BA (1977) The immobile cilia syndrome: a congenital ciliary abnormality as an etiologic factor in chronic airway infections and male sterility. New Engl J Med 297:1–6
Forti G, Krausz C (1998) Clinical review: evaluation and treatment of the infertile couple. J Clin Endocrinol Metab 83:4177–4188
Gardner RJM and Sutherland GR (1996) Chromosome abnormalities and genetic counselling. Oxford University Press, New York, pp 478
Giwercman A, Bruun E, Frimodt-Moller C, Skakkebaek NE (1989) Prevalence of carcinoma in situ and other histopathological abnormalities in testes of men with a history of cryptorchidism. J Urol 142:998–1002
Gottlieb B, Trifiro M, Lumbroso R, Vasiliou DM, Pinsky L (1996) The androgen receptor gene mutations database. Nucleic Acids Res 24:151–154
Harari O, Bourne H, Baker G, Gronow M, Johnston I (1995) High fertilization rate with intracytoplasmic sperm injection in mosaic Klinefelter's syndrome. Fertil Steril 63:182–184
Harding AE, Thomas PK, Baraitsen M, Bradbury PG, Morgan-Hughes JA, Ponsford JR (1982) X-linked recessive bulbospinal neuronopathy: a report of ten cases. J Neurol Neurosurg Psychiatry 45:1012–1019
Hargreave TB (1997) Varicocele: overview and commentary on the results of the World Health Organization varicocele trial. In: Waites GMH, Frick J, Baker GWH (eds) Current advances in andrology. Proc. VIth Int Congr of Andrology, Salzburg (Austria), May 25,1997
Hassold T, Chen N, Funkhouser J, Jooss T, Manuel B, Matsuura J, Matsuyama A, Wilson C, Yamane JA, Jacobs PA (1980) A cytogenetic study of 1000 spontaneous abortions. Ann Hum Genet 44:151–178

Hens L, Bonduelle M, Liebaers I (1988) Chromosome aberrations in 500 couples referred for in vitro fertilization or related infertility treatment. Hum Reprod 3:451–457

Huszar G, Vigue L, Morshedi M (1992) Sperm creatine phosphokinase M-isoform ratios and fertilizing potential of men: a blinded study of 84 couples treated with in vitro fertilization. Fertil Steril 57:882–888

Imbeaud S, Belville C, Messika-Zeitoun L, Rey R, di Clemente N, Josso N, Picard JY (1996) A 27 base-pair deletion of the anti-Mullerian type II receptor gene is the most common cause of the persistent Mullerian duct syndrome. Hum Mol Genet 5:1269–1277

Jackson MB (1988) The John Radcliff Hospital Cryptorchidism Research Group. The epidemiology of cryptorchidism. Horm Res 30:153–156

Kaplan H, Aspillaga M, Shelley TF, Gardner LI (1963) Possible fertility in Klinefelter's syndrome. Lancet 1:506

Kjessler B (1974) Chromosomal constitution and male reproductive failure. In: Mancini RE, Martini L (eds) Male fertility and sterility. Academic Press, New York, pp 231–247

Komori S, Kasumi H, Kanazawa R, Sakata K, Nakata Y, Kato H, Koyama K (1999) CAG repeat length in the androgen receptor gene of infertile Japanese males with oligozoospermia. Mol Hum Reprod 5:14–16

Koulischer L, Schoysman R (1975) Etude des chromosomes mitotiques et méiotiques chez les hommes infertiles. J Génét Hum 23:50–70

Koulischer L, Schoysman R, Gillerto Y (1982) Chromosomes meiotiques et infertilite masculine: evaluation des resultats. J Genet Hum 30:81–99

Krausz C, West K, Buckingham D, Aitken J (1992) Development of a technique for monitoring the contamination of human semen samples with leukocytes. Fertil Steril 57:1317–1325

Krausz C, Mills C, Rogers S, Tan SL, Aitken RJ (1994) Stimulation of oxidant generation by human sperm suspensions using phorbol esters and formyl peptides: relationship with motility and fertilization in vitro. Fertil Steril 62:599–605

Krausz C, Bonaccorsi L, Maggio P, Luconi M, Criscuoli L, Fuzzi B, Pellegrini S, Forti G, Baldi E (1996) Two functional assays of sperm responsiveness to progesterone and their predictive values in in vitro fertilization. Hum Reprod 11:1661–1667

Kumar TR, Wang Y, Lu N, Matzuk M (1997) Follicle stimulating hormone is required for ovarian follicle maturation but not male fertility. Nat Genet 15:201–204

Laron Z, Dickerman Z, Zamir R, Galatzer A (1982) Paternity in Klinefelter's syndrome – a case report. Arch Androl 8:149–151

La Spada AR, Roling DB, Harding AE, Warner CL, Spiegel R, Hausmanowa-Petrusewicz I, Yee WC, Fischbeck KH (1992) Meiotic stability and genotype-phenotype correlation of the trinucleotide repeat in X-linked spinal and bulbar muscular atrophy. Nature Genet 2:301–304

Ledbetter DH, Engel E (1995) Uniparental disomy in humans: development of an imprinting map and its implications for prenatal diagnosis. Hum Mol Genet 4:1757–1764

Leifke E, Simoni M, Kamischke A, Gromoll J, Bergmann M, Nieschlag E (1997) Does the gonadotropic axis play a role in the pathogenesis of Sertoli cell only syndrome? Int J Androl 20: 29–36

Lyon MF (1974) Sex chromosome activity in germ cells. In: Coutinho EM, Fuchs F (eds) Physiology and genetics of reproduction. Part A. Plenum Press, New York, pp 63–71

Martini E, Geraedts JP, Liebaers I, Land JA, Capitanio GL, Ramaekers FC, Hopman AH (1996) Constitution of semen samples from XYY and XXY males as analysed by in situ hybridization. Hum Reprod 11:1638–1643

Mau UA, Backert IT, Kaiser P, Kiesel L (1997) Chromosomal findings in 150 couples referred for genetic counselling prior to intracytoplasmic sperm injection. Hum Reprod 12:930–937

Meshede D, De Geyter C, Nieschlag E, Horst J (1995) Genetic risk in micromanipulative assisted reproduction. Human Reprod 10:2880–2886

Meshede D, Lemcke B, Exeler JR, de Geyter Ch, Behre HM, Nieschlag E, Horst J (1998) Chromosome abnormalities in 447 couples undergoing intracytoplasmic sperm injection – prevalence, types, sex distribution and reproductive relevance. Hum Reprod 13:576–582

Minegishi T, Nakamura K, Takakura Y, Ibuki Y, Igarishi M (1991) Cloning and sequencing of human FSh receptor cDNA. Biochem Biophys Res Commun 175:1125–1130

Narayan D, Krishnan SN, Upender M, Ravikumar TS, Mahoney MJ, Dolan TF Jr, Teebi AS, Haddad G (1994) Unusual inheritance of primary ciliary dyskinesia (Kartagener's syndrome). J Med Genet 31:493–496

Nashan D, Behre HM, Grunert JH, Nieschlag E. (1990) Diagnostic value of scrotal sonography in infertile men: report on 658 cases. Andrologia 22:387–395

Neilson J, Wohlert M (1991) Chromosome abnormalities found among 34 910 newborn children: results from a 13-year incidence study in Arhus, Denmark Hum Genet 70:81–83

Neugebauer DC, Neuwinger J, Jockenhovel F, Nieschlag E (1990) "9+0" axoneme in spermatozoa and some nasal cilia of a patient with totally immotile spermatozoa associated with thickened sheath and short midpiece. Hum Reprod 5:981–986

Nieschlag E, Hertle L, Fischedick A, Abshagen K, Behre HM (1998) Update on tttreatment of varicocele: counselling as effective as occlusion of the vena spermatica. Hum Reprod 13:2147–2150

Palermo GD, Colombero LT, Schattman GD, Davis OK, Rosenwaks Z (1996) Evolution of pregnancies and initial followup of newborns delivered after intra-cytoplasmic sperm injection. JAMA 276:1893–1897

Peschka B, Schwanitz G, van der Ven K (1996) Type and frequency of constitutional chromosome aberrations in couples undergoing ICSI. Hum Reprod 11:224–225

Retief AE, Van Zyl JA, Menkveld R, Fox MF, Kotze GM, Brusnicky J (1984) Chromosome studies in 496 infertile males with sperm count below 10 million/ml. Hum Genet 66:162–164

Rutland J, de Iongh RU (1990) Random ciliary orientation. A casue of respiratory tract disease. N Engl J Med 323:1681–1684

Samuel I (1987) Kartagener's syndrome with normal spermatozoa. (letter) JAMA 258:1329–1330

Schill WB (1991) Some disturbances of acrosomal development and function in human spermatozoa. Hum Reprod 6:969–978

Schwanzel-Fukuda M, Bick D, Pfaff DW (1989) Luteinizing hormone-releasing hormone (LHRH)-expressing cells do not migrate normally in an inherited hypogonadal (Kallmann) syndrome. Brain Res Mol Brain Res 6:311–326

Sharpe RM, Skakkebaek NE (1993) Are oestrogens involved in falling sperm counts and disorders of the male reproductive tract? Lancet 341:1392–95

Silber SJ, Van Sterteighem AC, Nagy Z , Liu J, Tournaye H, Devroy P (1996) Normal pregnancies resulting from testicular sperm extraction and intracytoplasmatic sperm injection for azoospermia for maturation arrest. Fertil Steril 66:110–117

Skakkebaek N, Philip J, Hammen R (1969) Meiotic chromosomes in Klinefelter's syndrome. Nature 221:1075–1076

Skakkebaek NE, Bryant JI, Philip J (1973) Studies on meiotic chromosomes in infertile men and controls with normal karyotypes. J Reprod Fertil 35:23–36

Steinberger E, Smith KD, Perloff WH (1965) Spermatogenesis in Klinefelter's syndrome. J Clin Endocrinol Metab 25:1325–1330

Sturgess JM, Chao J, Wong J, Aspin N, Turner JA (1979) Cilia with defective radial spokes; a cause of human respiratory disease. N Engl J Med 300:53–56

Tapainen JS, Aittomaki K, Min J, Vaskivuo T, Huhtaniemi I (1997) Men homozygous for an inactivating mutation of the follicle stimulating hormone (FSH) receptor gene present variable suppression of spermatogenesis and fertility. Nat Genet 15:205–210

Terzoli G, Lalatta F, Lobbiani A (1992) Fertility in a 47, XXY patient: assessment of biological paternity by deoxyribonucleic acid fingerprinting. Fertil Steril 58:821–822

Tesarik J, Mendoza C (1992) Defective function of a nongenomic progesterone receptor as a sole sperm anomaly in infertile patients. Fertil Steril 58:793–797

Tesarik J, Mendoza C (1996) Genomic imprinting abnormalities: a new potential risk of assisted reproduction. Mol Hum Reprod 2:295–298

Tiepolo L, Zuffardi O (1976) Localization of factors controlling spermatogenesis in the non-fluorescent portion of the human Y chromosome long arm. Hum Genet 34:119–124

Tiepolo L, Fraccaro M, Giarola A (1981) Chromosome abnormalities and male infertility. In: Frajese G, Hafez ESE, Conti C, Fabbrini A (eds) Oligozoospermia: recent progress in andrology. Raven Press, New York, pp.233–245

Tietze C (1956) Statistical contribution to the study of human fertility. Fertil Steril 7:88–95.

Tietze C (1968) Fertility after discontinuation of intrauterine and oral contraception. Int J Fertil 13:385–9

Tournaye H, Camus M, Liebaers I (1996) Testicular sperm recovery in nine 47,XXY Klinefelter patients. Hum Reprod 11:1644–1649

Tuerlings JHAM, Ligtenberg MJL, Kremer JAM, Siers M, Meuleman EJH, Braat DDHM, Hoefsloot LH, Merkus MWM, Brunner HG (1998) Screening male intracytoplasmic sperm injection candidates for mutations of the follicle stimulating hormone receptor gene. Hum Reprod 13:2098–2101

Tut TG, Ghadessy FJ, Trieiro MA, Pinsky L, Yong EL (1997) Long polyglutamine tracts in the androgen receptor are associated with reduced trans-activation, impaired sperm production, and male infertility. J Clin Endocrinol Metab 82:3777–3782

Van Opstal D, Los FJ, Ramlakhan S, Van Hemel JO, Van Den Ouweland AM, Brandenburg H, Pieters MH, Verhoeff A, Vermeer MC, Dhont M, In't Veld PA (1997) Determination of the parent of origin in nine cases of prenatally detected chromosome aberrations found after intracytoplasmic sperm injection. Hum Reprod 12:682–686

Van Steirteghem AC, Nagy Z, Joris H, Liu J, Staessen C, Smitz J, Wisanto A, Devroey P (1993) High fertilization rate and implantation rates after intracytoplasmic sperm injection. Hum Reprod 8:1061–1066

Van Steirteghem A, Nagy P, Joris H, Verheyen G, Smitz J, Camus M, Tournaye H, Ubaldi F, Bonduelle M, Silber S, Liebaers I, Devroey P (1996) The development of intracytoplasmic sperm injection. Hum Reprod 11 (Suppl 1):59–72

Vogt PH, Edelmann A, Kirsch S, Henegariu O, Hirschmann P, Kiesewetter F, Kohn FM, Schill WB, Farah S, Ramos C, Hartmann M, Hartschuh W, Meschede D, Behre HM, Castel A, Nieschlag E, Weidner W, Grone HJ, Jung A, Engel W, Haidl G (1996) Human Y chromosome azoospermia factors (AZF) mapped to different subregions in Yq11. Hum Mol Genet 5:933–943

Wang Q, Ghadessy J, Trounson A, de Kretser D, McLachlan R, Ng SC, Yong EL (1992) Azoospermia associated with a mutation in the ligand-binding domain of an androgen receptor displaying normal ligand binding, but defective trans-activation. J Clin Endocrinol Metab 83:4303–4309

Wennerholm UB, Bergh C, Hamberger L, Nilsson L, Reismer E, Wennergren M, Wikland M (1996) Obstetric and perinatal outcome of pregnancies following intracytoplasmic sperm injection. Hum Reprod 11:1113–1119

Willems PJ (1994) Dynamic mutations hit double figures. Nature Genet 8:213–215

Wilson JD, Harrod MJ, Goldstein JL, Hemsell DL, MacDonald PC (1974) Familial incomplete male pseudohermaphroditism, type 1: evidence for androgen resistance and variable clinical manifestations in a family with the Reifenstein syndrome. N Engl J Med 290:1097–1103

World Health Organization (1987) Towards more objectivity in diagnosis and management of male infertility. Int J Androl 7:1–53

World Health Organization (1992) WHO laboratory manual for the examination of human semen and sperm-cervical mucus interaction, 3rd edn. Cambridge University Press, Cambridge, UK pp 44–45

Yoshida A, Kazukiyo M, Shirai M (1996) Chromosome abnormalities and male infertility. Assist Reprod Rev 6:93–99

Zuhlke C, Thies U, Braulke I, Reis A, Schirren C (1994) Down's syndrome and male fertility: PCR derived fingerprinting, serological and andrological investigations. Clin Genet 46:324–326

The Cell Biology and Molecular Genetics of Testis Determination

Craig A. Smith and Andrew H. Sinclair

1 Introduction

The variety of sexual dimorphisms that distinguish males from females all stem from a single critical process: sex determination. In humans and other mammals, sex is determined at the time of fertilization by the inheritance of sex chromosomes. Individuals receiving two X chromosomes develop ovaries and become females, while those receiving one X and one Y chromosome develop testes and become males. These alternative sex chromosome constitutions must initiate different developmental programmes within the embryonic gonads. In mammalian embryos, testicular differentiation begins prior to ovarian differentiation. Consequently, female development has traditionally been regarded as a 'default' state that must be pre-empted by a male-determining signal if a testis is to form. Furthermore, Alfred Jost's classic castration experiments on rabbit embryos, conducted over 40 years ago, showed that the testis is necessary for male-specific development of the reproductive tract and external genitalia in eutherian mammals. In the absence of the gonads, both genetic sexes follow the female pathway of differentiation (Jost 1953, 1970; reviewed in Jost et al. 1973). The masculinizing effect of the testis is due to the hormones, testosterone and Anti-Müllerian hormone (AMH), (reviewed in George and Wilson 1988). Thus, the various processes that constitute sexual differentiation depend upon the presence/absence of a testis. Testis determination is therefore a decisive developmental event.

Recent years have seen some major advances in our understanding of testis determination at the molecular level. This chapter gives an account of these advances, beginning with a brief review of the chromosomal basis of human sex determination, followed by a description of gonadogenesis and the cell biology of testis differentiation. The testis-determining *SRY* and *SOX9* genes are then discussed, and other factors implicated in gonadogenesis are considered, particularly the orphan nuclear receptors, SF-1 and DAX-1. Finally, a developmental cascade controlling testis determination is presented. Since

Department of Paediatrics and Centre for Hormone Research, The University of Melbourne, Royal Children's Hospital, Melbourne, Victoria, 3052, Australia

Results and Problems in Cell Differentiation, Vol. 28
McElreavey (Ed.): The Genetic Basis of Male Infertility

somatic differentiation of the testis can occur in the absence of primordial germ cells (McLaren 1991), male germ cell development is not considered here. A discussion of the male germ line can be found elsewhere in this volume.

2
Human Sex Determination is Chromosomally Based

Sex in humans and other mammals is determined genetically by the sex chromosome constitution. The inheritance of two X chromosomes leads to female development, while an X and a Y chromosome result in male development. Human sex chromosomes are heteromorphic, the X chromosome being several times larger than the Y. The X carries major "house-keeping genes", the majority of which are irrelevant to sex determination. In females, most of these genes are subject to inactivation in one X homologue, thought to be necessary to equalise gene dosage with that of males (i.e., down-regulation of X-linked gene activity). The smaller Y chromosome carries far fewer genes than the X and is largely heterochromatic. Early studies on humans and mice with aberrant numbers of sex chromosomes showed that the Y chromosome is male-determining, regardless of the number of X chromosomes present. Thus, in humans, XY or XXY (Klinefelter's syndrome) individuals develop as males, while XX or XO (Turner's syndrome) individuals develop as females (Ford et al. 1959; Jacobs and Strong 1959). Studies in the 1960s and 1970s identified mutations in XX mice that resulted in female-to-male sex reversal (Sxr: Sex reversal; Cattanach et al. 1971). These XXSxr mice were subsequently shown to carry a small fragment of the Y chromosome translocated to the X chromosome (Evans et al. 1982; Singh and Jones 1982; reviewed in Goodfellow and Darling 1988). This finding supported the proposition that a master testis-determining factor is carried on the mammalian Y chromosome.

2.1
Sex reversal

Primary sex reversal occurs in humans when gonadal sex is inconsistent with chromosomal sex (reviewed in Schafer 1995). In the case of XY sex reversal, the individual has an XY sex chromosome constitution but the gonad fails to develop as a normal testis. Poorly differentiated ovarian tissue or a histologically amorphous "streak gonad" is present, while the ducts and external genitalia are female. This condition is known as 46XY complete gonadal dysgenesis. In other cases, 46XY partial gonadal dysgenesis exists, characterised by partial testicular tissue together with weakly or normally masculinized external genitalia (Berkovitz 1992). 46XX males have testes or partially masculinized gonads and normal male or ambiguous genitalia. Genetic analysis of such sex-reversed patients has been invaluable in defining the region of the Y chromosome that harbours the testis-determining factor (Page et al. 1987; Sinclair et al. 1990). Most (but not all) XX male patients with completely unambiguous external genitalia are known to carry a small fragment of the Y chromosome, due to unequal crossing over between the X and Y chromosomes during paternal meiosis (reviewed in McElreavey et al. 1995).

XY human females and XX human males are sterile. Two X chromosomes appear to be necessary for oogenesis in humans (absent in XY females), while two X chromosomes are incompatible with spermatogenesis in XX males (in addition, genes regulating spermatogenesis are carried on the Y chromosome, absent in XX males; Vogt et al. 1996). Spontaneous true hermaphrodites have both ovarian and testicular tissue, usually as ovotestes but less commonly as separate gonads. Chromosomally, they may be male (46XY), female (46XX) or mosaics (e.g; 45X/46XY or 46XX/46XY) (reviewed in Larsen 1993). Pseudohermaphroditism occurs when the phenotype of the external genitalia is inconsistent with gonadal sex. This condition reflects dysfunction at the hormonal level of sexual differentiation after testis or ovary formation. Male pseudohermaphrodites have a 46XY genotype and bilateral testes together with feminized genitalia, while female pseudohermaphrodites have a 46XX genotype, ovaries and masculinized genitalia. Male pseudohermaphroditism is attributed to anomalies in the production or functioning of androgen. In testicular feminization syndrome (Tfm), for example, testes develop normally and secrete androgens, however the androgen receptor is mutated, target tissues do not respond and external genitalia are therefore feminized (Warne et al. 1993).

3 Gonadal Sex Differentiation

In humans and other eutherian mammals, the first overt morphological sign of sex determination is differentiation of the embryonic gonads into testes or ovaries. Studies aimed at unravelling the mechanisms of sex determination have therefore focused upon the embryonic gonads, for it is here that major sex-determining pathways must operate. A description of gonadal sex differentiation will provide the necessary context for considering the molecular basis of testis determination.

3.1 Testicular and Ovarian Morphogenesis in Human Embryos

Development of the human gonad during embryogenesis reflects its future functions of gamete production and sex hormone synthesis. Gonadogenesis involves the enclosure of germ and supporting cells within membrane-bound compartments where gametogenesis can occur. Interstitial cells concerned purely with steroidogenesis differentiate outside these compartments. This fundamental organisation is common to both sexes. However, males and females differ in the structural design of the compartments (seminiferous tubules in the testis and follicles in the ovary), and in the timing of developmental events, such as meiosis.

Ovaries and testes have a common embryonic origin. They are derived from intermediate mesoderm, developing in close association with the mesonephric (embryonic) kidneys (Fig. 1). The first morphological sign of gonadogenesis occurs early in embryogenesis and is marked by a proliferation of coelomic epithelium (the "germinal epithelium") on the ventromedial surface of the mesonephros. Primordial germ cells (PGCs) are of extra-gonadal origin, they migrate from the yolk sac to the posterior body wall and enter the gonadal primordia. Mesenchymal cells accumulate beneath the coelomic epithelium and cords of cells - the primary sex cords - develop within the mesenchyme (Fig. 1). The exact origin of the sex cords is controversial (Wartenberg 1978; Fukuda et al. 1988); traditionally, germinal epithelial cells are thought to invade the underlying mesenchyme, giving rise to the cords. This view has been supported by recent cell labelling studies in male mouse embryos (Karl and Capel 1998), which show that cells of the germinal epithelium migrate into the interior of the gonad and can subsequently give rise to Sertoli cells within testicular cords.

At the early undifferentiated stage, the gonad is organised into two components; a cortex (thickened epithelium of somatic and germ cells) and underlying medulla (primary sex cords enclosing fewer germ cells together with surrounding mesenchyme; Fig. 1). This stage is evident by 40 days post ovulation (40dpo) in developing human embryos. During this time, the gonads of presumptive males and females are morphologically indistinguishable and are considered "indifferent" or "bipotential". Strands of epithelial cells, rete cords, extend through the core of the gonad at the cranial pole and join the mesonephros (= future vasa efferentia in males). Müllerian (paramesonephric) ducts arise from folds of epithelial and mesenchymal cells in close association with the Wolffian (mesonephric) ducts.

Between the sixth and seventh weeks of embryogenesis in humans (around 45 dpo), sex differences become apparent in the gonads as sexual differentiation begins in the male. In presumptive male embryos, Sertoli cells differentiate within the primary sex cords, giving rise to seminiferous tubules (Fig. 1). PGCs become enclosed in these tubules and enter mitotic arrest as prospermatogonia. The germinal epithelium regresses. A subpopulation of interstitial cells differentiates into androgen-producing Leydig cells soon after Sertoli cell differentiation. The Leydig cells are thought to be derived from mesenchymal cells that migrate into the gonad from the mesonephros. Maintenance of their differentiated state requires gonadotrophin from the pituitary gland, but their initial differentiation is believed to be induced by Sertoli cells. Other interstitial cells within the organising testis form peritubular myoid cells and connec-

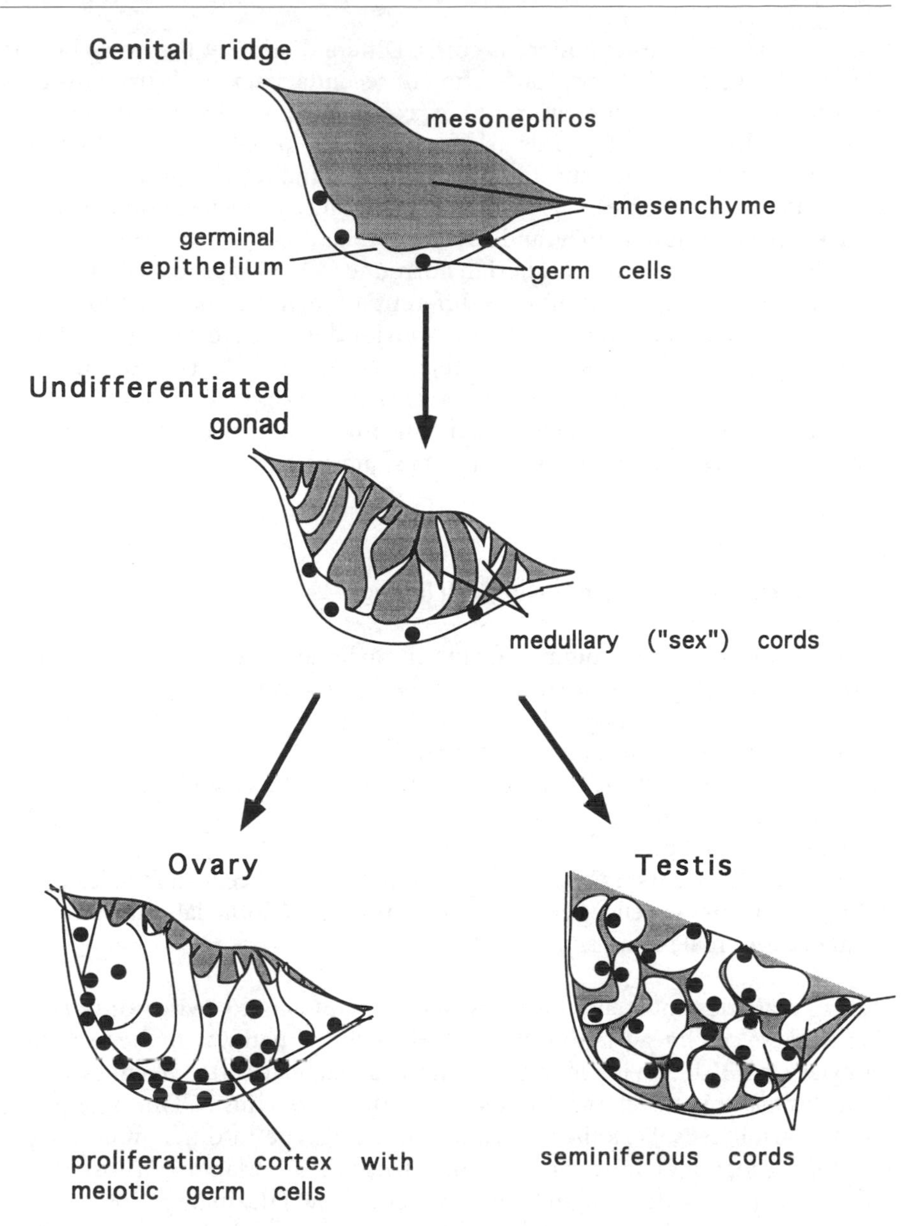

Fig. 1. Schematic representation of gonadogenesis in mammals. The genital ridge forms as a swelling of coloemic epithelium (germinal epithelium) on the ventromedial surface of the mesonephric kidney. Germ cells populate the germinal epithelium and mesenchyme proliferates beneath the epithelium. During the indifferent (bipotential) phase, primary (medullary) sex cords penetrate the inner part of the undifferetiated gonad. Ovarian differentiation is marked by regression of the primary cords in the medulla, and proliferation of secondary cords together with meiotic germ cells in the outer zone (cortex) of the gonad. Testis formation is marked by growth and differentiation of the primary cords to form seminiferous cords. Cortical germ cells are drawn into these cords, where they enter mitotic arrest. The germinal epithelium (cortex) regresses

tive tissue around the seminiferous cords. Differentiation of the ovary begins later and is characterised by proliferation of secondary sex cords from the cortex and regression of the primary sex cords (Fig. 1). Germ cells become enclosed in the secondary cords and enter meiotic prophase. The secondary sex cords ultimately become fragmented, surrounding the germ cells and undergoing folliculogenesis- granulosa and thecal cell development (reviewed in McCarrey and Abbott 1979, and Byskov, 1986).

In the male embryo, Anti-Müllerian hormone (AMH) is produced by Sertoli cells during the early stages of testis differentiation. It is responsible for disintegration of the Müllerian (paramesonephric) ducts. Testosterone produced by the Leydig cells induces the mesonephric (Wöllfian) ducts to become the epididymi and, posteriorly, the vas deferentia. In the absence of these two hormones, the mesonephric ducts degenerate in females, while the Müllerian ducts can differentiate into Fallopian tubes, uterus and the upper portion of the vagina.

3.2
The importance of the supporting cell lineage

Since the gonads have a common origin in males and females, they comprise a similar group of "bipotential" cell lineages. These cells must be directed to follow one of two divergent developmental pathways. Four cell lineages are recognised within the undifferentiated gonad:

1. Primordial germ cells (PGCs; presumptive spermatozoa and ova),
2. Supporting somatic cells (Sertoli cell precursors in males, and follicular (granulosa) cell precursors in females),
3. "Steroid cell precursors" (Leydig cells in males, thecal cells in females),
4. Connective tissue cells (peritubular myoid cells, endothelial cells and stromal cells in males, and similar cells in females).

During male development, the testis-determinant acts to switch differentiation of these cell lineages from the "default" ovarian pathway to the testicular pathway (Jost 1970; reviewed in Burgoyne and Palmer 1993). Current evidence indicates that this switch is activated within the supporting cell lineage, giving rise to Sertoli cells. Sertoli cells are the first cell type to differentiate in the developing mammalian testis (Merchant-Larios 1976; Magre and Jost 1991). Other cell lineages are thought to be channelled down the male pathway under the direction of the differentiating Sertoli cell population (reviewed in McLaren 1991; Lovell-Badge 1992a,b, and in Burgoyne and Palmer 1993). In the absence of the Y chromosome, cells differentiate in the female-specific mode (reviewed in McLaren 1991). As discussed below, germ cells are not necessary for somatic differentiation of the testis (sterile mutant mice have small but otherwise normal testes). In contrast, meiotic germ cells appear to be required for complete ovarian differentiation (reviewed in McLaren 1991).

McLaren (1991) has drawn together many lines of evidence showing that mammalian gonadal sex differentiation hinges upon the fate of the supporting cell lineage. Testis formation appears to depend upon sufficient numbers of Sertoli cells differentiating within a specific time frame of development. In cases where these conditions are not met, ovaries or ovotestes can differentiate. Under the direction of the Y chromosome, supporting cells differentiate into primordial Sertoli cells (pre-Sertoli cells). However, this cell lineage can differentiate into pre-follicular rather than pre-Sertoli cells if the testis-determining factor (TDF) is not present/not expressed (e.g., in normal XX females), if expression of the TDF is delayed (e.g., in crosses between certain mouse strains; Eicher and Washburn 1986; Palmer and Burgoyne 1991a; Taketo et al. 1991), or if few pre-Sertoli cells initially develop (e.g., in XX↔XY chimeric or X0/XY mosaic mouse gonads, with few XY supporting cells; Burgoyne et al. 1988b; Palmer and Burgoyne 1991b). In females, supporting cells normally begin differentiation as pre-follicular cells. However, their further development into follicle (granulosa) cells seems to require the presence of meiotic germ cells (Merchant 1975; reviewed in McLaren 1991). If oocytes are lost at the stage of folliculogenesis, developing follicle cells can apparently transdifferentiate into Sertoli cells (McLaren,1991). This can occur when embryonic mouse ovaries grown *in vitro* are exposed to AMH, which induces the degeneration of germ cells (Vigier et al. 1988; Charpentier and Magre, 1990).The freemartin effect in cattle, whereby a female embryo is masculinized in the presence of a male twin (see Jost 1953, 1970), may also depend upon the loss of germ cells. It has been suggested that AMH secretion from the male twin leads to germ cell loss and therefore seminiferous cord formation in the female twin (McLaren 1991).

Studies of embryonic mouse gonads add further support to the idea that XX supporting cells can differentiate into apparent Sertoli cells in the absence of any part of the Y chromosome. For example, XX gonadal primordia transplanted under the renal capsule of adult male or female mice can develop seminiferous tubules containing morphologically and functionally differentiated Sertoli cells (Taketo-Hosotani et al. 1985; Taketo et al. 1993). Similarly, in female mice expressing high levels of an AMH transgene, supporting cells can take on the characteristics of Sertoli cells (Behringer et al. 1994). Like the freemartin cattle mentioned above, both of these phenomena are associated with germ cell loss in the XX gonad, and are compatible with the idea that follicular cells can transdifferentiate into Sertoli cells upon removal of the germ cell lineage. Thus, both XX and XY supporting cells seem to have the *ability* to form Sertoli cells. Under normal conditions, a gene expressed in XX embryos must block this ability. This implies that the function of TDF would be to inhibit this gene, allowing Sertoli cell differentiation to proceed (see Sect. 4.3 below for further discussion).

3.3 The contribution of the mesonephros

The mesonephric kidney makes an important contribution to the developing testis in mammalian embryos. When the gonads of male mouse embryos are excised at the onset of sexual differentiation (11.5 days post coitum, dpc) and grown in vitro, Sertoli cells can differentiate in the absence of the mesonephric kidney (Merchant-Larios et al. 1993). However, the presence of the mesonephros appears to be necessary for the *organisation* of these cells into seminiferous cords (Buehr et al. 1993; Merchant-Larios et al. 1993). Studies on embryonic mouse gonads grown in culture indicate that the presence of the mesonephros promotes the deposition of basement membrane within the developing testis (Kanai et al. 1995). When the mesonephros is labelled with transgenic or radioactive markers, labelled mesonephric cells can be traced into the gonads following explantation at 11.5 dpc. Once in the gonad, these cells form peritubular myoid and other interstitial cells. These cells therefore appear to be the mesonephric input necessary for seminiferous cord organisation in male mouse gonads after 11.5 dpc (Buehr et al. 1993; Merchant-Larios et al. 1993). (There may be other mesonephric contributions to the developing gonad prior to 11.5 dpc.). Studies of adult rat testes indicate that myoid and Sertoli cells cooperate in laying down basement membrane components (laminin, fibronectin and collagen IV; Tung et al. 1987; Richardson et al. 1995). A similar interaction is likely to occur during fetal development, resulting in the basement membrane remodelling that characterises seminiferous cord formation. Indeed, cell-cell signalling between interstitial cells and differentiating Sertoli cells must be an integral part of testis formation.

3.4 Genes involved in formation of the gonadal primordium

Some genes have been identified that play a role in the formation of the gonadal primordium. One of these genes is *WT1*, originally identified as an oncogene involved in the paediatric renal cancer, Wilms' tumour (Call et al. 1990). The *WT1* gene encodes a zinc finger protein and is presumed to operate as a transcription factor. Heterozygous mutations in *WTI* occur in patients with Denys-Drash syndrome, characterised by kidney failure and genital abnormalities, including XY sex reversal (Pelletier et al. 1991a; reviewed in Schafer and Goodfellow 1996). The mouse homologue, *Wt1*, is expressed in the developing kidneys and in the undifferentiated gonads of both sexes (Pelletier et al. 1991b). Both of these organs fail to develop when *Wt1* is rendered non-functional by gene targeting in mice (Kreidberg et al. 1993), indicating that this gene is essential for formation of the urogenital system. The existence of Denys-Drash patients with sex-reversal and *WT1* mutations implies that the gene has an additional role in testis development downstream of its function in early function in kidney and gonad development. The human *WT1* gene comprises 10 exons and produces several different transcripts generated by alternative mRNA slicing and alternative translation start sites (Sharma et al. 1994). The different WT1 isoforms may perform a variety of regulatory functions, including DNA binding and RNA processing. WT1 is implicated in the regulation of inductive interactions bewteen mesenchyme and epithelia, which are likely to be important for gonadal differentiation (reviewed in Swain and Lovell-Badge 1999).

Lim1 (a homeo-paired box gene) also appears to be necessary for kidney and gonad development, as both organs are absent in *Lim1* knockout mice (Shawlot and Behringer 1995). Similarly, the orphan nuclear receptor, steroidogenic factor 1 (SF1) is essential for early gonad formation; SF1 knockout mice lack gonads. In the case of SF1, gonads initially form, but regress at the time of sexual differentiation (Luo et al. 1994) (see Sect. 6 for a more detailed discussion of SF1). Recently, the murine *M33* gene has been implicated in gonadal development. This gene is a homologue of the *Drosophila* polycomb group of genes (PcG), and *M33*-deficient mice show retarded gonadal development associated with varying degrees of XY sex reversal (Katoh-Fukui et al. 1998). In *Drosophila*, The PcG proteins act as multimers to regulate the co-ordinate expression of homeotic genes, and M33 may act in a similar way by targetting *Hox* gene expression within the mammalian urogenital system. However, while *M33*, *Wt1*, *Lim1* and *SF1* all appear to be important for gonadal formation, their precise roles are unknown.

4
The Testis-Determining Factor (TDF)

The morphological processes underlying testis formation must be finely orchestrated by a hierarchy of gene expression. Some of these genes are expected to encode regulatory (transcription) factors, while others must specify structural proteins. This hierarchy is initiated by a testis-determining factor (TDF) carried on the Y chromosome, now known as the *SRY* gene.

Despite its known whereabouts on the Y chromosome, isolation of TDF proved difficult, and its discovery came over 30 years after the sex chromosomes themselves were first identified. From the 1970s up until 1990, several different factors were proposed as candidate TDFs. These included the H-Y antigen (Wachtel 1983), *Bkm* sequences (Banded krait minor satellite; a group of tandem GATA or GACA repeats) (Jones and Singh 1981) and the *ZFY* gene, which encodes a zinc finger protein (Page et al. 1987; Page 1988). All of these candidates have since been rejected in favour of the Y-linked gene, *SRY* (*S*ex-determining *R*egion on the *Y* chromosome) (Sinclair et al. 1990). This gene is now regarded as the master testis determinant in humans and other mammals.

4.1 *SRY* is TDF

The discovery of human *SRY* (and its mouse homologue, *Sry*) was reported in 1990 (Gubbay et al. 1990; Sinclair et al. 1990). The human gene was isolated by sequence analysis of small Y chromosome fragments translocated to the X chromosome in XX sex-reversed patients. *SRY/Sry* has all the characteristics expected of a master sex-determining gene. In humans, *SRY* maps to the short arm of the Y chromosome, very close to the pseudoautosomal region (a small part of the Y that pairs with the X chromosome during meiosis). This region at the tip of the Y chromosome is the smallest fragment known to be necessary for testis development (Sinclair et al. 1990). Some (but not all) XY females have loss-of-function mutations in *SRY* (Berta et al. 1990; Jäger et al. 1990; summarised in Goodfellow et al. 1993, and in Cameron and Sinclair 1997). [Those XY females that have no obvious changes to the *SRY* gene may have mutations downstream in the testis-determining pathway (Pivnick et al. 1992).] *Sry* is carried by the smallest fragment of the Y chromosome capable of inducing sex reversal when translocated onto the X chromosome in XX*Sxr'* (sex-reversed) mice (Gubbay *et.* al 1990).

The human *SRY* gene comprises a single exon and encodes a protein of 204 amino acids, including a 79-residue conserved DNA-binding domain, the HMG-box (High Mobility Group) (Clepet et al. 1993). This suggests that the *SRY* gene product regulates gene expression. In male mouse embryos, *Sry* shows the spatial and temporal expression profiles expected of a testis determinant. Reverse transcription and the polymerase chain reaction (RT-PCR) and RNase protection assays have shown that *Sry* is briefly expressed within the somatic cells of male mouse gonads at the onset of testis differentiation (between 10.5 and 12.5 days dpc) (Koopman et al. 1990; Hacker et al. 1995). Furthermore, *Sry* is the only Y-linked gene necessary for testis development in mice. XX mice transgenic for a 14kb fragment containing only *Sry* can develop as males with (sterile) testes (Koopman et al. 1991). All of these findings indicate that *SRY/Sry* is *TDF.SRY* has been identified in all eutherians tested, and marsupial homologues have been isolated (reviewed in Graves et al. 1993). However, despite its presence on the Y chromosome of virtually all mammals examined, *SRY* has not been well conserved. Human and mouse SRY proteins have 71% homology within the conserved HMG-box (85% when conservative changes are taken into account) (Gubbay et al. 1990). Homology drops sharply outside the HMG-box. For example, the human and rabbit SRY gene products have 82% sequence identity within the HMG-box, but this falls to 54% outside the box (Sinclair et al. 1990).

Current evidence indicates that *SRY* is expressed in pre-Sertoli cells within the presumptive testis (Rossi et al. 1993; reviewed in Lovell-Badge 1992b). This evidence is consistent with two important observations: (1) Unlike other cell lineages, Sertoli cells are predominantly or exclusively XY in XX↔XY chimeric mouse gonads (Burgoyne et al. 1988a; Palmer and Burgoyne, 1991c; Patek et al. 1991).(2) Sertoli cells are the first cell type to differentiate in the male gonad (Magre and Jost, 1991; see Sect. 3.2 above). It is thought that *SRY* acts largely in a cell-autonomous fashion to induce Sertoli cell differentiation. However, data from XX↔XY chimeric mouse gonads show that some cells with an XX genotype can be recruited into the Sertoli cell population (Palmer and Burgoyne 1991c). This suggests that part of *Sry* action may involve an extra-cellular signalling mechanism. Altogether, the data indicate that the role of *SRY* is to act as a molecular switch, stimulating Sertoli cell formation and therefore testis determination. The expression data in the mouse indicate that the gene is not required for the maintenance of Sertoli cell differentiation, as *Sry* is down-regulated once differentiation has been initiated (Hacker et al. 1995).

Northern blot analysis shows that *Sry* expression also occurs in the testes of adult mice (Koopman et al. 1990). In adult mice, *Sry* is expressed at high levels in meiotic and post-meiotic germ cells and at lower levels in Sertoli cells (Rossi et al. 1993). The transcript in adult mice is circular, due to the presence of flanking inverted repeat sequences. In this form, the Sry transcript is thought not to be translated (Capel et al. 1993). Human *SRY* is expressed in a variety of non-gonadal embryonic and adult tissues (Clépet et al. 1993). Its function in these tissues is unclear, and indeed the transcript may be inactivated in non-gonadal sites as it appears to be in the adult mouse (but by mechanisms other than circularisation, which has not been reported in humans).

At present, the factors regulating SRY expression during testis differentiation are unknown. There is no definitive evidence as yet that genes involved in early gonad development directly activate *SRY*. Vilain et al. (1992) used primer extension, 5′RACE and RNase protection assays to define the 5′ flanking region of *SRY*. While the promoter region lacks an obvious TATA box, it is GC rich and contains putative regulatory response elements upstream of the transcription initiation site. Such elements include tandem recognition sites for the binding protein, Sp1, which may therefore participate in SRY activation. The sequence motif recognised by the SRY protein, AACAAAG, is also present in the region of the transcription start site of the *SRY* gene itself, implying possible autoregulation (Vilain et al. 1992).

4.2 The SRY Protein and Its Targets

Several lines of evidence indicate that testis differentiation requires a threshold level of *SRY* expression within a precise window of development. If this requirement is not met, ovarian differentiation can occur (reviewed in Burgoyne and Palmer 1993, and in Capel 1995). The exact mechanism whereby *SRY* controls Sertoli cell differentiation and therefore testis development is not known. It has been shown that the SRY protein is localised in the nucleus (Poulat et al. 1995), supporting the proposal that *SRY* provides a regulatory sig-

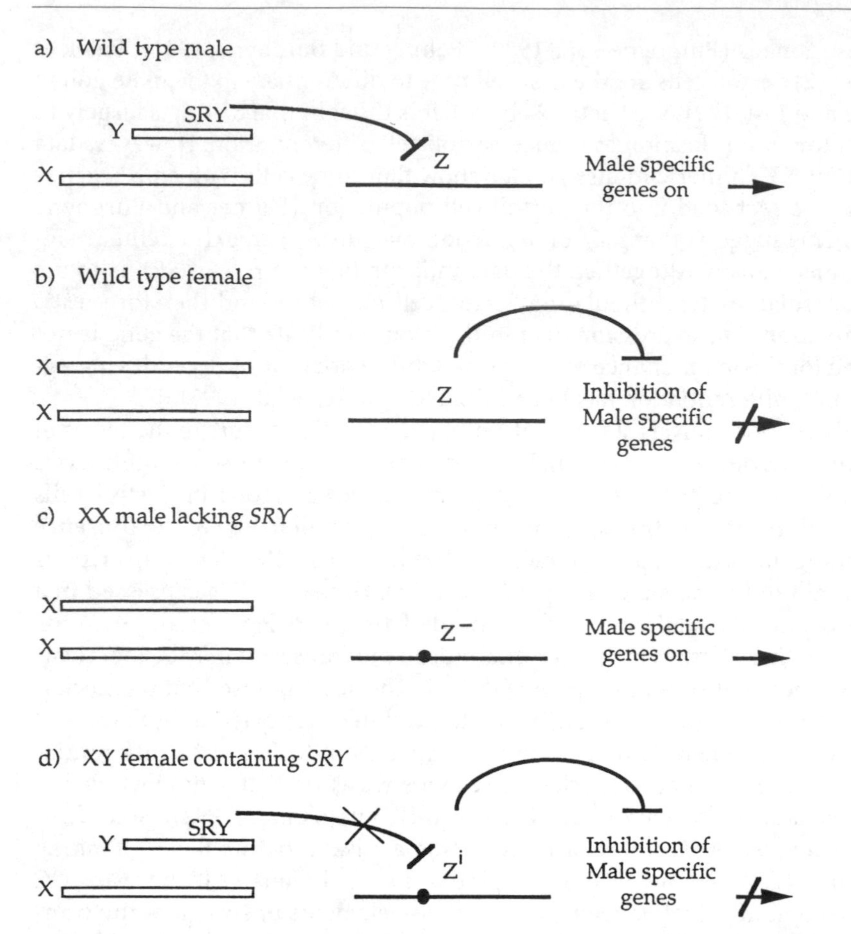

Fig. 2. Regulatory cascade hypothesis for mammalian sex determination. **a** In wild type males, functional *SRY* gene product is present and acts to inhibit *Z*, a repressor of male-specific genes, allowing male development to proceed. **b** In wild type females, *SRY* is absent, *Z* is activated, male-specific genes are inactivated by *Z* and female development occurs. **c** XX individuals lacking *SRY* may be homozygous for a Z mutation (Z^-), rendering it non-functional, it cannot inactivate male-specific genes and male development occurs. **d** In XY females carrying functional SRY, a Z^i mutation renders the *Z* locus insensitive to *SRY* inhibition. As a result, Z^i can repress male-specific genes and female development occurs. Redrawn from McElreveay et al. (1993)

nal that operates at the level of gene transcription. The HMG-box appears to be the critical feature necessary for this function. All but two reported cases of 46XY females with complete gonadal dysgenesis have mutations in the HMG box (see Fig. 2). (The exceptions are a deletion 5′ to the box; McElreavey et al. 1992, and a point mutation 3′ to the box, generating a truncated protein; Tajima et al. 1994). XY sex-reversed patients with mutations in the HMG box show reduced or abolished DNA-binding (Harley et al. 1992).

In vitro studies indicate that the SRY protein binds to linear DNA with sequence specificity (Harley et al. 1992, 1994). The HMG-box domain can induce dramatic bending in the structure of DNA (Ferrari et al. 1992; reviewed Harley and Goodfellow 1994). This may expose or block recognition sites for other regulatory proteins, thereby activating or inhibiting gene expression. It is possible that the HMG box of SRY acts to juxtapose normally distant regions of DNA, generating binding sites for other transcription factors (Giese et al. 1992). SRY may therefore function as a local regulator of chromatin structure rather than having a specific transactivating function typical of classical transcription factors. In some in vitro systems, SRY can act as a transcriptional enhancer (Cohen et al. 1994; Dubin and Oster 1994), but in vivo, it could have either an activating or an inhibiting function (or both). Intriguingly, mouse Sry contains a glutamine-rich putative transactivation domain that transgenic studies have shown to be essential for its sex-determining function, but this domain is lacking in the human protein (Bowles et al. 1999). This observation suggests that SRY, in addition to being poorly conserved, may act in different ways in different mammalian species.

Since the putative consensus recognition sequence for SRY binding is widely distributed throughout the genome, the identification of bona fide SRY target genes has been difficult. No direct in vivo targets for SRY have at present been identified. In vitro studies indicate the presence of SRY binding sites in the promoters of several genes, including *aromatase*, *Anti-Müllerian Hormone* (*AMH*) and *fra-1* (a component of the transcription factor, AP1; reviewed in Schafer and Goodfellow 1996). AMH is the first known product to be secreted from developing Sertoli cells, and in addition to its effect upon Müllerian ducts, it can masculinise embryonic ovaries in vitro (Vigier et al. 1998). AMH might therefore seem a good candidate for regulation by SRY. However, expression of *Sry* begins at least 20 hrs prior to the onset of *AMH* expression in embryonic mouse testis (Münsterberg and Lovell-Badge 1991; Hacker et al. 1995), while in vitro co-transfection experiments reveal SRY-dependent but indirect activation of the *AMH* promoter (Haqq et al. 1994). These observations imply the presence of an intervening gene/s between Sry and AMH in the pathway of male differentiation (see Sect. 6 below). In fact, AMH itself may not be part of the testis determining cascade per se. While AMH is clearly necessary for the downstream process of male sexual differentiation, testis formation is normal in human patients with loss-of-function AMH mutations (reviewed in McElreavey 1995) and in AMH knockout mice (Behringer et al. 1994).

4.3 Is SRY a Negative Regulator?

While *SRY* is clearly sufficient for testis determination, it is not necessary under all circumstances. Human 46XX males have been reported with testicular tissue but with no evidence of *SRY* (Berkovitz 1992, Vilain et al. 1994).

Similarly, in some species of rodents (voles and wood lemmings) males can have normal testes but lack *Sry* (Fredga 1994; Just et al. 1995). If *SRY/Sry* controls testis determination, how might these observations be explained? It has been suggested that *SRY* may repress an inhibitor of male development. McElreavey et al. (1993) postulate the presence of an autosomal locus, *Z*, that switches off male-determining genes. In normal males, *SRY* negatively regulates this locus, allowing activation of male-determining genes and testis development (Fig. 2a). In normal females, *Z* is activated, turning off male-specific genes, thereby permitting female development (Fig. 2b), making *Z*, in effect, an ovary-determining gene. This model could explain XX human males known to lack *SRY*, and XY human females that carry a normal copy of the gene (Pivnick et al. 1992; Vilain et al. 1994). XX males may have a recessive mutation in *Z*, so that it cannot switch off male-determining genes (Fig. 2c). In XY females, a different mutation in *Z* may render it insensitive to *SRY*, so that it inhibits male-determining genes despite the presence of a functional *SRY* gene product (Fig. 2d) (McElreavey et al. 1993). The idea that *SRY/Sry* has a repressor function might partly explain its low level of conservation between species. If the SRY protein inhibits a gene, and if it does so largely through physical mechanisms such as DNA bending, then its primary structure may be less critical and evolution of the gene less constrained.

5 The *SOX9* Gene and Testis Determination

While the isolation of SRY was significant in its own right, it also led to the discovery of an entirely new family of developmental genes, the *SOX* genes (Gubbay et al. 1990; Sinclair et al. 1990). The members of this multigene family share the conserved SRY-like HMG BOX (hence the term *SOX*), and all are thought to operate as regulators of gene expression. There are now over 20 known *SOX* genes, and studies in mice and humans have implicated these genes in a variety of developmental processes (reviewed in Prior and Walter 1996). One of these genes, *SOX9*, is implicated in testis determination.

5.1 Campomelic Dysplasia, Sex Reversal and *SOX9*

The role of *SOX9* in testis determination was discovered due to its association with the bone dysmorphic syndrome, Campomelic Dysplasia (CD). Patients with this syndrome have a number of skeletal and extraskeletal abnormalities, including bowing of the long bones, brachydactyly and micrognathia. While 46,XX individuals with CD have normal ovaries, 75% of 46,XY CD patients are sex-reversed females (Cooke et al. 1985). Tommerup and colleagues (1993) analysed CD patients with translocations of chromosome 17 and were able to

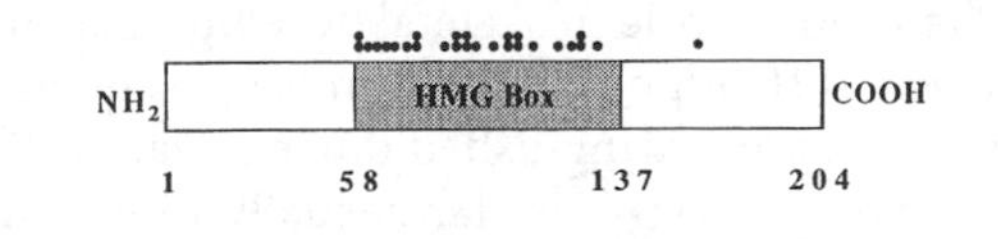

Fig. 3. Comparison between human SRY and SOX9 proteins, showing the positions of sex-reversing mutations (*closed circles*) and, for SOX9, mutations causing campomelic dysplasia but not sex reversal (*open circles*). For SRY, reported mutations are almost all confined to the HMG box, while mutations in SOX9 are more widely distributed. Note the presence of an activation domain in SOX9 but not in SRY. (Reproduced from Cameron and Sinclair, 1997, with permission)

map the the locus causing both CD and sex reversal to 17q24.3-25.1. This positional cloning approach subsequently led to the identification of *SOX9* within the critical region of 17q (Foster et al. 1994; Wagner et al. 1994). Sex-reversed patients with CD have mutations in *SOX9*, implying that this single gene has a role in both skeletal and testis development. In most cases only one allele is mutated, implying that haploinsufficency of *SOX9* is responsible (reviewed in Cameron and Sinclair 1997). As appears to be the case with *SRY*, gene dosage may be critical for normal *SOX9* function. Figure 3 compares the distribution of mutations reported for SRY and SOX9. In sharp contrast to *SRY* mutations, loss-of-function mutations in *SOX9* have been identified throughout the open reading frame of the gene. Most of these mutations result in a protein with a truncated C-terminus, implying that this region is critical for SOX9 function. Indeed, a conserved transactivation domain has been identified within amino acid residues 402 to 509 of the human SOX9 protein (Südbeck et al. 1996; Fig. 2). This suggests that *SOX9* specifies a classic transcription factor capable of activating other genes.

5.2 Embryonic expression of *Sox9*

Only male-to-female sex reversal has been observed in patients with campomelic dysplasia. This implies that mutations in *SOX9* prevent testicular but not ovarian development, and that *SOX9* lies within the testis-determining pathway. In mouse embryos, *Sox9* is expressed in mesenchymal precursors during chondrogenesis, consistent with its proposed role in skeletal development, and in the indifferent gonads of both sexes (10.5dpc). Expression is subsequently upregulated during testis differentiation and extinguished during ovary differentiation (Morais da Silva et al. 1996; Fig. 4). A similar sexually dimorphic expression profile is also seen in the embryonic gonads of chickens and in reptiles with temperature-sensitive sex determination (Kent et al. 1996; Spotila et al. 1998; Western et al. 1999), implying a conserved role for *Sox9* in vertebrate testis development. (In contrast, *SRY* has only been identified in mammals). Immunohistochemical staining has shown that the *Sox9* protein is localised in the nuclei of developing Sertoli cells in the male gonad (Morais da Silva et al. 1996). This suggests that *Sox9* acts as a regulator of Sertoli cell differentiation. Given the importance of Sertoli cells for testis formation, it is understandable that mutations in *SOX9* cause sex reversal.

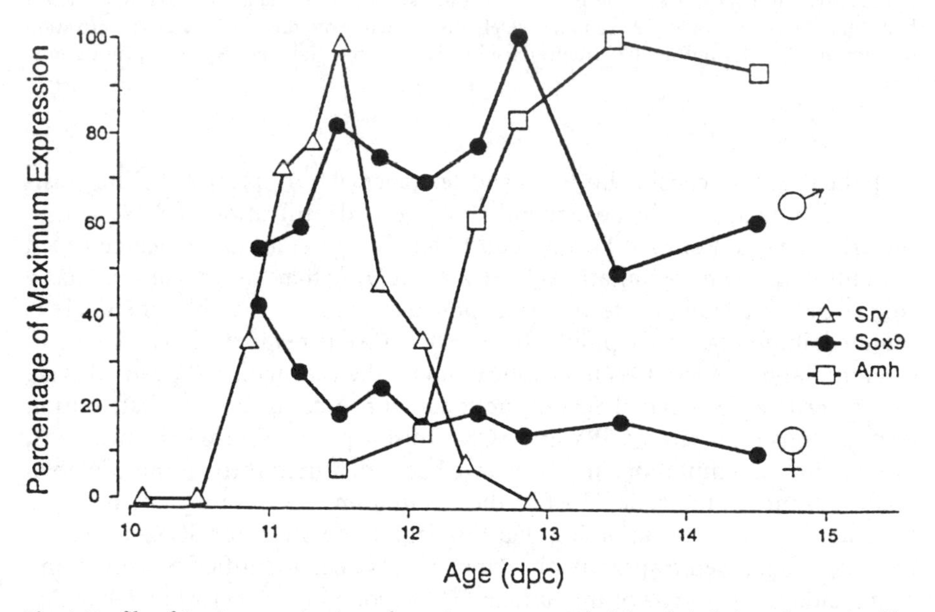

Fig. 4. Profile of *Sox9* expression in embryonic mouse gonads (*closed circles*), as determined by RNase protection assays. *Sox9* expression increases during male development and declines during female development. Expression profiles of *Sry* (*open triangles*) and *Amh* (*open squares*) during testis development are shown for comparison. Morphological differentiation of the testis begins between days 11.5 and 12.5 days *post coitum* (*dpc*). (Reproduced from Morias da Silva et al. 1996, with permission)

5.3 Where is SOX9 placed in the testis-determining cascade?

Both SRY and SOX9 are proteins containing an HMG box (Fig. 3), and both recognise the same binding motif in vitro. Like SRY, SOX9 is thought to act as a transcription factor, but unlike human SRY, SOX9 has an putative transactivation domain at its C-terminus (Fig. 2) (Südbeck et al. 1996; reviewed in Schafer and Goodfellow 1996). SRY could directly regulate *SOX9* expression in the Sertoli cell lineage. If so, *SOX9* could provide the missing link between *SRY* and *AMH*. A conserved SOX-binding sequence is present within the 180bp promoter region of the AMH gene. A binding site for steroidogenic factor 1 (SF1) is also present and recent in vitro and in vivo studies suggest that Sox9 interacts with steroidogenic factor-1 (SF1) to regulate *AMH* gene expression (DeSanta Barbara et al. 1998). In the developing mouse skeletal system, at least one gene known to be regulated by Sox9 is *collagen type 2a* (*Col2a*) (Bell et al. 1997, Ng et al. 1997). While type II collagen is not expressed in embryonic gonads, *Sox9* might regulate the expression of similar extracellular matrix proteins during testis formation (collagen type V, laminin, etc.).

6 Orphan Nuclear Receptors and Sex Determination

The nuclear receptors comprise an extensive family of transcription factors, including the sex steroid and thyroid hormone receptors, and the so-called orphan receptors, for which ligands have not been identified. One of the orphan nuclear receptors, steroidogenic factor 1 (SF1), regulates steroidogenic enzyme gene expression in the gonads and also plays a critical role in gonadogenesis. Another orphan receptor, DAX1, is implicated in dosage-sensitive sex reversal in humans.

6.1 Steroidogenic Factor 1 (SF1)

Steroidogenic factor-1 (SF1) was initially identified as a transcription factor controlling the expression of genes specifying steroidogenic enzymes. DNase-footprinting and gel mobility shift assays showed that SF1 binds to a conserved motif in the proximal promoter region of genes encoding the cytochrome P450 steroid hydroxylases (Morohashi et al. 1992; Lynch et al. 1993). SF1 is expressed in primary steroidogenic tissues, including the adrenal cortex, Leydig cells of the testis and ovarian follicles (reviewed in Parker and Schimmer 1997). Like other nuclear receptors, the human SF1 protein has a conserved zinc finger DNA-binding domain (Fig. 5). This domain shows 100% conservation in all mammals that have been examined. At the C-terminus,

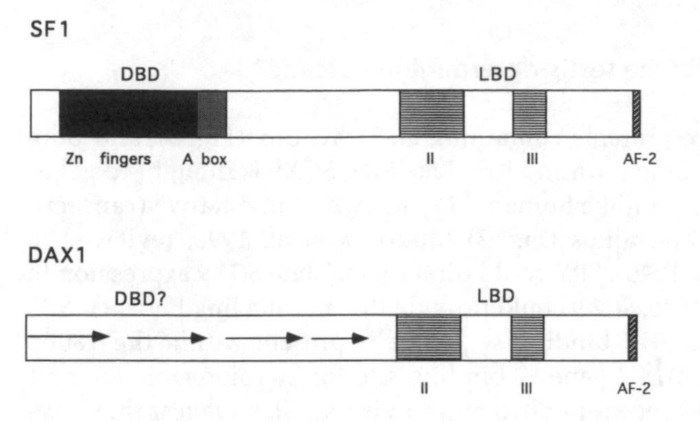

Fig. 5. Schematic representation of the SF1 and DAX1 orphan nuclear receptor proteins. Both proteins have a putative ligand binding domain (*LBD*) and a conserved transactivation domain at the carboxy terminus (*AF-2*). SF1 has a DNA-binding domain (*DBD*) at the amino terminus, comprising a zinc finger region and a non-zince finger A box. In contrast, DAX1 lacks a classic zinc finger region, but has a conserved 67–68 amino acid repeat motif that may represent a DNA-binding domain

there are regions that have homology to domains necessary for ligand binding in the ligand-activated nuclear receptors (Wong et al. 1996). A highly conserved putative transactivation domain called AF2 is present at the end of the C-terminus (Fig. 5). SFI is encoded by a homologue of the *Drosophila Ftz-F1* gene, which regulates expression of the homeobox segmentation gene, *fushi tarazu* (Lala et al. 1992). The mammalian *Ftz-F1* gene contains 8 exons and, in the mouse at least, produces different transcripts via alternative promoter usage and/or 3' exon splicing. These alternative transcipts, called ELPs, have various tissue-specific expression profiles. The SF1 isoform is predominantly expressed in the gonads and adrenal glands (reviewed in Parker and Schimmer 1997).

In addition to its function in mature gonads, SFI plays a role in gonadal differentiation during embryogenesis. Targeted disruption of the *Ftz-F1* gene encoding SF1 produces null mutant mice lacking both gonads and adrenal glands (Luo et al. 1994). In wild type mice, SF1 is first expressed in the undifferentiated gonad at the very earliest stages of urogenital development (9.5dpc) (see section 3.4 above). It continues to be expressed during testis differentiation, while its expression is down-regulated during ovary formation (Ikeda et al. 1994). In the testis, SF1 expression becomes localised in Sertoli and Leydig cells. These data indicate that SF1 has a role in both the formation of the gonadal primordium and in subsequent sexual differentiation. SF1 is also expressed at other sites in the developing embryo, including the hypothalamus and in pituitary gonadotropes, indicating that it functions at several levels of the reproductive axis (Ingraham et al. 1994; reviewed in Parker and Schimmer 1997).

It is unlikely that the importance of SF1 for gonadal development is due to its regulation of steroidogenesis alone, as sex steroids are thought not to be necessary for gonadal sex differentiation in mammals (see Pang et al. 1992). The precise role of SF1 during gonadogenesis remains to be defined, although it appears to act at several steps in the sex-determining cascade. One target for SF1 appears to be AMH. The two proteins are co-expressed in the Sertoli cell lineage during testis formation (Shen et al. 1994) and SF1 activates AMH expression by binding to a conserved regulatory element (Shen et al. 1994; Giuili et al. 1997). Recent in vitro studies have shown that SF1 can synergise with an isoform of WT1 lacking 3 amino acids (WT1-KTS) to activate AMH gene expression (Nachtigal et al. 1998). In these studies, WT1-KTS appeared to function as a coactivator of SF1-mediated *AMH* expression. However, like sex steroids, AMH is not essential for gonad formation per se; the essential role of SF1 in this process must be mediated by activation of other genes. Since it is first expressed prior to Sry, it is possible that SF1 activates Sry expression during testis determination. However, SF1 is also initially expressed in females, while Sry is not.

6.2
DAX1 and Gonadal Differentiation

The discovery of a single dominant male determinant carried by the Y chromosome seems to refute the idea that X chromosome dosage might play a role in mammalian sex determination (Chandra 1985). However, some sex-reversed human females with an XY genotype have a normal copy of *SRY*, but part of the X chromosome (Xp) is duplicated (reviewed in McElreavey et al. 1995). It has been hypothesised that, when duplicated, a gene/s on the X-chromosome interferes with testis differentiation (Ogata et al. 1992). Bardoni et al. (1994) examined a microduplication on the short arm of the X chromosome in a 46,XY female and were able to assign the dosage sensitive sex reversal (DSS) locus to a critical 160kb region at Xp21. The locus for X-linked adrenal hypoplasia congenita (AHC) also maps to this region. Positional cloning led to the identification of the gene responsible for AHC, called *DAX1* (DSS-AHC critical region on the X, gene 1) (Zanaria et al. 1994). While there may be other genes within the DSS region, *DAX1* appears to be responsible for DSS as well as AHC (Muscatelli et al. 1994; Zanaria et al. 1994; Swain et al. 1996). The *DAX1* gene encodes a novel orphan nuclear receptor. The protein is unusual; while it contains a putative ligand-binding domain similar to those of other nuclear receptors, it lacks the usual zinc finger DNA-binding motif (Fig. 5). Instead, it has an amino terminal domain comprising three and one half repeats of a 65–67 amino acid motif rich in alanine and glycine. No other known nuclear receptor has such an arrangement at the N-terminus.

Since deletions in *DAX1* do not affect male development in 46,XY individuals, while duplications do, it is thought that the additional dose of *DAX1* in some way disrupts testis determination. *DAX1* could represent an inhibitor of testis determination that must be repressed by SRY (the Z locus in Fig. 2 above). In this sense, *DAX1* would be an ovary-determining gene. A double dose of *DAX1* may overcome negative regulation by SRY, preventing testis differentiation. Alternatively, the SRY and DAX1 proteins may compete for common target genes, or act as antagonists, with SRY being outcompeted by DAX1 when the latter is present in two doses. The two proteins could compete in the regulation of *Sox9*, for example, Sry having a positive effect and Dax1 a negative effect. A role for *Dax1* in gonadal differentiation has been supported be expression studies in mice. *Dax1* is differentially expressed in male and female embryonic mouse gonads, being downregulated during testis development (Fig. 6; Swain et al. 1996). In males, expression is initially localised in Sertoli cells but later expression is seen in Leydig cells (Ikeda et al. 1996).

Transgenic mice carrying extra copies of *Dax1* can show male-to-female sex reversal, although only in the presence of weak alleles of *Sry* (Swain et al. 1998). Sex reversal has not been achieved in the presence of a "normal" copy of *Sry*. In humans, *DAX1* is likely to be subject to inactivation on the X chromosome, as XXY individuals (Klinefelter's syndrome) are males, not sex-reversed females. Thus, normal males and females could each be expected to have one functional copy of *DAX1*. Furthermore, the marsupial *DAX1* orthologue is

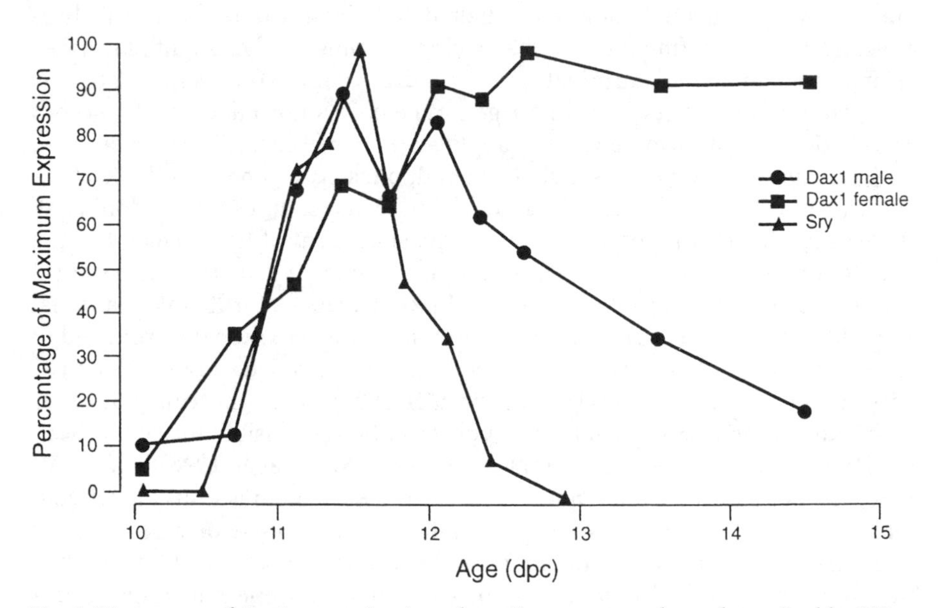

Fig. 6. Timecourse of *Dax1* expression in embryonic mouse gonads, as determined by RNase protection assays. *Dax1* expression declines during male development (*closed circles*) and increases during female development (*closed squares*). Sry expression in males (*closed triangles*) is shown for comparison. Age is given as days *post coitum* (dpc). (Reproduced from Swain et al. 1996, with permission)

autosomal, and therefore does not show gene dosage differences between the sexes. These two observations imply that *DAX1* does not participate in gonadal sex differentiation by a gene dosage mechanism under normal conditions. However, it has been found that facultative X chromosome inactivation occurs in embryonic male mouse gonads at the time of *Sry* expression and testis organisation (Jamieson et al. 1997). The functional significance of this finding is unclear, but it could provide a mechanism of achieving unequal doses of *Dax1* between the sexes, and/or accounting for the downregulation of *Dax1* in male gonads.

The observation that *Dax1* expression is down-regulated during male mouse development (but maintained in females) adds further weight to the proposal that the gene has an anti-testis function or is ovary-determining. However, recent *Dax1* knockout studies in mice have shown that it is not necessary for somatic differentiation of either the testis or the ovary (Yu et al. 1998). Spermatogenesis was imparied in males, however, revealing a role for *Dax1* in the male germ line. Subtle changes in the morphology of steroidogenic cell lineages in both sexes of knockout mice also point to a role for *Dax1* in endocrine function (Yu et al. 1998). Since it is expressed in Sertoli cells, while knockouts effect germ cells, *Dax1* may be necessary for proper Sertoli-germ cell signalling.

Some putative target genes for DAX1 have been defined. The Dax1 protein localises to the nucleus and, like other nuclear receptors, is expected to regulate the expression of specific target genes. Zanaria et al. (1994) found that DAX1 binds retinoic acid response elements (RERs) in vitro, mediating retinoic acid transactivation. This relationship to RERs may or may not play a role during gonadogenesis. If DAX1 inhibits a testis-determining gene, as the human studies indicate, one candidate target might be SF1. Dax1 and SF1 have very similar spatial and temporal expression patterns during embryogenesis (gonads, adrenals, hypothalamus; Ikeda et al. 1996). in vitro co-expression studies have shown that DAX1 can inhibit SF1-mediated transactivation (Ito et al. 1997; reviewed in Swain and Lovell-Badge, 1999). However, the onset of *Dax1* expression occurs after *SF1* expression in embryonic mouse gonads, implying that the former does not directly activate the latter. Could SF1 regulate *Dax1*? A binding site for SF1 has been identified in the 5' flanking region of the *DAX1* gene (Burris et al. 1995). *Dax1* expression is maintained, however, after targetted disruption of the gene encoding SF1 in mice (Ikeda et al. 1996). These observations argue against a direct hierarchical link between the two genes during gonadogenesis. It is possible, however, that the gene products interact; the two receptors could form heterodimers or interact in other ways that influence their putative testis- and ovary-determining functions. It has been proposed that SF1 and Dax1 may interact antagonistically during gonadal development, particularly within endocrine cells, via protein-protein interactions. Nachtigal et al. (1998) have carried out a number of detailed in vitro studies to show that human DAX1 can antagonise the synergistic stimulation of the AMH promoter by SF1 and the –KTS isoform of WT1. Further studies are necessary to confirm the precise role of *DAX1* during gonadal development.

7 Summary: A Genetic Cascade for Testis Determination

Testis determination must involve an ordered cascade of gene expression within the developing gonad. Several components of this cascade have now been identified and their relative positions are shown in Fig. 7. WT1 and SF1 operate early in the cascade, being required for development of the undifferentiated urogenital ridge from intermediate mesoderm. Downstream of these genes, *SRY* and *SOX9* act at a similar level in the pathway to direct Sertoli cell differentiation in XY gonads, hence initiating testis differentiation. When either of these genes is mutated, gonadal sex differentiation is diverted to the female pathway. It is unclear at present whether SRY and SOX9 directly interact or whether they are sequential in the cascade. In addition to its earlier role, SF1 also participates in testis determination. SF1 regulates *AMH* expression in developing Sertoli cells. In doing so, SF1 appears to interact with other factors, possibly WT1 and/or SOX9. SF1 also stimlutes testosterone production in differentiating Leydig cells, and probably has other as yet undefined functions. In the XX gonad, SRY is absent, SOX9 and SF1 are downregulated at the time of ovarian differentiation, while (at least in the mouse) *DAX1* expression is maintained. In mammals, *Dax1* may have an "anti-testis" role in antagonising both SRY and SF1, but it may also be involved in the endocrine function of both sexes and in spermatogenesis.

Gene dosage appears to be an important feature of the sex-determining pathway. A sufficient level of SRY expression within a precise window of time is necessary to trigger testis differentiation. Similarly, the functions of both *SOX9* and the DSS gene (*DAX1*) appear to be dose-related. *SOX9* mutations leading to female sex reversal result from haploinsufficency, while the DSS locus can block SRY activity when duplicated. These observations are reminiscent of the sex-determining systems in the invertebrate models, *Drosophila melanogaster* and *Caenorhabditis elegans*. In these organisms, the dose (number) of X chromosomes relative to autosomes provides the sex-determining switch (reviewed in Cline and Meyer 1996). An ancestral sex-determining system in mammals may have relied primarily upon gene dosage phenomena, but has now come under control of a dominant determinant, *SRY*.

The cascade presented in Fig. 7 is by no means complete. Additional components of the pathway await discovery. A number of human syndromes of unknown etiology include gonadal dysgenesis, raising the possibility that the gene/s involved has a role in the sex-determining pathway. XY sex reversal occurs in Frasier syndrome, for example, and in some cases of Smith-Lemli-

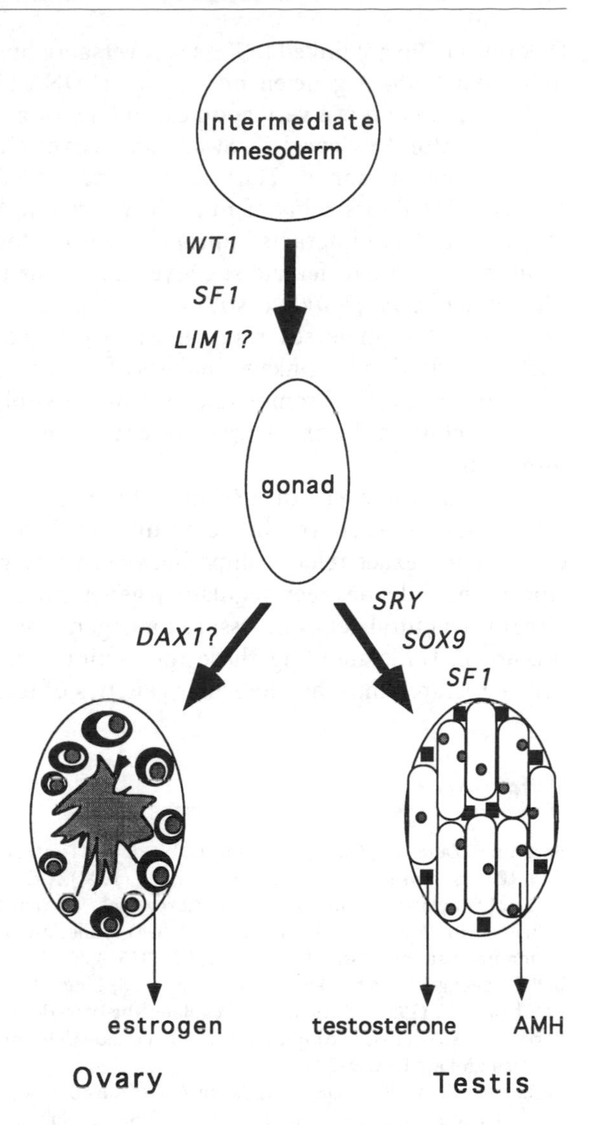

Fig. 7. Gene hierarchy controlling gonadal sex differentiation in mammals. Development of the undifferentiated from intermediate mesoderm requires *WT1* and *SF1* genes, and probably *LIM1*. Testis differentiation involves *SRY*, *SOX9* and *SF1* expression. AMH is then expressed from developing Sertoli cells and testosterone synthesis occurs in interstitial Leydig cells. In females (and in males), *DAX1* may influence endocrine function of the developing gonad. During ovary formation, estrogen is synthesised and released from developing follicles

Opitz syndromes [the former appears to be due to mutations in the *WT1* gene]. While gonadal dysgenesis in these syndromes may be secondary to other developmental abnormalities, it is also possible that mutations in unknown testis-determining genes are involved. One syndrome involving male-to-female sex reversal, atypical ATR-X (α-thalassemia and mental retardation) has recently been ascribed to a novel mutation in a putative DNA helicase, *XH2*. This X-linked gene may therefore participate in the testis-determining cascade (Ion et al. 1996). Several other chromosomes have been identified in humans and mice that appear to carry sex-determining genes.

Deletions in 9p are linked to XY sex reversal in humans, for example (reviewed in Schafer 1995). A gene encoding a novel DNA-binding protein, *DMT1*, maps to this region and is a very good candidate for a dose-dependent testis-determining factor (reviewed in Swain and Lovell-Badge, 1999). In mice, testis-determining autosomal (*Tda*) loci have been identified by analysis of crosses between different strains of mice, in which the Y chromosome (specifically, *Sry*) of one strain interacts improperly with autosomal loci in another strain, resulting in male-to-female sex reversal. In particular, the Y chromosome of *Mus musculus Poschiavinus* strain, $Y^{Pos,}$ causes ovotestis development or complete male-to-female sex reversal when crossed onto certain other genetic backgrounds. Genetic linkage analysis of such crosses shows a high correlation of sex reversal with chromosomes 2, 4 and possibly 5 (Eicher et al. 1996). These findings point to the existence of several autosomal loci involved in sex determination.

Since the discovery of *SRY* in 1990, major advances have been made in uncovering genes controlling testis differentiation and gonadogenesis in general. Yet the exact relationships between these genes remain poorly understood. The links between regulatory genes such as *SRY* and *SOX9* and downstream structural genes necessary for organising the developing testis are also unknown. Understanding these interactions will provide the challenge for future research into the molecular genetics of testis determination.

References

Bardoni B, Zanarai E, Guioli S, Floridia G, Worley K, Tonini G, Ferrante E, Chiumello G, McCabe ERB, Fraccaro M, Zuffardi O, Camerino G (1994) A dosage sensitive locus at chromosome Xp21 is involved in male to female sex reversal. Nat Genet 7:497–501

Behringer RR, Finegold MJ and Cate RL (1994) Müllerian inhibiting substance function during mammalian sexual development. Cell 79:415–425

Bell D, Leung KH, Wheatley SC, Ng LJ, Zhou S, Ling KW, Sham MH, Koopman P, Tam PPL, Cheah KSE (1997) SOX9 directly regulates the type-II collagen gene. Nat Genet 16:174–178

Berkovitz GD (1992) Abnormalities of gonad determination and differentiation. Sem Perinatology 16:289–298

Berta P, Hawkins JR, Sinclair AH, Taylor A, Griffiths B, Goodfellow PN, Fellous M (1990) Genetic evidence equating SRY and the testis determining factor. Nature 348:448–450

Bowles J, Cooper L, Berkman J, Koopman P (1999) Mouse Sry requires a CAG repeat domain for male sex determination. Human Genome Meeting 1999. Brisbane, Australia (Abstr)

Buehr M, Gu S, McLaren A (1993) Mesonephric contribution to testis differentiation in the fetal mouse. Development 117:273–281

Burgoyne PS, Palmer SJ (1993) Cellular basis of sex determination and sex reversal in mammals. In: Hillier SG (ed) Gonadal development and function. Raven Press, New York, pp 17–30

Burgoyne PS, Buehr M, Koopman P, Rossant J, McLaren A (1988a) Cell-autonomous action of the testis-determining gene: Sertoli cells are exclusively XY in XX↔XY chimaeric mouse testes. Development 102:443–450

Burgoyne PS, Buehr M, McLaren A (1988b) XY follicle cells in ovaries of XX↔XY female mouse chimaeras. Development 104:683–688

Burris TP, Guo W, Le T, McCade ERB (1995) Identification of a putative steroidogenic factor-1 response element in the DAX-1 promoter. Biochem Biophys Res Com 214:576–581

Byskov AG (1986) Differentiation of mammalian embryonic gonad. Physiol Rev 66:71–117

Call KM, Glaser T, Ito CY, Buckler AJ, Pelletier J, Haber DA, Rose EA, Kral A, Yeger H, Lewis WH, Jones C, Houssman DE (1990) Isolation and characterisation of a zinc finger polypeptide gene at the human chromosome 11 Wilm's tumor locus. Cell 60:509–520

Cameron F, Sinclair AH (1997) Mutations in *SRY* and *SOX9*: testis-determining genes. Hum Mutat 9:388–395

Capel (1995) New bedfellows in the mammalian sex determination affair. Trends Genet 11: 161–163

Capel B Swain A, Nicolis S, Hacker A, Walter M, Koopman P, Goodfellow PN, Lovell-Badge R (1993) Circular transcripts of the testis-determining gene *Sry* in adult mouse testis. Cell 73:1019–1030

Cattanach BM, Pollard CE, Hawkes SG (1971) Sex-reversed mice: XX and XO males. Cytogenetics 10:318–337

Chandra HS (1985) Is human X chromosome inactivation a sex-determining device? Proc Natl Acad Sci USA 82:6947–6949

Charpentier G, Magre S (1990) Masculinising effect of testes on developing rat ovaries in organ culture. Development 440:839–849

ClineT, Meyer B (1996) Vivé la difference: males vs females in flies vs worms. Ann Rev Genetics 30:637–702

Clépet C, Schafer AJ, Sinclair AH, Palmer MS, Lovell-Badge R, Goodfellow PN (1993) The human SRY transcript. Human Mol Genet 2:2007–2012

Cohen D, Sinclair A, McGovern J (1994) SRY protein enhances transcription of Fos-related antigen 1 promoter constructs. Proc Natl Acad Sci USA 91:4372–4376

Cooke CT, Mulcahy MT, Cullity GJ, Watson M, Sprague P (1985) Campomelic dysplasia with sex reversal; morphological and cytogenetic studies of a case. Pathology 17:526–529

De Santa Barbara P, Bonneaud N, Boizat B, Desclozeaux M, Moniot B, Sudbeck P, Scherer G, Poulat F, Berta P (1998) Direct interaction of SRY-related protein SOX9 and steroidogenic factor 1 regulates transcription of the human Anti-Müllerian hormone gene. Mol Cell Endocrinol 18:6653–6665

Dubin R, Ostrer H (1994) Sry is a transcriptional activator. Molec Endocrinol 8:1182–1192

Eicher EM, Washburn LL (1986) Genetic control of primary sex determination in mice. Annu Rev Genet 20:327–360

Eicher EM, Washburn LL, Schork NJ, Lee BK, Shown EP, Xu X, Dredge RD, Jodeane Pringle M, Page DC (1996) Sex-determining genes on mouse autosomes identified by linkage analysis of C57BL/6J-Y^{POS} sex reversal. Nat Genet 14:206–209

Evans EP, Burtenshaw M, Cattanach BM (1982) Cytological evidence for meiotic crossing-over between the X and Y chromosomes of male mice carrying the sex reversing (Sxr) factor. Nature 300:443–445

Ferrari S, Harley VR, Pontiggia A, Goodfellow PN, Lovell-Badge R, Bianchi ME (1992) SRY, like HMG1, recognises sharp angles in DNA. EMBO J 11:4497–4506

Ford CE, Jones KW, Polani PE, De Almeida JC, Briggs JH (1959) A sex chromosome anomaly in a case of gonadal dysgenesis (Turner's syndrome). Lancet 1:711–713

Foster JW, Dominguez-Steglich MA, Guioli S, Kwok C, Weller PA, Stevanovic M, Weissenbach J, Mansour S, Young ID, Goodfellow PN, Brook JD, Schafer AJ (1994) Campomelic dysplasia and autosomal sex reversal caused by mutations in an SRY-related gene. Nature 372: 525–530

Fredga K (1994) Bizarre mammalian sex-determining mechanisms. In: Short RV, Balaban E (eds) The differences between the sexes. Cambridge University Press, Cambridge, pp 419–432

Fukuda O, Miyayama Y, Fujimoto T, Okamura H (1988) Electron microscopic study in the gonadal development in early human embryos. Okajimas Folia Anat Japon 64:363–384

George FW, Wilson JD (1988) Sex determination and differentiation. In: Knobil E, Neill J, Ewing L, Markert L, Greenwald GS, Pfaff D (eds) The physiology of reproduction. Raven Press, New York, pp 3–26

Giese K, Cox J, Grosschedl R (1992) The HMG domain of lymphoid enhancer factor 1 bends DNA and facilitates assembly of functional nucleoprotein structures. Cell 69: 185–195

Giuili G, Shen W-H, Ingraham HA (1997) The nuclear receptor SF-1 mediates dimorphic expression of Müllerian inhibiting substance, in vivo. Development 124: 1799–1807

Goodfellow PN, Koopman P, Sinclair AH, Harley VR, Hawkins JR, Lovell-Badge R (1993) Identification and characterisation of the mammalian testis-determining factor. In: Reed KC, Graves JAM (eds) Sex chromosomes and sex-determining genes, Harwood Academic Publ, Chur, Switzerland, pp 309–324

Goodfellow P, Darling SM (1988) Genetics of sex determination in man and mouse. Development 102: 251–258

Graves, JAM, Foster, JW, Hampikian, GK, Brennan, FE (1993) Marsupial SRY-related genes and the evolution of the testis-determining factor. In: Reed KC, Graves JAM (eds) Sex chromosomes and sex-determining genes. Harwood Academic Publ, Chur, Switzerland, pp 325–336

Gubbay J, Collignon J, Koopman P, Capel B, Economou A, Münsterberg A, Vivian N, Goodfellow PN, Lovell-Badge R (1990) A gene mapping to the sex determining region of the mouse Y chromosome is a member of a novel family of embryonically expressed genes. Nature 346: 245–250

Hacker A, Capel B, Goodfellow P, Lovell-Badge R (1995) Expression of *Sry*, the mouse sex determining gene. Development 121: 1603–1614

Haqq CM, King C-Y, Ukiyama E, Falsafi S, Haqq TN, Donahoe PK, Weiss MA (1994) Molecular basis of mammalian sexual determination: activation of Müllerin inhibiting substance gene expression by SRY. Science 266: 1494–1500

Harley VR, Goodfellow PN (1994) The biochemical role of *SRY* in sex determination. Mol Reprod Devel 39: 184–193

Harley VR, Jackson DI, Hextall PJ, Hawkins JR, Berkovitz GS, Sockanathan S, Lovell-Badge R, Goodfellow PN (1992) DNA binding activity of recombinant SRY from normal males and XY females. Science 255: 453–456

Harley VR, Lovell-badge R, Goodfellow PN (1994) Definition of a consensus DNA binding site for SRY. Nucleic Acid Res 22: 1500–1501

Ikeda Y, Shen W-H, Ingraham HA, Parker KL (1994) Developmental expression of mouse steroidogenic factor 1, an essential regulator of steroid hydroxylases. Mol Endocrinol 8: 654–662

Ikeda Y, Swain A, Weber TJ, Hentges KE, Zanaria E, Lalli E, Tamai KT, Sassone-Corsi P, Lovell-Badge R, Camerino G, Parker KL (1996) Steroidogenic factor 1 and Dax 1 co-localise in multiple cell lineages: potentil links in endocrine development. Mol Endocrinol 10: 1261–1272

Ingraham H, Lala DS, Ikeda Y, Luo X, Shen W-H, Nachtigal MW, Abbud R, Nilson JH, Parker KL (1994) The nuclear receptor steroidogenic factor 1 acts at multiple levels of the reproductive axis. Genes Dev 8: 2302–2312

Ion A, Telvi L, Chaussain JL, Galacteros F, Valayer J, Fellous M, McElreavey K (1996) A novel mutation in the putative DNA helicase XH2 is responsible for male-to-female sex reversal associated with an atypical form of the ATR-X syndrome. Am J Hum Genet 58: 1185–1191

Ito M, Yu R, Jameson JL (1997) DAX-1 inhibits SF-1-mediated transactivation via carboxy-terminal domain that is deleted in adrenal hypoplasia congenita Molec Cell Biol 17: 1476–1483

Jacobs PA, Strong JA (1959) A case of human intersexuality having a possible XXY sex determining mechanism. Nature 183: 302–303

Jäger RJ, Anvret M, Hall K, Scherer G (1990) A human XY female with a frameshift mutation in the candidate testis determining gene SRY. Nature 210: 352–354

Jamieson RV, Zhou SX, Tan SS, Tam PPL (1997) X-chromosme inactivation during development of the male urogenital ridge of the mouse. Int J Dev Biol 41: 49–55

Jones KW, Singh L (1981) Conserved repeated DNA sequences in vertebrate sex chromosomes. Human Genet 58 46–53

Jost A (1953) Problems of fetal endocrinology: the gonadal and hypophyseal hormones. Rec Prog Horm Res 8:379–418

Jost A (1970) Hormonal factors in the sex differentiation of the mammalian foetus. Philos Trans R Soc Lond 259:119–130

Jost A, Vigier B, Prepin J, Perchellett JP (1973) Studies on sex differentiation in mammals. Rec Prog Horm Res 29:1–41

Just W, Rau W, Vogel W, Akhverdian M, Fredga K, Graves JAM, Lyapanova E (1995) Absence od SRY in species of the vole *Ellobius*. Nat Genet 11:117–118

Kanai Y, Kanai-Azuma M, Kurohmaru M, Yazaki K, Hayashi Y (1995) Effects of extracellular matrix od differentiation of mouse fetal goands in the absence of mesonephros in vitro. Microsc Res Techn 3:437–448

Karl J, Capel B (1998) Sertoli cells of the mouse testis originate from the coelomic epithelium. Dev Biol 203:232–333

Katoh-Fukui Y, Tsuchiya R, Shiroishi T, Nakahara Y, Hashimoto N, Noguchi K, Higashinakagawa T (1998) Male-to-female sex reversal in *M33* mutant mice. Nature 393:688–692

Kent J, Wheatley SC, Andrews JE, Sinclair AH, Koopman P (1996) A male-specific role for *SOX9* in vertebrate sex determination. Development 122:2813–2822

Koopman P, Münsterberg A, Capel B, Vivian N, Lovell-Badge R (1990) Expression of a candidate sex-determining gene during mouse testis determination. Nature 348:450–452

Koopman P, Gubbay J, Vivian N, Goodfellow PN, Lovell-Badge R (1991) Male development of chromosomally female mice transgenic for *Sry*. Nature 351:117–121

Kreidberg J A, Sariola H, Loring J M, Maeda M, Pelletier J, Houssman D, Jaenisch R (1993) WT-1 is required for early kidney development. Cell 74:679–691

Lala DS, Rice DA, Parker KL (1992) Steroidogenic factor 1, a key regulator of steroidogenic enzyme expression, is the mouse homologue of fushi tarazu-factor 1. Mol Endocrinol 6: 1249–1258

Larsen WJ (1993) Human embryology. Churchill Livingstone, New York

Lovell-Badge R (1992a) Testis determination: soft talk and kinky sex. Current Opin Genet Dev 2:596–601

Lovell-Badge R (1992) The role of *SRY* in mammalian sex determination. In: Chadwick DJ, Marsh J (eds) Post-implantation development in the mouse. John Wiley, New York, pp 162–182

Luo X, Ikeda Y, Parker KL (1994) A cell-specific nuclear receptor is essential for adrenal and gonadal development and sexual differentiation. Cell 77:481–490

Lynch JP, Lala DS, Peluso JJ, Luo W, Parker KL, White BA (1993) Steroidogenic factor1, and orphan nuclear receptor, regulates the expression of the rat aromatase gene in goandal tissues. Mol Endocrinol 7:776–786

Magre S, Jost A (1991) Sertoli cells and testicular differentiation in the rat fetus. J Electron Microsc Techn 19:172–188

McCarrey JR, Abbott UK (1979) Mechanisms of genetic sex determination, gonadal sex differentiation, and germ-cell development in animals. Adv Genet 20:217–290

McElreavey K, Vilain E, Abbas N, Herskowitz I, Fellous M (1992) XY sex reversal associated with a deletion 5′ to the SRY HMG box in the testis-determining region. Proc Natl Acad Sci USA 89:11016–11020

McElreavey K, Vilain E, Abbas N, Herskowitz I, Fellous M (1993) A regulatory cascade hypothesis for mammalian sex determination: SRY represses a negative regulator of male development. Proc Natl Acad Sci USA 90:3368– 3372

McElreavey K, Barbaux S, Ion A, Fellous M (1995) The genetic basis of murine and human sex determination: a review. Heredity 75:599–611

McLaren A (1991) Development of the mammalian gonad: the fate of the supporting cell lineage. BioEssays 13:151–156

Merchant H (1975) Rat gonadal and ovarian organogenesis with and without germ cells. An ultrastructural study. Dev Biol 44:1–21

Merchant-Larios H (1976) The onset of testicular differentiation in the rat: an ultrastructural study. Am J Anat 145:319–330
Merchant-Larios H, Moreno-Mendoza N, Buehr M (1993) The role of the mesonephros in cell differentiation and morphogenesis of the mouse fetal testis. Int J Develop Biol 37:407–415
Morais da Silva S, Hacker A, Harley V, Goodfellow P, Swain A, Lovell-Badge R (1996) *Sox9* expression during gonadal development implies a conserved role for the ene in testis differentiation in mammals and birds. Nat Genet 14:62–68
Morohashi K, Honda S, Inomata Y, Handa H, Omura T (1992) A common trans-acting factor, Ad4-binding protein, to the promoters of steroidogenic P-450s. J Biol Chem 267: 17913–17919
Münsterberg A, Lovell-Badge R (1991) Expression of the mouse Anti-Müllerian hormone gene suggests a role in both male anf female sexual differentiation. Development 113:613–624
Muscatelli F, Strom TM, Walker AP, Zanaria E, Récans D, Meindl A, Bardoni B, Guioli S, Zehetner G, Rabi W, Schwarz HP, Kaplans J-C, Camerino G, Meitinger T, Monaco AP (1994) Mutations in the *DAX-1* gene give rise to both X-linked adrenal hypoplasia congenita and hypogonadotrophic hypogonadism. Nature 372:672–676
Ng L-J, Wheatley S, Muscat GEO, Conway-Campbell J, Bowles J, Wright E, Bell DM, Tam PPL, Cheah KSE, Koopman P (1997) SOX9 binds DNA, activates transcription, and coexpresses with type II collagen during chondrogenesis in the mouse. Develop Biol 183:108–121
Ogata T, Hawkins JR, Taylor A, Matsuo N, Hata J, Goodfellow PN (1992) Sex reversal in a child with a 46,X,Yp+ karyotype: support for the existence of a gene(s) located in distal Xp, involved in testis determination. J Med Genet 29:226–230
Page DC, Mosher R, Simpson EM, Fisher EMC, Mardon G, Pollack J, McGillivray B, De La Chapelle A, Brown LG (1987) The sex-determining region of the human Y chromosome encodes a finger protein. Cell 51:1091–1104
Page DC (1988) Is *ZFY* the sex-determining gene on the human Y chromosome? Phil Trans R Soc Lond 322:155–157
Palmer SJ, Burgoyne PS (1991a) The *Mus musculus domesticus Tdy* allele acts later than the *Mus musculus musculus Tdy* allele: a basis for XY sex-reversal in C57BL/6-Y^{POS} mice. Development 113:709–714
Palmer SJ, Burgoyne PS (1991b) XY follicle cells in the ovaries of XO/XY and XO/XY/XYY mosaic mice. Development 111:1017–1020
Palmer SJ, Burgoyne PS (1991c) In-situ analysis of fetal, prepubertal and adult XX↔XY chimaeric mouse testes: Sertoli cells are predominantly, but not exclusively, XY. Development 112: 265–268
Pang S, Yang X, Wang M, Tissot R, Nino M, Manaligod J, Bullock LP, Mason JI (1992) Inherited congneital adrenal hyperplasia in the rabbit: absent cholesterol side-chain cleavage enzyme cytochrome P450 gene expression. Endocrinology 131:181–186
Parker KL, Schimmer BP (1997) Steroidogenic factor 1; A key determinant of endocrine development and function. Endocrine Rev 18:361–377
Patek CE, Kerr JB, Gosden RG, Jones KW, Hardy K, MUggleton -Harris AL, Handyside AH, Whittingham DG, Hooper ML (1993) Sex chimerism, fertility and sex determination in the mouse. Development 113:311–325
Pelletier J, Bruening W, Kashtan CE, Mauer SM, Manivel JC, Stiegel JE, Houghton DC, Junien C, Habib R, Fouser L, Fine RN, Silverman BL, Haber D, Housman D (1991a) Germline mutations in the Wilms' tumor suppressor gene are associated with abnormal urogenital development in Denys-Drash syndrome. Cell 67:437–447
Pelletier J, Schalling M, Buckler A, Rogers A, Haber DA, Housman D (1991b) Expression of the Wilms' tumor gene wt-1 in the murine urogenital system. Genes Dev 5:1345–1356
Pivnick EK, Wachtel S, Woods D, Simpson JL, Bishop CE (1992) Mutations in the conserved domain of *SRY* are uncommon in XY gonadal dysgenesis. Human Genet 90:308–310
Poulat F et al (1995) Nuclear localisation of the testis determining gene product SRY. J Cell Biol 128:737–748

Prior HM, Walter MA (1996) Sox genes: architects of development. Mol Med 2:405–412
Richardson L, Kleinman HK, Dym M (1995) Basement membrane gene expression by Sertoli and peritubular myoid cells in vitro in the rat. Biol Reprod 52:320–330
Rossi P, Dolci S, Albanesi C, Grimaldi P, Geremia R (1993) Direct evidence that the mouse sex-determining gene *Sry* is expressed in the somatic cells of male fetal gonads and in the germ line in the adult testis. Mol Reprod Dev 34:369–373
Schafer A J (1995) Sex determination and its pathology in man. In: Hall JC, Dunlap JC (eds) Advances in Genetics, vol 33. Academic Press, San Diego, pp 275–329
Schafer A, Goodfellow P N (1996) Sex determination in humans. BioEssays 18:955–963
Sharma P, Bowman M, Madden S, Rauscher III FJ, Sukumar S (1994) RNA editing in the Wilm's tumour susceptibility gene. Genes Dev 8:720–731
Shawlot W, Behringer R (1995) Requirement for *Lim1* in head organsier function. Nature 374: 425–430
Shen W-H, Moore CCD, Ikeda Y, Parker KL, Ingraham HA (1994) Nuclear receptor steroidogenic factor 1 regulates the Mullerian inhibiting substance gene: a link to the sex determination cascade. Cell 77:651–661
Sinclair AH (1994) The cloning of *SRY*. In: Wachtel SS (ed) Molecular genetics of sex determination. Academic Press, San Diego, pp 23–41
Sinclair AH, Berta P, Palmer MS, Hawkins JR, Griffiths BL, Smith MJ, Foster JW, Frischauf A-M, Lovell-Badge R, Goodfellow PN (1990) A gene from the human sex-determining region encodes a protein with homology to a conserved DNA-binding motif. Nature 346: 240–244
Singh L, Jones K W (1982) Sex reversal in the mouse (*Mus musculus*) is caused by a recurrent nonreciprocal crossover involving the X and an aberrant Y chromosome. Cell 28:205–216
Spotila LD, Spotila JR, Hall SE (1998) Sequence and expression analysis of WT1 and Sox9 in the red-eared slider turtle, Trachemys scripta. J Exp Zool 281:417–427
Südbeck P, Leinhard Schmitz M, Baeuerle PA, Scherer G (1996) Sex reversal by loss of the C-terminal transactivation domain of human SOX9. Nat Genet 13:230–232
Swain A, Zanaria E, Hacker A, Lovell-badge R, Camerino G (1996) Mouse *Dax1* expression is consistent with a role in sex determination as well as in adrenal and hypothalamus function. Nat Genet 12:404–409
Swain A, Narvaez V, Burgoyne P, Camerino G, Lovell-Badge R (1998) Dax1 antagonises Sry action in mammalian sex determination. Nature 391:761–767
Swain A, Lovell-Badge R (1999) Mammalian sex determination: a molecular drama. Genes Dev 13:755–767
Tajima T, Nakae J, Shinohara N, Fujieda K (1994) A novel mutation localised in the 3′ non-HMG box region of the SRY gene in 46,XY gonadal dysgenesis. Hum Mol Genet 3:1187–1189
Taketo T, Saeed F, Nishioka Y, Donahoe PK (1991) Delay of testicular differentiation in the B6.Y^{DOM} ovotestis demonstrated by immunocytochemical staining for Müllerian inhibiting substance. Dev Biol 146:386–395
Taketo T, Saeed J, Manganaro T, Takahashi M, Donahoe PK (1993) Müllerian Inhibiting Substance production associated with loss of oocytes and testicular differentiation in the transplanted mouse XX gonadal primordium. Biol Reprod 49:13–23
Taketo-Hosotani T, Merchant-Larios H, Thau RB, Koide SS (1985) Testicular cell differentiation in fetal mouse ovaries following transplantation into adult male mice. J Exp Zool 236: 229–237
Tommerup N, Schempp W, Meinecke P, Pedersen S, Bolund L, Brandt C, Goodpasture C, Guldberg P, Held KR, Reinwein H, Saugstad OD, Scherer G, Skjeldal O, Toder R, Westvik J, van der Hagen CB, Wolf U (1993) Assignment of an autosomal sex reversal locus (*SRA1*) and campomelic dysplasia (CMPD1) to 17q24.3–q25.1. Nat Genet 4:170–173
Tung PS, Skinner MK, Fritz IB (1987) Cooperativity between Sertoli cells and peritubular myoid cells in the formation of the basal lamina in the seminiferous tubule Ann NY Acad Sci 438: 435–446

Vigier B, Charpentier G, Josso N (1988) Anti-Müllerian hormone sex-reverses the fetal ovary. In: Imura H, Shizume K, Yoshida S (eds) Progress in endocrinology, Vol. 1. Excerpta Medica, Amsterdam, pp 679–684

Vilain E, Fellous M, McElreavey K (1992) Characterisation and sequence of the 5′ flanking region of the human testis-determining factor *SRY*. Methods Molec Cell Biol 3:128–134

Vilain E, le Fiblec B, Morichon-Delvallez N, Brauner R, Dommergues M, Dumez Y, Jaubet F, Boucekkine C, McElreavey K, Vekemans M, Fellous M (1994) SRY-negative XX fetus with complete male phenotype. Lancet 343:240–241

Vogt P, Edelmann A, Kirch S, Henegariu O, Hirschmann P, Kiesewetter F, Köhn F M, Schill WB, Farah S, Ramos C, Hartmann M, Hartschuh W, Meschede D, Behre HM, Castel A, Nieschlag E, Weidner W, Gröneh J, Jung A, Engel W, Haidl G (1996) Human Y chromosome azoospermia factors (AZF) mapped to different subregions in Yq11. Hum Mol Genet 5:933–943

Wachtel SS (1983) H-Y Antigen and the biology of sex determination. Grune and Stratton, New York

Wagner T, Wirth J, Meyer J, Zabel B, Held M, Zimmer JJP, Bricarelli FD, Keutal J, Hustert E, Wolf U, Tommerup N, Schempp W, Scherer G (1994) Autosomal sex reversal and campomelic dysplasia are caused by mutations in and around the SRY-related gene SOX9. Cell 79:1111–1120

Warne GL, Maclean HE, Zajac JD (1993) Genetic disorders of human sex differentiation. In: Reed KC, Graves JAM (eds) Sex chromosomes and sex-determining Genes. Harwood Academic Publ, Chur, Switzerland, pp 57–68

Wartenberg H (1978) Human testicular development and the role of the mesonephros in the origin of the dual Sertoli cell system Andrologia 10:1–21

Wartenberg H, Rodemer-Lenz E, Viebahn C (1989) The dual Sertoli cell system and its role in testicular development and in early germ cell differentiation (prespermatogenesis). Reprod Biol Med 1989:44–57

Western PS, Harry JL, Graves JA, Sinclair AH (1999) Temperature-dependent sex determination; up-regulation of Sox9 expression after commitment to male development. Dev Dynamics 214:171–177

Wong M, Ramayya MS, Chrousos GP, Driggers PH, Parker KL (1996) Cloning and sequence analysis of the human gene encoding steroidogenic factor 1. J Mol Endocrinol 17:139–147

Yu R, Ito M, Saunders T L, Camper SA, Jameson JL (1998) Role of Ahch in gonadal development and gametogenesis. Nat Genet 20:353–356

Zanaria E, Muscatelli F, Bardoni B, Strom TM, Guioli S, Guo W, Lalli E, Moser C, Walker AP, McCabe RRB, Meitinger T, Monaco AP, Sassone-Corsi P, Camerino G (1994) An unusual member of the nuclear hormone receptor superfamily responsible for X-linked adrenal hypoplasia congenita. Nature 372:635–641

The Sertoli Cell-Germ Cell Interactions and the Seminiferous Tubule Interleukin-1 and Interleukin-6 System

B. Jégou, J. P. Stéphan, C. Cudicini, E. Gomez, F. Bauché, C. Piquet-Pellorce, and A.M. Touzalin

1 Organisation of the Testis

The testis consists of convoluted seminiferous tubules embedded in a connective tissue matrix, called the interstitium. The seminiferous tubule, in which spermatozoa are produced, comprises the germ cells, the non-proliferating Sertoli cells and the peritubular cells, which surround the Sertoli and germ cells. The seminiferous epithelium consists of continually dividing germ cells and Sertoli cells, which stop dividing during puberty. The interstitium contains a mixture of blood and lymph vessels, nerves, fibroblast cells, macrophages, lymphocytes, occasionally mast cells, and Leydig cells. The last produce androgens crucial for differentiation of the embryonic male reproductive organs, sexualisation of the hypothalamus and brain, and for the formation and control of accessory sex organs, secondary sexual characteristics and spermatogenesis.

1.1 Spermatogenesis

Spermatogenesis is the process by which spermatozoa are formed from the most immature germ cells. This process involves three major developmental phases: (1) the proliferative or mitotic phase in which the primitive germ cells, the spermatogonia undergo several mitotic divisions; (2) the meiotic phase in which spermatocytes, undergo two consecutive divisions (meiosis) to produce the haploid spermatids; and (3) the spermiogenic phase, or spermiogenesis, involving the conversion of spermatids into spermatozoa. Germ cells of the same stage of differentiation are connected by intercellular bridges during each phase of spermatogenesis. This makes possible the exchange of signals and the equilibration of expressed genes, post-meiotic genes in particular. New spermatogonial divisions are initiated before any one round of spermatogenesis is completed, so these three phases of spermatogenesis occur simul-

GERM-INSERM U435, Université de Rennes I, Campus de Beaulieu, 35042 Rennes, Bretagne, France

Results and Problems in Cell Differentiation, Vol. 28
McElreavey (Ed.): The Genetic Basis of Male Infertility

taneously in the tubules, leading to the daily production of tens of millions to billions of spermatozoa, depending on the mammalian species.

In any given segment of the tubule, several generations of germ cells, developing simultaneously, succeed one another in contact with the Sertoli cells, from the base to the apex of the epithelium. The development of each generation of germ cells is strictly synchronized with that of the others. This leads to the formation of defined cell associations called stages of the seminiferous cycle. The succession of a complete series of these stages in a given area of the tubule constitutes the seminiferous epithelial cycle. Fourteen stages have been characterized in the rat (Leblond and Clermont 1952). In contrast with most mammals, including rat, multiple stages of the cycle can be seen in a single tubule cross section in humans and a few other primates (Clermont 1963; Chowdhury and Marshall 1980). In humans, six stages of the seminiferous epithelial cycle have been defined (Clermont 1963). The various stages of the cycle occur in succession along the tubule, either linearly, in most mammals (Courot et al. 1970), or helically, in several primates including humans (Schulze and Rehder 1984). This progression of stages is known as the spermatogenetic wave. The helical nature of the human spermatogenetic wave has recently been called into question (Johnson 1994).

1.2 The Sertoli Cell

The Sertoli cell extends from the innermost layer of the basement membrane lining the seminiferous tubule towards the lumen. It also has cytoplasmic processes which envelop the associated germ cells (de Kretser and Kerr 1988). Tight junctions are formed between Sertoli cells, baso-laterally, during postnatal development of the testis. This is the principal anatomical sign that the Sertoli cells are potentially mature (Jégou 1992, 1993). Due to its position in the seminiferous epithelium, the Sertoli cell is unique in being able to communicate with all germ cell generations, the myoid cells, and also, via its base, with the cells of the interstitium (Fawcett 1975; Jégou 1992).

1.3 The Interstitial Tissue

There are various patterns of interstitium organisation in mammalian species (Fawcett 1973). However, the principal location and function of this tissue is the same for all mammalian species. The general organisation of the interstitial tissue, with its extensive vascularization and Leydig cell association with blood and lymphatic vessels that facilitate the diffusion of steroids, is consistent with it having an endocrine function (de Kretser and Kerr 1988). The location of the interstitium is also consistent with a paracrine function because the products of the interstitial cells (primarily Leydig cells) have access to the seminiferous tubules, either via long spaces in the lymph vessels, as in rat, or via diffusion through the loose connective tissue encountered in other species such as humans, in which the lymph vessels are smaller and located at some distance from the tubule wall.

2 Endocrine Regulation of Testicular Function

The endocrine control of testicular function involves interactions between the central nervous system, particularly the hypothalamus and the anterior pituitary gland, and the testis itself. Endocrine regulation of the testis is mediated primarily by two pituitary hormones under the control of the hypothalamic gonadotropin-releasing hormone (GnRH): the luteinising hormone (LH), which stimulates Leydig cells, and the follicle-stimulating hormone (FSH), that acts on Sertoli cells. Leydig cells produce testosterone which inhibits LH, whereas Sertoli cells secrete inhibin/follistatin which inhibit FSH and activin which stimulates FSH. Both FSH and testosterone are required for normal qualitative and quantitative spermatogenesis (Sharpe 1994; Weinbauer and Nieschlag 1997).

3 Paracrine Regulation of Testicular Function

The existence of communication between cells within the testis was suggested by the first histological observations of the testis (Jégou et al. 1992), Sertoli (1865) himself suggested that Sertoli cells were nursing cells for germ cells. Bouin and Ancel (1903) studied the testes of cryptorchid pigs and postulated the production of "nutritive materials" by the "interstitial gland" (interstitial cells) destined to act on the "seminal gland" (seminiferous tubules). Subsequently, testosterone was the first intratesticular regulatory factor to be identified, when it was shown, in the 1940s, to maintain spermatogenesis in hypophysectomised animals (Steinberger 1971). Apart from the role and action of testosterone, it was not until the end of the 1970s and the beginning of the 1980s, that the concept of paracrine regulation of spermatogenesis acting together with endocrine regulation became accepted. However, although we can clearly represent the endocrine system, we do not yet have enough information to draw a clear picture of the testicular paracrine system of spermatogenesis regulation. Little is known of the paracrine system because anatomy of the testis is extremely complicated, rendering the exploration of the cell to cell communication network extremely difficult. Furthermore, unlike the hormones of the endocrine system, many molecules are involved in the paracrine system and they are pleiotropic and often have redundant activity and interactions among themselves.

Extensive reviews have been published in recent years on the paracrine regulation of testicular function (Skinner 1991; Jégou 1993; Jégou and Sharpe 1993; Parvinen 1993; Sharpe 1993). Therefore, rather than producing another general review, we will focus on the seminiferous tubule interleukin-1 (IL-1) interleukin-6 (IL-6) system, which demonstrates the way in which scientists have approached the study of new paracrine molecules, their targets, regulation and putative functions.

3.1 The Place of the IL-1/IL-6 System in the Sertoli Cell-Germ Cell Communication Network

3.1.1 *IL-1 and IL-6*

IL-1 is a family of polypeptides with a wide range of biological activities produced after infection, injury, or antigenic challenge (Dinarello 1991; Cavaillon 1996a). IL-1 was originally detected in culture media of blood macrophages, but is synthetised by a large variety of haematopoietic and non-haematopoietic cells. When reaching the circulation, IL-1 acts like an hormone, inducing a broad spectrum of systemic changes in neurological, metabolic, haematological and endocrine systems (Dinarello 1991). When produced locally, within a given organ, it may act as a paracrine factor, also eliciting a broad range of cellular responses, including regulation of cell division and differentiation.

The term IL-1 encompasses at least three molecules: the biologically active IL-1α and IL-1β, both with a relative molecular mass (Mr) of 17 kDa and the natural antagonist of IL-1 receptors, IL-1Ra, which has a Mr of 23–26 kDa. There are two types of receptors, type I (IL-1RI) and type II (IL-1RII). The IL-1RII receptor is considered to be a decoy receptor, the function of which is to prevent unwanted effects of IL-1. IL-1α and IL-1β recognise both forms of the receptors, but IL-1α, like IL-1Ra, preferentially binds to IL-1RI, which is responsible for signal transduction, with a much higher affinity than IL-1RII. In contrast, IL-1β binds to IL-1RII with a higher affinity than to IL-1RI (Martin and Falk 1997).

Interleukin 6 (IL-6), like IL-1, is a multifunctional cytokine with a Mr of 25–30 kDa, causing growth stimulation or inhibition and cell differentiation in various organs (Yamasaki et al. 1989; Cavaillon 1996b). IL-6 also acts on a large number of cell types. An IL-6 receptor (IL-6R) is expressed by many types of cells consistent with the pleiotropic actions of IL-6. This receptor consists of two structures, an 80 kDa (gp80) protein which binds IL-6, and a 130 kDa (gp130) protein, common to other cytokine receptors of the IL-6/LIF family. In the absence of binding to gp80 signal transduction is impossible (Cavaillon 1996b).

3.2 Sertoli Cell-Germ Cell Interactions

Both structural and functional aspects are involved in the dialogue between Sertoli cells and germ cells, the two major cellular components of the seminiferous tubule. Three types of structure are involved (Jégou 1993): (1) those involved in cell attachment, movement and shaping, such as the desmosome-like and ectoplasmic specialisations that interconnect Sertoli cells with peritubular cells, and with germ cells; (2) those that fulfill these functions but are also involved in the transfer of molecules and materials from Sertoli cells to germ cells and vice versa (spermatogonial processes, tubulobulbar complexes, spermatid processes and gap junctions belong to this category); and (3) those strictly involved in transfer, such as residual bodies, which are the cytoplasm of elongated spermatids that detach at spermiation and undergo phagocytosis by Sertoli cells.

3.2.1 *The Role of the Sertoli Cell*

The location of the Sertoli cells within the tubule is such that they can supply each germ cell generation with the physical and chemical signals required for division, differentiation and metabolism. It is thought that, in addition to synchronising the development of germ cells across the tubule, Sertoli cells are involved in the maintenance of the wave of spermatogenesis along the length of the tubule (Jégou 1993).

An essential function of the Sertoli cell is to control the entry of substances within the seminiferous tubule. This control is partly mediated by specialised junction complexes, the tight or occluding junctions between Sertoli cells, which constitute the Sertoli barrier, a major component of the blood-testis barrier (Ploën and Setchell 1992). The Sertoli cell barrier forms during postnatal testicular development and one of its functions is to prevent a number of substances present in testicular blood and lymph from penetrating the seminiferous epithelium. This barrier also separates the seminiferous epithelium into a basal compartment containing spermatogonia and early primary spermatocytes, and an adluminal compartment containing differentiated primary spermatocytes, secondary spermatocytes and the various steps of haploid spermatids. The Sertoli cell barrier creates a unique microenvironment essential for normal meiosis and spermiogenesis. The reciprocal control of Sertoli and germ cells is illustrated by the simultaneous breakdown and reconstruction of the Sertoli cell barrier, during the meiosis phase, in which clusters of spermatocytes migrate from the basal to the adluminal compartment of the tubule (de Kretser and Kerr 1988; Jégou 1992).

Sertoli cells also secrete the tubule fluid and a large number of peptides, proteins and steroids required by germ cells for the various phases of spermatogenesis (Bardin et al. 1988; Jégou 1992). The tubule fluid is essential for the nutrition of germ cells, and the transport of signals across the seminiferous epithelium, and along the tubule. It is also required for the transport of spermatozoa within tubules and from the tubules to the rete testis and thence to the caput epididymis. The various Sertoli cell substances can be classified as: factors involved in germ cell division, differentiation and metabolism; transport or binding proteins; proteases; components of the extracellular matrix; energy metabolites; and components of the Sertoli cell membrane and junction complexes (Jégou 1992, 1993). The compartmentalisation of the seminiferous epithelium results in the bidirectional secretion of these products (Byers et al. 1993a). The Sertoli cell substances preferentially secreted towards the apex of the cell and the tubule lumen are thought to be particularly required by haploid cells, spermatozoa and epithelial cells. The other substances secreted toward the basal compartment and the interstitium may be directed towards early germ cells, myoid and Leydig cells.

3.2.2
The Action of Germ Cells

It is clear that Sertoli cells assist germ cells thoughout their development. However, it is also now recognised that changes in the composition of germ cell complement, germ cell size and shape, movements of the germ cells in the tubule, and the production of various factors by these cells also markedly affect Sertoli cell morphology and function by feedback regulation (Jégou 1993).

Various pathways exist by which germ cells exert their control on Sertoli cells. The production of germ cell factors that bind to specific receptors on Sertoli cells is the best known example. Other pathways, like those for morphoregulatory mechanisms and membrane molecules also exist. Changes in Sertoli cell shape caused by permanent changes in the nature and composition of the germ cell population physically regulate transcription, via changes in the contacts between the cytoskeleton and the nuclear matrix of the Sertoli cell (Jégou et al. 1992; Byers et al. 1993b). Furthermore, membrane molecules such as Sertoli cell testins (Zong et al. 1992), which are components of the junction complexes, and liver regulating protein (Gérard et al. 1994), which is produced by early primary spermatocytes and Sertoli cells are also involved in germ cell-Sertoli cell cross-talk. There is also transfer of germ cell materials, such as cellular portions from spermatogonia and late spermatids to Sertoli cells (Jégou 1993). The effect of the spermatogonial processes on Sertoli cell function is unknown. However, there is evidence that the phagocytosis of residual bodies may trigger a cascade of events resulting in changes in the spermatogenic process (see Sect. 3.6.1).

3.3 Sources of Seminiferous-Tubule IL-1 and IL-6

3.3.1 *Tubular IL-1*

In 1987, Khan et al. discovered that homogenates of whole rat testis, isolated seminiferous tubules, testicular cytosol, and culture media conditioned by seminiferous tubules, contain high concentrations of IL-1 with Mr of 17 to 20 kDa. This cytokine had an isoelectric point of 5–6, suggesting that it was IL-1α. Sertoli cells or germ cells were proposed as the most likely source of the cytokine (Khan et al. 1987). Sertoli cells were also suggested to be a source of IL-1α by Syed et al. (1988). Interleukin-1-like factors were detected in human testis cytosol (Khan et al. 1988). Attempts were made to determine the cellular source of IL-1α within the tubule, by isolating and culturing Sertoli cells, peritubular cells and germ cells. No IL-1 activity was detected in peritubular cell and germ cell culture media. However, bioactive IL-1 was detected in media from Sertoli cells isolated from 35 and 45-day-old rats (Gérard et al. 1991), but not in Sertoli cell media from 10 or 20-day-old animals (Khan et al. 1987; Gérard et al. 1991). The Sertoli cell IL-1 activity was specifically abolished by an IL-1α antiserum (Gérard et al. 1991). Approximately twice as much Sertoli cell IL-1α was produced in vitro at 45 days than at 35 days. The age-related increase in IL-1 production observed in cultured Sertoli cells by Gérard et al. (1991) was also observed in cultured seminiferous tubules by Syed et al. (1988). IL-1 production in rats begins at the onset of spermatogenesis and reaches a plateau by day 60 to 90, when spermatogenesis is fully developed and the animals are fertile. These early studies implicated Sertoli cells as the origin of IL-1α in the seminiferous tubule. However, Haugen et al. (1994), found that germ cells are also involved in the production of this cytokine. They found that moderate levels of IL-1α mRNA (2.2 kb) are present in Sertoli cells from 35 and 45-day-old rats, whereas high levels of this transcript are present in both pachytene spermatocytes and round spermatids from 32-day-old rats. In 44-day-old animals, high levels of the IL-1α transcript were still detected in pachytene spermatocytes, but markedly lower levels were detected in early spermatids. Furthermore, immunoblot analysis showed that IL-1α proteins of 17 kDa (mature form) and 31 kDa (precursor form) in size are present in pachytene spermatocytes, and in early and late spermatids collected from 35-day-old rats. The same proteins were found in the same cells of 44-day-old animals, but at lower concentrations. Therefore, there may be two sources of IL-1α in the tubule: the Sertoli cells and the germ cells. Sertoli cell IL-1 production increases with age, whereas that of germ cells seems to decline. Consistent with this, Jonsson et al. (1996, 1998) used in situ approaches to clearly demonstrate that Sertoli cells are by far the most essential source of IL-1α in the seminiferous tubules of adult rats. In humans, Sertoli cells also synthesise IL-1α mRNA and produce immunoreactive and biologically active IL-1 protein (Cudicini et al. 1997a). We have also found that isolated human germ cells produce IL-1α (C. Cudicini and B. Jégou, unpubl. results). Leydig cells also produce IL-1 in mice, rats and humans, but they essentially produce IL-1β (Wang et al. 1991; Lin et al. 1993; Xiong and Hales 1994; Cudicini et al. 1997a). Testicular macrophages have also been shown to produce IL-1β (Kern et al. 1995; Hayes et al. 1996).

3.3.2
Sertoli Cell IL-6

IL-1 induces IL-6 production in monocytes and the pituitary (Tosato and Jones 1990; Spangelo et al. 1991, respectively). Therefore, the presence of IL-6 in the testes was expected. Syed et al. (1993) used a specific bioassay and a specific monoclonal anti-IL-6 antibody to demonstrate that rat Sertoli cells, but not germ cells or peritubular cells, produce this cytokine. IL-6 mRNA has also been found in rat Sertoli cells (Okuda et al. 1994) and both IL-6 mRNA and protein were discovered in human Sertoli cells, but not in human germ cells (Cudicini et al. 1997a). Furthermore, Leydig cells synthesise this cytokine in the rat (Bookfor et al. 1994; Okuda et al. 1994) and in humans (Cudicini et al. 1997a).

3.4
Testicular IL-1 and IL-6 Receptors.

Attempts have been made to identify the potential cellular targets of IL-1 and IL-6 within the testis by looking for testicular receptors for these cytokines. An autoradiography technique was used to show that high affinity binding sites for ^{125}I-recombinant human IL-1α were present in mouse testicular interstitial tissue (Takao et al. 1990). IL-1α receptors were also detected in this tissue in mouse by Cunningham et al. (1992) and by Gomez et al. (1997). The mRNA encoding the type I IL-1 receptor (IL-1RI), was detected using in situ hybridisation with ^{35}S-labelled antisense RNA probes. RT-PCR has been used to detect IL-1-RI mRNA and IL-I RII mRNA in isolated rat macrophages and Leydig cells and in human Leydig cells (Gomez et al. 1997). The IL-1RI and IL-1RII mRNAs have also been detected in rat and mouse peritubular cells, Sertoli cells, pachytene spermatocytes and early spermatids, but neither in mouse late spermatids, nor in human germ cells (Gomez et al. 1997). Confirmation of the presence of IL-1-RI mRNA in rat Sertoli cells was recently provided by Wang et al. (1998), who showed that IL-1 stimulated its expression.

IL-6 R transcript have also been detected in rat Leydig cells and Sertoli cells, but not in pachytene spermatocytes or early spermatids, and Sertoli cell IL-6 R mRNA synthesis is stimulated by IL-1, IL-6 and FSH (Okuda and Morris 1994).

3.5
Testicular Effects of IL-1 and IL-6

High concentrations of IL-1-RI are found on the Leydig cell surface (see Sect. 3.4), and IL-1 has been shown to affect Leydig cell proliferation (Khan et al. 1992) and Leydig cell function (Verhoeven et al. 1988; Calkins et al. 1988; Fauser et al. 1989; Warren et al. 1990; Moore and Moger 1991; Hales et al. 1992; Lin et al. 1992; Wang et al. 1995). IL-1, in the interstitial compartment, causes acute inflammation-like changes in the testicular microcirculation of adult rats (Bergh and Söder 1990; Bergh et al. 1996). Furthermore, Sertoli cell œstradiol and transferrin production is regulated in vitro by IL-1 and IL-6 (Bookfor and Schwarz 1991; Khan and Nieschlag 1991; Karzai and Wright 1992; Hoeben et al. 1996), and IL-1 stimulates Sertoli cell IL-6 secretion in a dose-dependent manner (Syed et al. 1993; Riccioli et al. 1995; Cudicini et al. 1997b). IL-6 strongly increases the amount of the intercellular adhesion molecule-1 (ICAM-1) and the vascular adhesion molecule 1 (VCAM2; Riccioli et al. 1995) produced at the Sertoli cell surface.

In addition to its autocrine effects on various testicular cells, IL-1 also exerts strong paracrine effects. For example, IL-1 concentrations in rats are lowest at stage VII of the seminiferous epithelium cycle, and markedly increase at stage VIII, when there is a high level of DNA replication in preleptotene spermatocytes. These data were interpreted by Söder et al. (1991) as an indirect indication that IL-1 is a spermatogonial growth factor. Pöllänen et al. (1989) suggested that IL-1 may be responsible for an increase in DNA synthesis in the spermatogonia of hypophysectomised adult rats. Parvinen et al. (1991) demonstrated in vitro that IL-1 is involved in the stimulation of spermatogonial and preleptotene spermatocyte DNA synthesis. Using the same approach, we found that IL-6 inhibits, in a dose-dependent manner, meiotic DNA synthesis in preleptotene spermatocytes and, to a lesser extent, DNA synthesis in advanced (A3–B) spermatogonia (Hakovirta et al. 1995).

However, male mice lacking a functional IL-1RI were recently found to function normally sexually (Cohen and Pollard 1997). This probably reflects the existence of alternative or rescue systems involving other cytokines which, like IL-1, are implicated in the paracrine control of testicular function.

3.6
The Regulation of Sertoli Cell IL-1 and IL-6 by Germ Cells and the Synchronisation of the Seminiferous Epithelium Cycle

The levels of most Sertoli cell products vary over the seminiferous epithelial cycle, and this has been interpreted as evidence that germ cells control Sertoli cell activity (Parvinen 1982; Jégou 1993). IL-1 levels also differ greatly according to the stage of the cycle considered (Parvinen et al. 1991). Jonsson et al. (1998) have demonstrated that the production of IL-1α mRNA and protein in rat Sertoli cells is dependent on there being a normal interaction between germ cells and Sertoli cells. The same pattern has been observed for IL-6: IL-6 production was minimal at stages VII and VIII and maximal at stages XIII–XIV and I-VI (Syed et al. 1993; Hakovirta et al. 1995). Therefore, germ cells probably exert feedback control over Sertoli cell IL-6 production. However, the precise mechanisms involved in this control are not fully understood. The two possible routes by which germ cells control Sertoli cell IL-1 and IL-6 production are the formation of residual bodies and the secretion of soluble factors.

3.6.1
Residual Bodies and Production of Sertoli Cell IL-1 and IL-6

Sertoli cells from 20-day-old rats contain IL-1α mRNA (N. Gérard and B. Jégou, unpubl. results) but not detectable levels of bioactive IL-1 (Gérard et al. 1991). Small amounts of IL-1 may be released spontaneously by various cell types, but most normal cells, like Sertoli cells from 20-day-old-rats, only produce IL-1 in response to exogenous stimulants. We carried out a series of experiments which demonstrated that IL-1 and IL-6 production is induced or stimulated when Sertoli cells prepared from 20-day-old-rats or older animals are exposed to classical exogenous activators of macrophages, such as the bacterial surface component lipopolysaccharide (LPS), the yeast extract zymosan and silica (Gérard et al. 1992; Syed et al. 1995; Cudicini et al. 1997a,b; Stéphan et al. 1997). IL-1 is also secreted by macrophages during phagocytosis of bacteria or latex beads so we incubated Sertoli cells with latex beads and showed that phagocytosis strongly induces both IL-1 and IL-6 release (Gérard et al. 1992; Syed et al. 1993).

A unique feature of the communication system within the seminiferous tubule is the transfer of cellular portions, residual bodies from late spermatids, to Sertoli cells. These residual bodies, packed with spermatid organelles and rejected by the mature sperm at spermiation, undergo rapid phagocytosis by Sertoli cells (Roosen-Runge 1952; Fawcett 1975; Jégou 1993). Co-culture of Sertoli cells with pachytene spermatocytes and early spermatids which, in previous experiments, were shown to stimulate or inhibit the production of several Sertoli cell factors (Jégou 1993), did not affect IL-1 or IL-6 production. Residual bodies, which had not been reported to have any effect on Sertoli cell activity (ABP, transferin, inhibin, œstradiol) in previous studies, stimulated the production of these two cytokines in a dose-dependent manner (Gérard et al. 1992; Syed et al. 1993). As IL-1 levels rise in the tubule at the same time as residual bodies are formed and undergo phagocytosis (stages VIII-IX; Söder et al. 1991), we have suggested that residual bodies are key inducers of IL-1 production by the Sertoli cell during spermatogenesis (Syed et al. 1993 1995). Further evidence for this has been provided by Wang et al. (1998) who showed that the exposure of Sertoli cells to residual bodies greatly increases Sertoli cell

IL-1 mRNA levels. We have shown that when LPS or residual bodies are incubated with Sertoli cells, the prevention of an increase in IL-1 activity by a specific anti-IL-1 antibody totally abolished the induction of IL-6 observed in the absence of the antibody (Syed et al. 1995). This demonstrates that it is the autocrine action of IL-1 that is responsible for the LPS- or residual body-induced production of IL-6. We have also demonstrated that the IL-1-induced stimulation of Sertoli cell IL-6 production is mediated by the arachidonic acid pathway and, more specifically, via the induction of a lipoxygenase and, therefore, by production of leukotrienes (Syed et al. 1995).

We constructed a model to explain how the seminiferous epithelium cycle may be synchronised in the rat, in which spermiation occurs at stage VIII, when A1 spermatogonia start their first mitotic division, and meiosis begins with the transformation of preleptotene spermatocytes into leptotene spermatocytes (Roosen-Runge 1952). In this model, phagocytosis of residual bodies triggers Sertoli cell IL-1 release which, in turn, stimulates IL-6 production. As IL-1 and IL-6 may both stimulate and inhibit mitotic and meiotic DNA replication (see Sect. 3.5), this sperm release-induced mechanism may regulate germ cell homeostasis during each cycle within the seminiferous epithelium.

3.6.2 *Control of Sertoli Cell IL-1 and IL-6 Production by Germ Cell Cytokines*

Sertoli cell IL-1 and IL-6 regulation by germ cells was further investigated by culturing rat Sertoli cells with cytokines only produced in the seminiferous tubule by germ cells (Stéphan et al. 1997). The cytokines tested were the tumour necrosis factor α (TNFα; De et al. 1993), interferon γ (IFNγ; Dejucq et al. 1995) and nerve growth factor β (NGF β; Ayer-Lelievre et al. 1988). These cytokines had different effects on IL-1 and IL-6 production. NGFβ stimulates IL-6 production, but has no effect on IL-1 production, whereas TNFα stimulates the production of both cytokines. Fifty to 100 U/ml IFNγ stimulates IL-1α, whereas this cytokine has an inhibitory effect on IL-6 production at concentrations of 200 to 400 U/ml (Stéphan et al. 1997).Thus, germ cells, which require IL-1 and IL-6 to regulate their division, may, in turn, regulate Sertoli cell activity via the production of other cytokines.

4 Conclusion

The establishment of intricate systems of cell to cell communication is crucial for the development of all organisms, from the simplest bacterium (Kaiser and Losick 1994), to the most complex. The more elaborate the function of an organ, the more sophisticated its anatomical organization, and the more complex the cell types comprising it are likely to be. Therefore, the extraordinarily sophisticated succession of coordinated cellular events constituting spermatogenesis are founded on an extremely complex intercellular communication network. This is reflected in the unique complexity of the organization of the seminiferous epithelium (Fawcett 1975). The autocrine and paracrine mechanisms regulating testicular function in vitro lead to the formation of quantitatively and qualitatively normal gametes, essential for the reproduction of individuals and the perpetuation of the species.

The complexity of this intercellular system is a major obstacle to the understanding of the physiology and the physiopathology of the male reproductive system. Significant progress has been made over the last 20 years in elucidating germ cell-Sertoli cell interactions. Our understanding of involvement of the IL-1 and IL-6 systems is one of the more important developments. However, much remains to be done to complete our understanding of the intratesticular control of spermatogenesis. Technological progress is required to extend our knowledge further. Unfortunately, experimental gene knockout experiments with IL-1, IL-1 RI and IL-6 have not generated any apparent negative phenotypes in the testis, presumably due to the major redundancies in the action of cytokines. Targeted gene knockout approaches may now be useful to check whether the results obtained in culture and co-culture systems, presented in this review, are relevant to the cellular communications within the intact testis. Other breakthroughs, such as the recent success in transferring germ stem cells from donor rodents to infertile recipient mice, leading to the production of spermatozoa (Brinster and Zimmerman 1994; Avarbock et al. 1996), will certainly provide new approaches for investigating the germ cell-Sertoli cell interaction, increasing our understanding of testicular physiology and pathology.

Acknowledgements. This work was supported by Institut National de la Santé et de la Recherche Médicale, the Ministère de l'Education Nationale, de l'Enseignement Supérieur et de la Recherche, the Fondation Langlois and Région Bretagne.

References

Avarbock MR, Brinster CJ, Brinster RL (1996) Reconstitution of spermatogenesis from frozen spermatogonial stem cells. Nat Med 2:693–696

Ayer-Lelievre C, Olson L, Ebendal T, Hallböök F, Persson H (1988) Nerve growth factor mRNA and protein in the testis and epididymis of mouse and rat. Proc Natl Acad Sci USA 85: 2628–2632

Bardin CW, Cheng CY, Musto NA, Gunsalus GL (1988) The Sertoli cell. In: Knobil E, Neill J (eds) The physiology of reproduction. Raven Press, New York, pp 933–974

Bergh A, Söder O (1990) Interleukin-1β, but not interleukin-1α, induces acute inflammation- like changes in the testicular microcirculation of adult rats. J Reprod Immunol 17: 155–165

Bergh A, Damber JE, Hjertkvist M (1996) Human chorionic gonadotrophin-induced testicular inflammation may be related to an increased sensitivity to interleukin-1. Int J Androl 19: 229–236
Bookfor FR, Schwarz LK (1991) Effects of interleukin-6, interleukin-2 and tumor necrosis factor α on transferrin release from Sertoli cells in culture. Endocrinology 129:256–262
Bookfor FR, Wang D, Lin T, Nagpal ML, Spangelo BL (1994) Interleukin-6 secretion from rat Leydig cells in culture. Endocrinology 134:2150–2155
Bouin P, Ancel P (1903) Recherches sur les cellules interstitielles du testicule des mammifères. Arch Zool Exp Gen 1:437–523
Brinster RL, Zimmerman JW (1994) Spermatogenesis following male germ-cell transplantation. Proc Natl Acad Sci USA 91:11298–11302
Byers S, Pelletier RM, Suarez-Quian C (1993a) Sertoli cell junctions and the seminiferous epithelium barrier. In: Russell LD, Griswold MD (eds) The Sertoli cell. Cache River Press, Clearwater, Florida pp 431–446
Byers S, Jégou B, MacCalman C, Blaschuk O (1993b) Sertoli cell adhesion molecules and the collective organization of the testis. In: Russell LD, Griswold MD (eds) The Sertoli cell. Cache River Press, Clearwater, Florida, pp 461–476
Calkins JH, Sigel MM, Nankin HR, Lin T (1988) Interleukin-1 inhibits Leydig cell steroidogenesis in primary culture. Endocrinology 123:1605–1610
Cavaillon JM (1996a) Interleukin-1. In: Cavaillon JM (ed) Les cytokines. 2ème édn. Masson, Paris, pp 93–117
Cavaillon JM (1996b) Interleukin-6. In: Cavaillon JM (ed) Les Cytokines. 2ème édn. Masson, Paris, pp 183–199
Chowdhury AK, Marshall G (1980) Irregular pattern of spermatogenesis in the baboon (*Papio anubis*) and its possible mechanism. In: Steinberger A, Steinberger E (eds) Testicular development, structure and function. Raven, New York, pp 129–137
Clermont Y (1963) The cycle of the seminiferous epithelium in man. Am J Anat 11:35–51
Cohen PE, Pollard JW (1997) Normal sexual function in male mice lacking a functional type I interleukin-1 (IL-1) receptor. Endocrinology 139:815–818.
Courot M, Hochereau-de Reviers MT, Ortavant R (1970) Spermatogenesis. In: Johnson AD, Gomes WR, Vandemark NL (eds) The testis. I. Development, anatomy and physiology. Academic Press, New York, pp 339–432
Cudicini C, Lejeune H, Gomez E, Bosmans E, Ballet F, Saez J, Jégou B (1997a) Human Leydig cells and Sertoli cells are producers of interleukin-1 and-6. J Clin Endocrinol Metab 82: 1426–1433
Cudicini C, Kercret H, Touzalin AM, Ballet F, Jégou B (1997b) Vectorial production of interleukin 1 and interleukin 6 by rat Sertoli cells cultured in a dual culture compartment system. Endocrinology 138:2863–2870
Cunningham ET Jr, Wada E, Carter DB, Tracey DE, Battey JF, De Souza EB (1992) Distribution of type I interleukin-1 receptor messenger RNA in testis: an in situ histochemical study in the mouse. Neuroendocrinology 56:94–99
De SK, Chen HL, Pace JL, Hunt JS, Terranova PF, Enders GC (1993) Expression of tumor necrosis factor-α in mouse spermatogenic cells. Endocrinology 133:389–396
Dejucq N, Dugast I, Ruffault A, van der Meide PH, Jégou B (1995) Interferon-α and-γ expression in the rat testis. Endocrinology 136:4925–4931
de Kretser DM, Kerr JB (1988) The cytology of the testis. In: Knobil E, Neill J (eds) The physiology of reproduction. Raven Press, New York, pp 837–932
Dinarello CA (1991) Interleukin-1. In: Thomson A (ed) The cytokines handbook. Academic Press, New York, pp 47–82
Fauser BCJM, Galway AB, Hsueh AJW (1989) Inhibitory action of interleukin-1β on steroidogenesis in primary cultures of neonatal rat testicular cells. Acta Endocrinol 120:401–408
Fawcett DW (1973) Observations on the organization of the interstitial tissue of the testis and on the occluding cell junctions in the seminiferous epithelium. Adv Biosci 10:83–99

Fawcett DW (1975) Ultrastructure and function of the Sertoli cell. In: Hamilton DW, Greep RO (eds) Handbook of physiology, vol 5. Williams and Wilkins, Baltimore, pp 21–55

Gérard N, Syed V, Bardin W, Genete N, Jégou B (1991) Sertoli cells are the site of interleukin-1α synthesis in rat testis. Mol Cell Endocrinol 82:R13–R16

Gérard N, Syed V, Jégou B (1992) Lipopolysaccharide, latex beads and residual bodies are potent activators of Sertoli cell interleukin-1α production. Biochem Biophys Res Commun 185: 154–161

Gérard N, Cornu A, Kneip B, Kercret H, Rissel M, Guguen-Guillouzo C, Jégou B (1994) Liver regulating protein (LRP) is a plasma membrane protein involved in cell contact mediated regulation of Sertoli cell function by primary spermatocytes. J Cell Sci 108:917–925

Gomez E, Morel G, Cavalier A, Liénard MO, Haour F, Courtens JL, Jégou B (1997) Type I and type II interleukin-1 receptor expression in rat, mouse, and human testes. Biol Reprod 56:1513–1526

Hakovirta H, Syed V, Jégou B, Parvinen M (1995) Function of interleukin-6 as an inhibitor of meiotic DNA synthesis in the rat seminiferous epithelium. Mol Cell Endocrinol 108: 193–198

Hales DB, Xiong Y, Tur-Kaspa I (1992) The role of cytokines in the regulation of Leydig cell P450c17 gene expression. J Steroid Biochem Mol Biol 43:907–917

Haugen TB, Landmark BF, Josefen GM, Hansson V, Högset A (1994) The mature form of interleukin-1α is constitutively expressed in immature male germ cells from rats. Mol Cell Endocrinol 105:R19–R23

Hayes R, Chalmers SA, Nikolic-Paterson DJ, Atkins RC, Hedger MP (1996) Secretion of bioactive interleukin-1 by rat testicular macrophages in vitro. J Androl 17:41–49

Hoeben E, Van Damme J, Put W, Swinnen JV, Verhoeven G (1996) Cytokines derived from activated human mononuclear cells markedly stimulate transferrin secretion by cultured Sertoli cells. Endocrinology 137:514–521

Hoeben E, Wuyts A, Proost P, Van Damme J, Verhoeven G (1997) Identification of IL-6 as one of the important cytokines responsible for the ability of mononuclear cells to stimulate Sertoli cell functions. Mol Cell Endocrinol 132:149–160

Jégou B (1993) The Sertoli-germ cell communication network in mammals. Int Rev Cytol 147: 25–96

Jégou B, Sharpe RM (1993) Paracrine mechanisms in testicular control. In: de Kretser (ed) The molecular biology of the male reproductive system, chapt 8, Academic Press, New York, pp 271–310

Jégou B, Syed V, Sourdaine P, Byers S, Gérard N, Velez de la Calle JF, Pineau C, Garnier DH, Bauché F (1992) The dialogue between late spermatids and Sertoli cells in vertebrates: a century of research. In: Nieschlag E, Habenicht UF (eds) Spermatogenesis-Fertilisation-Contraception. Schering Foundation Series. Springer, Berlin Heidelberg New York, pp 57–95

Jégou B (1992) The Sertoli cell. In: de Kretser DM (ed) The testis.: Baillière's Clinical Endocrinology and Metabolism, vol 6. Baillière Tindall, London Philadelphia Sydney Tokyo Toronto, pp 273–311

Jégou B (1993) The Sertoli-germ cell communication network in mammals. Int Rev Cytol 147: 25–96

Johnson L (1994) A new approach to study the architectural arrangement of spermatogenic stages revealed little evidence of a partial wave along the length of human seminiferous tubules. J Androl 15:435–441

Jonsson C, Johansson O, Parvinen M, Zetterström RM, Söder O (1996) Expression of IL-1α mRNA and protein in rat Sertoli cell. Miniposter book of the 9th European Testis Workshop on Molecular and Cellular Endocrinology. Miniposter F8, Geilo, Norway

Jonsson C, Fröysa B, Zetterström RH, Söder O (1998) Neighboring germ cells are required for proper interleukin-1α expression in rat Sertoli cells. Miniposter book of the 10th European Testis Workshop on Molecular and Cellular Endocrinology of the Testis. Miniposter C22, Capri, Italy

Kaiser D, Losick R (1994) How and why bacteria talk to each other. Cell 73:873–885
Karzai AW, Wright WW (1992) Regulation of the synthesis and secretion of transferrin and cyclic protein-2/cathepsin L by mature rat Sertoli cells in culture. Biol Reprod 47:823–831
Kern S, Robertson SA, Mau VJ, Maddocks S (1995) Cytokine secretion by macrophages in the rat testis. Biol Reprod 53:1407–1416
Khan SA, Nieschlag E (1991) Interleukin-1 inhibits follitropin-induced aromatase activity in immature rat Sertoli cells in vitro. Mol Cell Endocrinol 75:1–7
Khan SA, Söder O, Syed V, Gustafsson K, Lindh M, Ritzèn E (1987) The rat testis produces large amounts of an interleukin-1-like factor. Int J Androl 10:495–503
Khan SA, Schmidt K, Hallin P, Di Pauli R, De Geyter CH, Nieschlag E (1988) Human testis cytosol and ovarian follicular fluid contain high amounts of interleukin-1-like factor(s). Mol Cell Endocrinol 58:221–230
Khan SA, Khan SJ, Dorrington JH (1992) Interleukin-1 stimulates deoxyribonucleic acid synthesis in immature rat Leydig cells in vitro. Endocrinology 131:1853–1857
Leblond CP, Clermont Y (1952) Definition of the stages of the cycle of the seminiferous epithelium in the rat. Ann N Y Acad Sci 55:548–573
Lin T, Wang D, Nagpal ML, Chang W, Calkins JH (1992) Down-regulation of Leydig cell insulin-like growth factor-I-gene expression by interleukin-1. Endocrinology 130:1217–1224
Lin T, Wang D, Nagpal ML (1993) Human chorionic gonadotropin induces interleukin-1 gene expression in rat Leydig cells in vivo. Mol Cell Endocrinol 95:139–145
Martin MU, Falk W (1997) The interleukin-1 receptor complex and interleukin-1 signal transduction. Eur Cytokine Netw 8:5–17
Moore C, Moger WH (1991) Interleukin-1α-induced changes in androgen and cyclic adenosine 3′,5′-monophosphate release in adult rat Leydig cells in culture. J Endocrinol 129:381–390
Ogawa T, Aréchaga JM, Avarbock MR, Brinster RL (1997) Transplantation of testis germinal cells into mouse seminiferous tubules. Int J Dev Biol 41:111–122
Okuda Y, Morris PL (1994) Identification of interleukin-6 receptor (IL-6R) mRNA in isolated Sertoli and Leydig cells: regulation by gonadotrophin and interleukins in vitro. Endocrine 2: 1163–1168
Okuda Y, Sun XR, Morris PL (1994) Interleukin-6 (IL-6) mRNAs expressed in Leydig and Sertoli cells are regulated by cytokines, gonadotropins and neuropeptides. Endocrine 2:617–624
Parvinen M, Söder O, Mali P, Fröysa B, Ritzén EM (1991) In vitro stimulation of stage-specific deoxyribonucleic acid synthesis in rat seminiferous tubule segments by interleukin-1α. Endocrinology 129:1614–1620
Parvinen M (1982) Regulation of the seminiferous epithelium. Endocr Rev 3:404–417
Parvinen M (1993) Cyclic functions of Sertoli cells. In: Russel LD, Griswold MD (eds) The Sertoli cell. Cache River Press, Clearwater, Florida, pp331–347
Ploën L, Setchell BP (1992) Blood-testis barriers revisited. A homage to Lennart Nicander. Int J Androl 15:1–4
Pöllänen P, Söder O, Parvinen M (1989) Interleukin-1α stimulation of spermatogonial proliferation in vivo. Reprod Fertil Dev 1:85–87
Riccioli A, Filippini A, De Cesaris P, Barbaccia E, Stefanini M, Starace G, Ziparo E (1995) Inflammatory mediators increase surface expression of integrin ligands, adhesion to lymphocytes, and secretion of interleukin-6 in mouse Sertoli cells. Proc Natl Acad Sci USA 92: 5808–5812
Roosen-Runge EC (1952) Kinetics of spermatogenesis in mammals. Ann NY Acad Sci 55:574–584
Schulze W, Rehder U (1984) Organization and morphogenesis of the human seminiferous epithelium. Cell Tissue Res 237:395–407
Sertoli E (1865) Dell'esistenza di particolari cellule ramificate nei canalicoli seminiferi del testicolo umano. Morgagni 7:31–40
Sharpe RM (1993) Experimental evidence for Sertoli-germ cell and Sertoli-Leydig cell interactions. In: Russell LD, Griswold MD (eds) The Sertoli cell. Cache River Press, Clearwater, Florida, pp 391–418

Sharpe RM (1994) Regulation of spermatogenesis. In: Knobil E, Neil JD (eds) The physiology of reproduction. Raven Press, New York, pp 1363–1434

Skinner MK (1991) Cell-cell interactions in the testis. Endocr Rev 12:45–77

Söder O, Syed V, Callard GV, Toppari J, Pöllanen P, Parvinen M, Fröysa B, Ritzén EM (1991) Production and secretion of an interleukin-1-like factor is stage-dependent and correlates with spermatogonial DNA synthesis in the rat seminiferous epithelium. Int J Androl 14: 223–231

Spangelo BL, Jarvis WD, Judd AM, MacLeod RM (1991) Induction of interleukin-6 release by interleukin-1 in rat anterior pituitary cells in vitro: evidence for an eicosanoid-dependent mechanism. Endocrinology 129:2886–2894

Steinberger E (1971) Hormonal control of mammalian spermatogenesis. Physiol Rev 51:1–22

Stéphan JP, Syed V, Jégou B (1997) Regulation of Sertoli cell IL-1 and IL-6 production in vitro. Mol Cell Endocrinol 134:109–118

Syed V, Söder O, Arver S, Lindh M, Khan S, Ritzén EM (1988) Ontogeny and cellular origin of an interleukin-1-like factor in the reproductive tract of the male rat. Int J Androl 11:437–447

Syed V, Gérard N, Kaipia A, Bardin CW, Parvinen M, Jégou B (1993) Identification, ontogeny, and regulation of an interleukin-6-like factor in the rat seminiferous tubule. Endocrinology 132:293–299

Syed V, Stéphan JP, Gérard N, Legrand A, Parvinen M, Bardin CW, Jégou B (1995) Residual bodies activate Sertoli cell IL-1α release which triggers IL-6 production by an autocrine mechanism, through the lipoxygenase pathway. Endocrinology 136:3070–3078

Takao T, Mitchell WM, Tracey DE, De Souza EB (1990) Identification of interleukin-1 receptors in mouse testis. Endocrinology 127:251–258

Tosato G, Jones KD (1990) Interleukin-1 induces interleukin-6 production in peripheral blood monocytes. Blood 75:1305–1310

Verhoeven G, Cailleau J, Van Damme J, Billiau A (1988) Interleukin-1 stimulates steroidogenesis in cultured rat Leydig cells. Mol Cell Endocrinol 57:51–60

Wang D, Nagpal ML, Calkins JH, Chang W, Sigel MM, Lin T (1991) Interleukin-1β induces interleukin-1α messenger ribonucleic acid expression in primary cultures of Leydig cells. Endocrinology 129:2862–2866

Wang D, Nagpal ML, Shimasaki S, Ling N, Lin T (1995) Interleukin-1 induces insulin-like growth factor binding protein-3 gene expression and protein production by Leydig cells. Endocrinology 136:4049–4055

Wang JE, Josefsen GM, Hansson V, Haugen TB (1998) Residual bodies and IL-1alpha stimulate expression of mRNA for IL-1alpha and IL-1 receptor type I in cultured rat Sertoli cells. Mol Cell Endocrinol 137:139–144

Warren W, Pasupuleti V, Lu Y, Platler BW, Horton R (1990) Tumor necrosis factor and interleukin-1 stimulate testosterone secretion in adult male rat Leydig cells in vitro. J Androl 11: 353–360

Weinbauer GF, Nieschlag E (1997) Endocrine control of germ cell proliferation in the primate testis. What do we really know? Adv Exp Med Biol 424:51–58

Xiong Y, Hales DB (1994) Immune-endocrine interaction in the mouse testis: cytokine-mediated inhibition of Leydig cell steroidogenesis. Endocrine J 2:223–228

Yamasaki K, Taga T, Matsuka T, Suematsu S, Hirano T, Kishimoto T (1989) Interleukin-6 and its receptor. In: Fradelizi D, Bertoglio J (eds) Lymphokine receptor interactions. Colloque INSERM/John Libbey Eurotext Ltd, vol 179, pp143–150

Zong SD, Bardin CW, Phillips D, Cheng CY (1992) Testins are localized to the junctional complexes of rat Sertoli and epididymal cells. Biol Reprod 47:568–572

Leydig Cell Function and Its Regulation

M. P. Hedger and D. M. de Kretser

1 Introduction

The Leydig cell is the source of the male sex steroids, or androgens, which are essential for development and maintenance of the male phenotype, the male reproductive tract, and spermatogenesis. Disorders of androgen production or action, although relatively rare, have a profound impact upon development and fertility. In recent years, it has become evident that these steroids also impinge upon the function of other organ systems. Consequently, failure of Leydig cell function can have implications not only for reproductive health, but also for many aspects of general health. Beyond its traditional role as the primary source of androgens, Leydig cells produce other steroids and many non-steroidal factors. Increasingly, it is recognized that these other products play important roles in male reproduction, and have been implicated in vascular and immunological control within the environment of the testis. The non-androgenic functions of the Leydig cell have received comparatively little attention in the past, but are likely to excite far more interest as our understanding of the cellular and molecular environment of the testis develops.

The following review comprises a broad outline of our current knowledge of this essential testicular cell type. It is important to note that most studies on Leydig cell function have been carried out in laboratory rodents, in the boar, which possesses particularly large numbers of Leydig cells, and in Leydig tumor cell lines. Nonetheless, the data obtained from experimental animals and cell lines are largely consistent with the available human data, and where there are substantial differences this will be indicated.

Monash University Institute of Reproduction and Development, Monash Medical Centre, Clayton, Victoria 3168, Australia

Results and Problems in Cell Differentiation, Vol. 28
McElreavey (Ed.): The Genetic Basis of Male Infertility

2 Leydig Cell Morphology and Endocrine Function

2.1 Morphology of the Leydig Cell

The Leydig cell was first described by Franz Leydig (Leydig 1850). In all mammalian species, Leydig cells are found within the interstitial tissue of the testis between the seminiferous tubules, and are characterized as round or ovoid cells with extensive smooth endoplasmic reticulum, mitochondria with tubular cristae, variable numbers of lipid droplets, and a large round centrally-located nucleus with dense peripheral nuclear heterochromatin and a prominent nucleolus (de Kretser 1967; Christensen and Gillim 1969). The cytoskeleton of the Leydig cell comprises microtubules, actin microfilaments, and vimentin intermediate filaments which are characteristic of cells of mesenchymal origin (van Vorstenboch et al. 1984; Russell et al. 1987). The cytoplasm of human Leydig cells frequently contains variable numbers of large crystalloid structures, called Reinke crystals (de Kretser 1967). The morphological features of this cell are characteristic of an active steroid-secreting cell, although confirmation that the Leydig cell itself is the principal source of androgens was not obtained until more than 100 years after its discovery (Christensen and Mason 1965; Hall et al. 1969).

Across species, Leydig cell morphology differs substantially only in terms of variation in the number and volume of cytoplasmic lipid droplets. However, the arrangement and number of Leydig cells in the testis varies dramatically between species: in the human and laboratory rodents there are relatively few clusters of cells, sparsely scattered throughout the loose connective tissue of the intertubular tissue, while the testicular intertubular tissue is almost completely filled by Leydig cells in species such as pigs and many marsupials (Fawcett et al. 1973). The reasons for these remarkable species differences in Leydig cell number and organization remain unknown.

2.2 The Hypothalamo-Pituitary-Leydig Cell Axis

2.2.1 *Hypothalamo-Pituitary Activity and Androgen Secretion*

The development and maintenance of the Leydig cell is under the primary control of the anterior pituitary gonadotropin, luteinizing hormone (LH; Fig. 1). Leydig cells have an absolute requirement for this hormone during development (Mancini et al. 1963; Burger and Baker 1984; Teerds et al. 1989a). In the adult, withdrawal of LH by hypophysectomy, treatment with antagonistic ana-

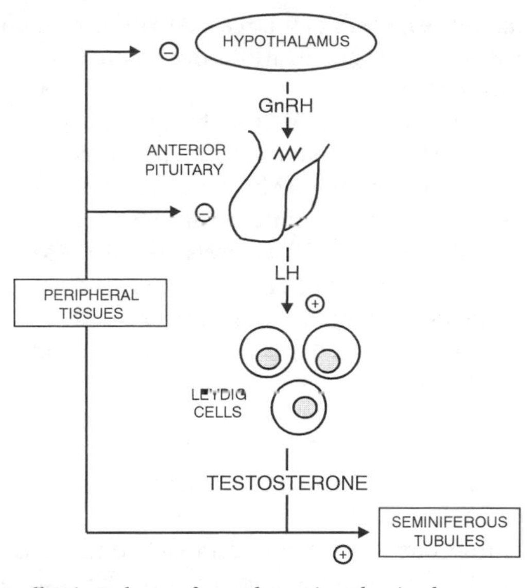

Fig. 1. The hypothalamo-pituitary-Leydig cell axis. Release of gonadotropin-releasing hormone (*GnRH*) in a pulsatile manner from the hypothalamus stimulates release of the anterior pituitary gonadotropin luteinizing hormone (*LH*). The secreted LH travels via the circulation to the testis where it stimulates Leydig cell function and the production of androgens, predominantly testosterone. Testosterone acts locally upon the seminiferous tubules to maintain spermatogenesis, and leaves the testis via the testicular efferent ducts and venous drainage to act upon the male reproductive tract and peripheral androgen-responsive tissues. This system is under negative feedback control via testosterone, which inhibits both GnRH release by the hypothalamus, and LH synthesis and secretion by the anterior pituitary

logues of the hypothalamic releasing hormone which regulates LH-release (gonadotropin-releasing hormone; GnRH), or immunization against GnRH, leads to regression of the Leydig cells, loss of the cellular machinery necessary for steroidogenesis, and undetectable serum androgens (Wing et al. 1985; Keeney et al. 1988, 1990; Duckett et al. 1997a). This regression is reversed by administration of LH, but not by other pituitary hormones.

The primary product of the Leydig cell is testosterone, which is the principal androgen in all mammalian species. Testosterone concentrations in the blood are not static, and measurements taken at different time-points from a single individual will vary considerably. Under the influence of GnRH, LH is released into the blood in a pulsatile fashion (approximately 60–90 min. apart), with corresponding stimulation of testosterone pulses (Santen and Bardin 1973; Boyar et al. 1974; Talbot et al. 1990; Chase et al. 1992). This pulsatility is extremely important, since continuous infusion of GnRH agonists causes a rapid down-regulation of pituitary sensitivity and LH release (Belchetz et al. 1978), and the onset of episodic LH and testosterone in the blood is an early sign of puberty in boys (Boyar et al. 1974). In addition to these regular pulses, there is a circadian rhythm in circulating testosterone levels, with the highest concentrations occurring in the morning (Resko and Eik-Nes 1966; Mock et al. 1978).

Testosterone acts directly upon specific androgen receptors in the Sertoli cells and peritubular cells of the seminiferous tubules to regulate spermatogenic development (Mulder et al. 1975; Skinner and Fritz 1985; Tan et al. 1988; Sun et al. 1990; Bremner et al. 1994; Fig. 1). Within the testis, the concentration of testosterone reaches levels 50- to 100-fold greater than those found in blood plasma. This comparatively high intratesticular testosterone concentration appears to be many times greater than that necessary to maintain spermatogenesis, for reasons that remain to be elucidated (Cunningham and Huckins 1979; Weinbauer et al. 1988).

2.2.2 *The Role of the Testicular Vasculature*

Numerous studies have established that treatments that alter testis blood flow or the formation and volume of the testicular interstitial fluid have serious negative consequences for testosterone production and serum testosterone levels (recently reviewed by Bergh and Damber 1993; Setchell et al. 1994; Desjardins 1996). These consequences are presumed to be due to restriction of the transport of LH, nutrients and other chemical agents essential for Leydig cell function, and to alterations in testosterone secretion via the venous, lymphatic or epididymal drainage. Consequently, there exist complex mechanisms responsible for controlling the microcirculation and fluid dynamics within the testis and the Leydig cell is central to this regulation, having both direct and indirect effects on the pattern of testicular blood flow, formation of interstitial fluid, and microvessel permeability (Setchell and Sharpe 1981; Damber et al. 1987; Maddocks and Sharpe 1989; Sharpe et al. 1991; Collin et al. 1993; Collin and Bergh 1996).

2.2.3 *Androgen Metabolism, Action and Negative Feedback Regulation*

In most androgen-responsive organs, testosterone is converted by the action of 5α-reductase into 5α-dihydrotestosterone (DHT), which has a much greater affinity for the androgen receptor than testosterone itself. The androgen receptor belongs to a family of ligand-activated transcription factors, which includes all steroid hormone receptors, the retinoic acid receptor and the thyroid hormone receptor (Lubahn et al. 1989; Tilley et al. 1989). Androgens act through this receptor to regulate not only spermatogenesis, but also brain development, anabolic effects on tissue development, hair growth and the development of other male characteristics (Bardin and Catterall 1981). In peripheral tissues, the conversion of testosterone to DHT mediates many of the androgen effects, particularly during development when circulating testosterone levels are low (Russell and Wilson 1994). Within the testis, conversion of testosterone to DHT does not appear to be necessary, presumably because of the very high local testosterone concentrations. Nonetheless, the role of 5α-reductase and DHT within the testis, particularly in conditions where androgen production is deficient, remains under investigation (O'Donnell et al. 1996).

Regulation of the hypothalamo-pituitary-Leydig cell axis involves a negative feedback loop whereby androgen inhibits GnRH and gonadotropin secretion, at the hypothalamus and anterior pituitary level, respectively (Fig. 1). There is evidence that this inhibition involves conversion of androgens to estrogens by the aromatizing enzyme complex (aromatase) in addition to androgens (Ryan et al. 1972; Naftolin and Feder 1973). Studies in men have shown that estradiol-17β (E_2) inhibits LH pulse amplitude but not frequency, while non-aromatizable 5α-reduced androgens such as DHT inhibit pulse frequency but not amplitude (Santen 1975, 1977). These observations suggested that testosterone may be converted to E_2 in order to exert its effects via pituitary estrogen receptors, while the effects of androgens on the hypothalamus and GnRH release are mediated via androgen receptors. The importance of aromatization of testosterone to E_2 in the negative feedback regulation of gonadotropins in the male, especially at the pituitary level, has been confirmed recently by studies in mice with targeted disruption of the principal form of estrogen receptor and of the aromatase gene (Fisher et al. 1998; Lindzey et al. 1998).

2.3
Leydig Cell Steroidogenesis

The pathway for androgen synthesis by the Leydig cell has been well-characterized, although some details remain unresolved (reviewed by Hall 1994; Fig. 2). Normally, cholesterol is synthesized de novo from acetate within the smooth endoplasmic reticulum of the Leydig cell (Morris and Chaikoff 1959). Cholesterol also may be derived from circulating lipoproteins, particularly under conditions of elevated or extended steroid production (Quinn et al. 1981; Hedger and Risbridger 1992). The cholesterol may be stored in lipid droplets, primarily as cholesterol esters until needed, or is transported into the mitochondria. Within the inner mitochondrial membrane, the cholesterol side-chain is removed by the cholesterol side-chain cleavage complex, which is an electron transport system comprising ferredoxin, ferredoxin reductase and a cytochrome P450-containing enzyme (P450CSCC). The resulting steroid product is pregnenolone, and the production of pregnenolone is the rate-limiting step of androgen synthesis in the Leydig cell.

Pregnenolone is transported back to the cytosol for conversion to testosterone within the smooth endoplasmic reticulum. Biosynthesis of testosterone

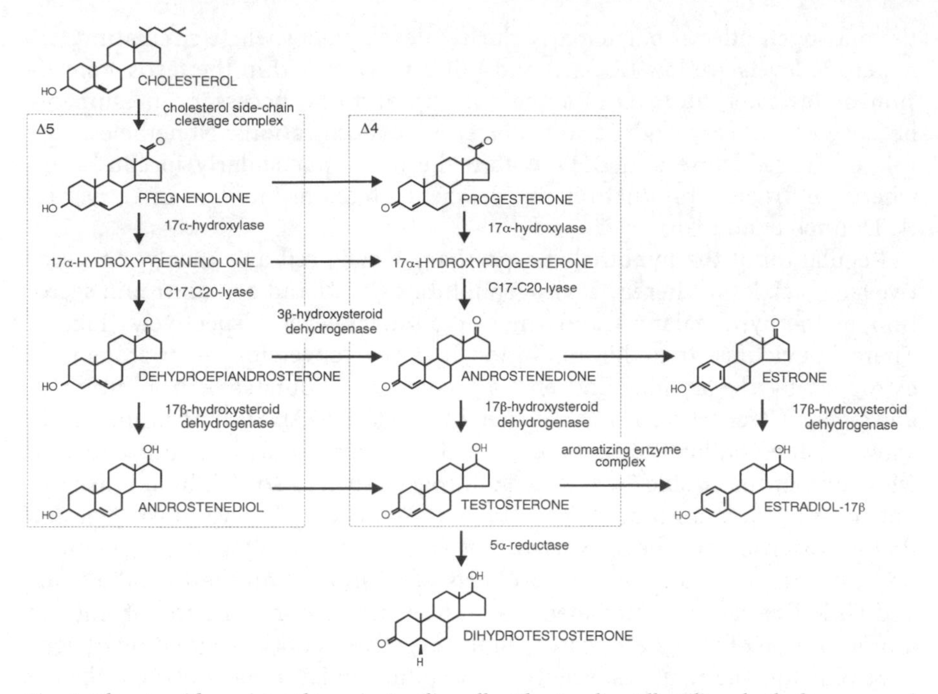

Fig. 2. The steroidogenic pathway in Leydig cells. The Leydig cell utilizes both de novo and stored cholesterol to synthesize androgens via a sequence of steroidogenic enzyme complexes within the smooth endoplasmic reticulum and mitochondria. Adult Leydig cells primarily produce testosterone, which is converted to the 5α-reduced androgen, dihydrotestosterone, in peripheral tissues. The Leydig cell also converts some androgen to estrogens via the aromatizing enzyme complex, although the extent of estrogen production varies widely between species

proceeds via Δ^5 (double bond in the B ring) intermediates (pregnenolone → 17α-hydroxy pregnenolone → dehydroepiandrosterone → 5-androstene-3β,17β-diol) or Δ^4 (double bond in the A ring) intermediates (progesterone → 17α-hydroxy progesterone → 4-androstene-3,17-dione), but in the human the Δ^5 pathway appears to be more important (Huhtaniemi 1977). The conversion of the less active Δ^5 precursors (pregnenolone, dehydroepiandrosterone and 5-androstenediol) to the more active Δ^4 steroids involves oxidation of the 3β-hydroxyl group to a ketone and rearrangement of the double bond by Δ^5-Δ^4 isomerase activity by the action of 3β-hydroxysteroid dehydrogenase (3βHSD; Lorence et al. 1991; Payne and Sha 1991). Studies on the microsomal enzymes have indicated that a single cytochrome P450-containing enzyme complex (P450c17), possessing both 17α-hydroxylase and C17–C20 lyase activity, catalyzes the conversion of pregnenolone/progesterone to their 17α-hydroxy forms and then to dehydroepiandrosterone or androstenedione, respectively (Nakajin et al. 1981). The enzyme 17β-hydroxysteroid dehydrogenase (17βHSD) catalyzes the conversion to androstenediol (Δ^5) and testosterone (Δ^4).

In the adult testis, the Leydig cell also produces estrogens, principally E_2 in the human, via the action of the aromatizing enzyme system (Payne et al. 1976). While estrogen production is relatively minor in the human testis, in some species such as horses and pigs, estrogen levels in the male can be comparatively high (Bedrak and Samuels 1969; Raeside and Renaud 1983; Eisenhauer et al. 1994). The importance of estrogens in male reproductive function, however, remains unclear. Disruption of the principal estrogen receptor isoform causes progressive testicular failure in adult mice, possibly as a result of inadequate fluid resorption in the epididymis (Eddy et al. 1996), however, aromatase 'knock-out' mice display relatively normal testicular function apart from elevated serum testosterone and gonadotropin levels (Fisher et al. 1998). These data suggest that estrogen is involved in regulating fluid dynamics in the male reproductive tract, and in the negative feedback control of pituitary gonadotropin secretion, but is not essential for maintaining reproductive function.

Conversion of androgens to the more potent 5α-reduced androgens, such as DHT, occurs through the action of 5α-reductases at many of its known sites of action such as the male accessory glands, muscle, liver, adipose tissue and in skin. Two isoforms of the 5α-reductase enzyme (type 1 and type 2) have been identified in the human and rat (Russell and Wilson 1994).

2.4 Non-Steroidal Products of the Leydig Cell

In addition to steroids, the Leydig cell produces many non-steroidal factors. Corticotropin-releasing factor (CRF), angiotensin II, serotonin and arginine vasopressin have been implicated in potential short-loop autoregulatory feedback mechanisms controlling LH action (Kasson et al. 1986a; Dufau et al. 1989; Tinajero et al. 1993). In addition, Leydig cells produce the pro-opiomelanocortin (POMC) peptides, including β-endorphin, which indirectly inhibits steroidogenesis via actions on the Sertoli cells (Cicero et al. 1989; He et al. 1991; Chandrashekar and Bartke 1992). Interestingly, Leydig cells are able to secrete the inflammatory cytokines interleukin-1α (IL-1α), IL-1β, and interleukin-6 (IL-6), and production of these cytokines is stimulated by IL-1β, which indicates a role for the Leydig cells in mediating local vascular and inflammatory responses (Wang et al. 1991a; Lin et al. 1993; Boockfor et al. 1994; Okuda et al. 1995; Cudicini et al. 1997). Another inflammatory cytokine that is secreted constitutively by the adult Leydig cell is macrophage-migration-inhibiting factor (MIF; Meinhardt et al. 1996). Curiously, this cytokine does not appear to be under hormonal control, and its functional role in the testis remains unresolved.

Studies have suggested that the heterodimeric gonadal hormone inhibin is produced by the adult Leydig cells in the human and rat (Risbridger et al. 1989; Vliegen et al. 1993), but others have proposed that the adult Leydig cell secretes only an inhibin α-subunit product without the corresponding β-subunit (de Winter et al. 1992). Nonetheless, there is strong evidence that the β-subunit of inhibin is produced by immature Leydig cells and Leydig tumor cell lines, in the form of the β-subunit dimer, activin (Lee et al. 1989; de Winter et al. 1992). Since activin is a regulatory cytokine involved in haematopoiesis and immunomodulation, this implicates yet another Leydig cell product in inflammation and immune responses (Hedger and Clarke 1993; Brosh et al. 1995; Yu and Dolter 1997). Surprisingly, the Leydig cells also produce a number of neural-specific proteins, such as substance P and neural cell adhesion molecules, which has led to the suggestion that at least some Leydig cells may have an ectodermal rather than mesenchymal origin (Schulze et al. 1987; Mayerhofer et al. 1992; Holash et al. 1993).

3 Leydig Cell Development

3.1 Species Variation in Leydig Cell Development

The pattern of Leydig cell development varies significantly between species. In all mammalian species studied to date, there is a fetal generation of Leydig cells, which appears to exert an influence on sexual differentiation (Pelliniemi and Niemi 1969; Reyes et al. 1974; Clements et al. 1976; Huhtaniemi 1977; Waters and Trainer 1996). In those species that have a significant temporal separation between birth and puberty, such as the human, monkey, and to a lesser extent, domestic species such as sheep, pigs and cattle, there is a perinatal period of development of Leydig cells and a commensurate increase in circulating testosterone, which in the human occurs over the first six months of life (Mancini et al. 1963; Faiman and Winter 1971; Kaplan et al. 1976; Nistal et al. 1986). The Leydig cells then regress to redevelop under the influence of LH secretion at puberty from ill-defined mesenchymal precursors. In the rat, the species that has been studied most extensively, spermatogenesis commences shortly after birth with spermatogonial division, and this species has essentially no "prepubertal" period (Lording and de Kretser 1972). In this species, the fetal and adult generations of Leydig cells are poorly separated.

3.2 Fetal, Perinatal and Prepubertal Development of the Leydig Cell in the Human

Development of the fetal testis is genetically determined by sex-determining genes located on the Y-chromosome, in particular the SRY gene (Harley 1993). As the testis develops, hormonal influences take over this process. The development of the fetal testis is primarily directed by androgens, derived from the fetal Leydig cell population. In the male, the indifferent gonad begins to develop into a recognizable testis around the seventh week of gestation, with the formation of the testicular cords containing the primordial germ cells and Sertoli cells. At this time, identifiable Leydig cells differentiate from mesenchymal precursors and fill the tissue between the cords (Pelliniemi and Niemi 1969). These fetal Leydig cells begin to secrete androgens, reaching a peak of activity in the human at 12–14 weeks of gestation. This peak of testosterone production is stimulated by chorionic gonadotropin (hCG) from the placenta and the pattern of testosterone secretion temporally reflects the pattern of hCG in the maternal circulation. Through its action on the LH-receptor, hCG is believed to be the major stimulus of the early differentiation of the Leydig cell and androgen production in the fetus (Reyes et al. 1974; Clements et al. 1976), although fetal LH is present and may also be involved, particularly later in gestation (Hagen and McNeilly 1975; Kaplan et al. 1976; Clark et al. 1984).

After this early period in the human, fetal pituitary activity declines and the Leydig cells regress (Mancini et al. 1963; Nistal et al. 1986; Waters and Trainer 1996). At the time of birth, there is a brief period of activation of the pituitary-testis axis, with an increase in Leydig cell activity and testosterone secretion (Faiman and Winter 1971; Kaplan et al. 1976). After several months, however, gonadotropins and testosterone production return to fetal levels and, until puberty, Leydig cells are only occasionally observed in the testis (Mancini et al. 1963). At puberty, increasing gonadotropin levels, indicated by a gradual increase in the pulses of serum LH (Boyar et al. 1974), bring about the maturation of the Leydig cells once more, resulting in a dramatic increase in testosterone release and the development of the accessory sexual organs and adult male characteristics.

3.3 Pubertal Development of the Leydig Cell

Since the vast majority of studies on the regulation of adult Leydig cell development have been performed in the rat, it is important to outline briefly the postnatal developmental process of the Leydig cell in this rodent species (Fig. 3). The fetal population of Leydig cells persists in the rat testis until after birth, and there is no postnatal surge of development as there is in the human (Kerr and Knell 1988). The fetal Leydig cells are characterized by their appearance, typically possessing large numbers of steroid droplets, and limited steroidogenic capacity. These cells express LH receptors, but produce relatively little testosterone (Habert and Picon 1984; Zhang et al. 1994). The appearance of an adult population of Leydig cells begins around 20 days of age in the rat and there is a period of overlap of the fetal and adult populations (Lording and de Kretser 1972; Kerr and Knell 1988). Although there is evidence that a very small number of fetal Leydig cells may persist in the adult testis (Kerr and

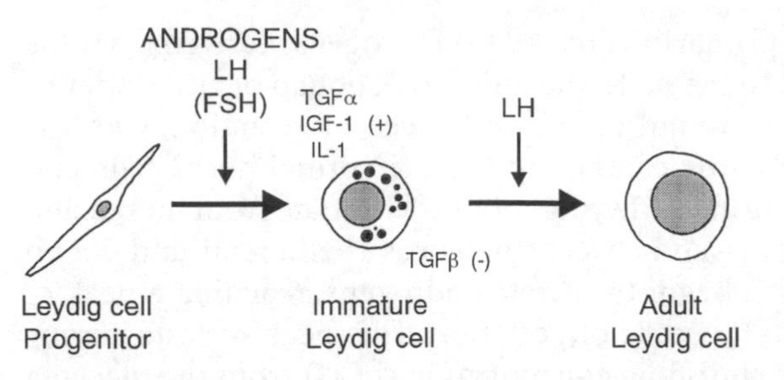

Fig. 3. Development of the Leydig cell and its regulation in the pubertal rat testis. Leydig cells are derived from a pool of luteinizing hormone (*LH*)-responsive mesenchymal-like progenitor cells. Under stimulation by LH, increasing androgen production, and local factors involving the resident testicular macrophages and follicle-stimulating hormone (*FSH*)-regulated Sertoli cell functions, these progenitors differentiate into steroidogenically-active immature Leydig cells. Proliferation of the immature Leydig cells is stimulated by transforming growth factor-α (*TGFα*), insulin-like growth factor-1 (*IGF-1*) and interleukin-1 (*IL-1*), while transforming growth factor-β (*TGFβ*) inhibits the proliferation and development of these cells. Eventually the immature Leydig cells differentiate under the influence of LH into adult-type Leydig cells which have a high steroidogenic capacity, and do not normally undergo further cell division. The general features of this sequence of events are replicated in the recovering adult testis after treatment with the Leydig cell cytotoxin, ethane dimethane sulfonate (EDS)

Knell 1988), it appears unlikely that the fetal Leydig cells contribute in any major way to the adult Leydig cell population (Mendis-Handagama et al. 1987; Hardy et al. 1989).

In the rat and in the human, the mesenchymal progenitor cells of the Leydig cells are characterized by some Leydig cell morphological features, and a limited steroidogenic capacity (Chemes et al. 1985; Hardy et al. 1990; Shan and Hardy 1992). These cells are spindle-shaped, or fibroblastic, in appearance, possess LH receptors (Hardy et al. 1990; Shan and Hardy 1992; Shan et al. 1995), and are particularly numerous in the prepubertal testis (between 14 and 28 days in the rat; Hardy et al. 1989). These progenitors gradually give rise to the immature-type Leydig cells of the pubertal testis, which have proliferative activity and primarily secrete 5α-reduced androgen metabolites (Cochran et al. 1979; Hardy et al. 1990). Division and differentiation of these cells eventually produces the final adult Leydig cell population by 60 days of age in the rat (Hardy et al. 1989). A small number of the progenitor cells persist in the adult testis, presumably to provide a stable stem cell reservoir for the maintenance of the adult Leydig cell population (Jackson et al. 1986; Kerr et al. 1987), since adult Leydig cells, unlike the immature Leydig cells of the pubertal testis, do not normally display significant mitotic activity (Christensen and Gillim 1969; Amat et al. 1986; Keeney et al. 1990). The adult Leydig cell population normally persists throughout life, although in the ageing human testis a gradual decline in testosterone production is associated with progressive atrophy, and possibly loss, of Leydig cells (Kaler and Neaves 1978; Paniagua et al. 1986).

3.4 The Ethane Dimethane Sulfonate Recovery Model

In recent years, many studies have employed the Leydig cell-specific cytotoxin, ethane dimethane sulfonate (EDS), to study Leydig cell development in adult rats (Morris et al. 1986; Jackson et al. 1986; Molenaar et al. 1986; O'Leary et al. 1986; Kerr et al. 1987; Teerds et al. 1989b,c; Gaytan et al. 1994a, b). At a particular sub-lethal dose, the alkylating agent EDS causes destruction of adult Leydig cells by a highly-selective mechanism that still remains poorly understood (Morris et al. 1997). Within 3 days after EDS treatment, all mature Leydig cells are destroyed and there is a period during which the testis is devoid of morphologically identifiable Leydig cells. By about 14–21 days after treatment however, there is a dramatic proliferation of perivascular and peritubular mesenchymal Leydig cell progenitors, which gradually develop into Leydig cells with the general characteristics of fetal-like or immature Leydig cells before transforming by 6 weeks, to an adult-type Leydig cell population. Analysis of this sequence of events has clearly established the fundamental similarities of this regeneration model with the normal process of Leydig cell development (Fig. 3).

3.5 Hormonal Regulation of Adult Leydig Cell Development

Most studies on the hormonal regulation of the development of adult Leydig cells have employed either the pubertal rat testis, or the EDS-treated regenerating rat testis (Fig. 3). Clearly, LH is crucial in both rat models (Molenaar et al. 1986; Teerds et al. 1989a, b), as it appears to be in the human (Mancini et al. 1963; Chemes et al. 1985). While LH may be of only marginal importance in stimulating Leydig cell proliferation during development (Teerds et al. 1989c; Khan et al. 1992a; Gaytan et al. 1994c), it is essential for Leydig cell differentiation from progenitor cells to mature Leydig cells (Chemes et al. 1985; Teerds et al. 1989a; Hardy et al. 1990; Benton et al. 1995). Androgens also appear to be involved in differentiation of the Leydig cell progenitors, possibly by stimulating LH receptor expression and responsiveness to LH (Hardy et al. 1990; Shan and Hardy 1992; Misro et al. 1993; Shan et al. 1995). Hormone replacement studies, following LH-withdrawal by hypophysectomy or other means, and treatment with androgen antagonists, have confirmed that both LH and testosterone are important in maintaining the differentiated Leydig cell population in the adult rat testis as well (Wing et al. 1985; Keeney et al. 1990; Duckett et al. 1997a, b).

In the immature hypophysectomized rat, follicle-stimulating hormone (FSH) increases Leydig cell numbers by stimulating proliferation and differentiation of their progenitor cells, an action which is most likely mediated through the Sertoli cells, as Leydig cells do not have FSH receptors (Kerr and

Sharpe 1985a, b; Teerds et al. 1989c). The effect of FSH on Leydig cell development is evidence that local influences are involved in this development. In contrast to the developing rat testis, FSH is not required for recovery of the Leydig cells after EDS-treatment in mature rats (Molenaar et al. 1986). Nonetheless, local influences appear to be important in the EDS-treated model, since the recovery of the Leydig cell population after EDS-treatment occurs more rapidly in rats in which spermatogenesis has been damaged by the experimental induction of cryptorchidism (O'Leary et al. 1986). Finally, recent data have established the importance of thyroid hormones in controlling the differentiation of the testis during the pubertal period. Delaying testicular development by experimental induction of hypothyroidism during early postnatal life leads to an extended period of Leydig cell proliferation, and a larger number of Leydig cells in the adult testis (Cooke and Meisami 1991; Hardy et al. 1993, 1996).

3.6 Local Factors and Leydig Cell Development

A number of locally produced growth factors have been implicated in Leydig cell development (Fig. 3). Insulin-like growth factor-1 (IGF-1), in particular, appears to be required during pubertal development, in order to stimulate the proliferation of the mesenchymal Leydig cell progenitors and immature Leydig cells (Khan et al. 1992a; Moore and Morris 1993), and the differentiation of Leydig cells from the immature to the mature form (Perrard-Sapori et al. 1987; Chatelain et al. 1991; Gelber et al. 1992). This responsiveness to IGF-1 is stimulated by LH (Lin et al. 1988). Other growth factors implicated in stimulating Leydig cell development are IL-1 and transforming growth factor-α (TGFα; Teerds et al. 1990; Khan et al. 1992a, b). Transforming growth factor-β (TGFβ), on the other hand, inhibits Leydig cell growth in vitro, and expression of the TGFβ family in the testis has been shown to be down-regulated during development within the testis (Mullaney and Skinner 1993; Teerds and Dorrington 1993; Gautier et al. 1994). In general, these growth factors are produced, or are assumed to be produced, by the Leydig cells themselves and/or the adjacent Sertoli cells (Wang et al. 1991a; Lin et al. 1992,1993).

A recent unexpected advance in the study of local regulation of Leydig cell development was the observation that testicular macrophages are also involved. Differentiation and maturation of Leydig cells is inhibited in the testes of pubertal and EDS-treated rats which have been depleted of macrophages by treatment with the liposome-encapsulated cytotoxin, dichloromethylene diphosphonate (Gaytan et al. 1994a, b,d;1995). In the adult testis, the Leydig cell and macrophage are closely associated in clusters and close structural specializations develop between the two cell types during pubertal development (Miller et al. 1983; Hutson 1992). The function of these specializations remains uncertain, but they provide a physical linkage between the two cells types within the interstitial tissue, and potentially mediate and/or facilitate direct cell-to-cell communication (see reviews by Hutson 1994; Hedger 1997). The mechanisms involved in this control remain to be discovered. Although both IL-1β and TGFβ are produced by the macrophage, this usually occurs only during inflammation and the resident macrophages of the testis are poorly responsive to inflammatory stimulation (Kern et al. 1995; Hayes et al. 1996).

4
Molecular Regulation of the Leydig Cell

4.1
Luteinizing Hormone and the LH Receptor

In the adult, LH acts at multiple levels to stimulate steroidogenesis and to maintain normal Leydig cell function. In vitro, LH exerts immediate effects on protein synthesis, protein phosphorylation, and steroid synthesis, and has longer-term effects on transcription of the steroidogenic enzymes and the intracellular structures important for steroidogenesis (Hall 1994; Payne and Youngblood 1995; Dufau et al. 1997). In the absence of LH in vivo, there is a rapid decline in testosterone secretion by the Leydig cell and a gradual regression of the Leydig cells with loss of cytoplasmic volume and the intracellular structures associated with steroidogenesis, although Leydig cell numbers are only marginally affected (Keeney et al. 1988; Keeney and Ewing 1990; Russell et al. 1992; Duckett et al. 1997a). The LH-receptor is located on the Leydig cell surface within the plasma membrane and binding of LH to its receptor leads to aggregation and rapid internalization of the receptor-ligand complex (de Kretser et al. 1969; Clayton 1996). More recent data have implicated the extracellular matrix, and proteoglycans in particular, in the process of LH-receptor interaction at the cell surface. Treatments which disrupt proteoglycan formation or activity inhibit LH-stimulated testosterone production by adult Leydig cells, but have no effect on basal production, while addition of the proteoglycan heparin to cultured Leydig cells stimulates steroidogenic activity (McFarlane et al. 1996).

The LH receptor belongs to the seven-transmembrane helix-G protein coupled receptor family and possesses a very long (64 kDa) extracellular domain which comprises the LH binding site (McFarland et al. 1989). The action of this hormone involves activation of a membrane G protein complex coupled to adenyl cyclase and protein kinase A-mediated protein phosphorylation (Fig. 4); however, other signal transduction pathways also appear to be involved, including Ca^{2+}, phospholipids and protein kinase C, and arachidonic acid metabolites (Nikula and Huhtaniemi 1989; Cooke et al. 1992). The key steps of the steroidogenic pathway which are acutely regulated by LH action

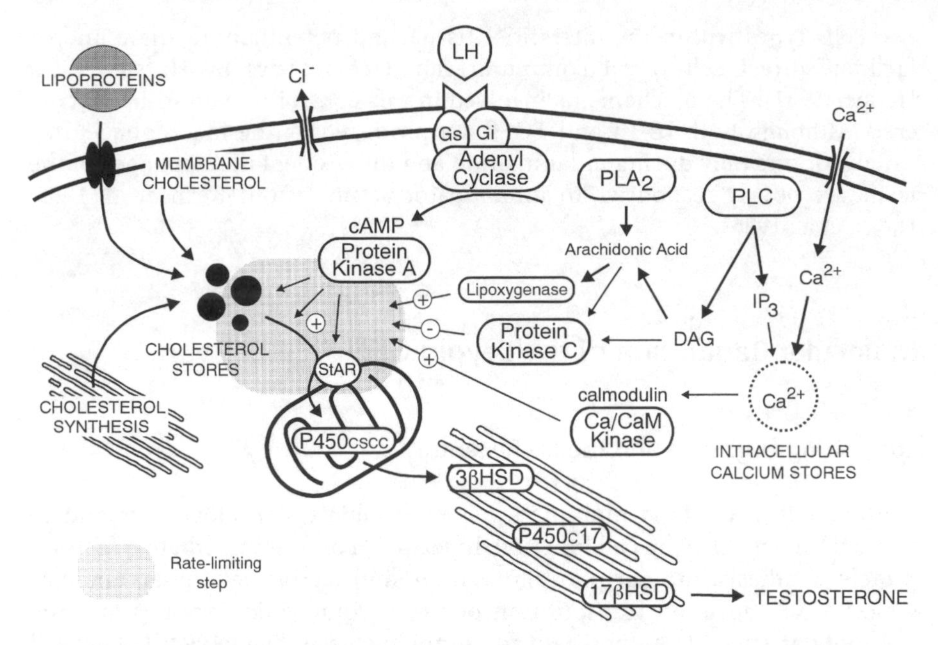

Fig. 4. Acute regulation of Leydig cell steroidogenesis by luteinizing hormone (*LH*). LH acts upon its specific surface receptor linked to a G-protein regulated adenyl cyclase, which produces cAMP, and activates protein kinase A-induced protein phosphorylation. This leads to mobilization of cholesterol from the plasma membrane or by hydrolysis of stored cholesterol esters (which may be synthesized de novo in the smooth endoplasmic reticulum, or derived from circulating lipoprotein), transport of cholesterol to the mitochondria via the cytoskeleton, and synthesis of the steroidogenesis acute regulatory protein (*StAR*). In turn, StAR transfers the cholesterol to the inner mitochondrial membrane where it is converted to pregnenolone by the cholesterol side-chain cleavage complex (*P450CSCC*). This sequence of events is the rate-limiting step in androgen production. Once produced, the pregnenolone passes to the smooth endoplasmic reticulum and is converted to testosterone via the actions of the microsomal enzymes 3β-hydroxysteroid dehydrogenase (*3βHSD*), 17α-hydroxylase/C17–C20 lyase (*P450C17*), and 17β-hydroxysteroid dehydrogenase (*17βHSD*). The response to LH also involves an increase in cytosolic Ca^{2+} via influx through plasma membrane channels and the release of intracellular Ca^{2+} stores, as well as efflux of chloride ions. The mechanism of Ca^{2+} action at least in part involves a calcium/calmodulin (*Ca/CaM*) protein kinase. There is evidence that activation of phospholipase C (*PLC*) leads to production of inositol triphosphate (IP_3), which stimulates the release of stored intracellular Ca^{2+}, and diacylglycerol (*DAG*). The latter either directly, or after conversion to arachidonic acid, activates protein kinase C, which has an inhibitory effect on testosterone production. Activation of phospholipase A_2 (PLA_2), also produces arachidonic acid, but conversion of arachidonic acid via the lipoxygenase pathway to leukotriene intermediates has a stimulatory effect on steroidogenesis. Note that many of the details of this complex regulation remain incompletely understood, or are subject to ongoing discussion. Moreover, in addition to these acute actions on steroidogenesis, LH also regulates the expression of the steroidogenic enzymes, and maintains the cellular machinery essential to steroid production

are the mobilization of stored cholesterol, transport of cholesterol into mitochondria and the resulting activity of the cholesterol side-chain cleavage complex (Hall and Young 1968).

The LH receptor is expressed by the Leydig cells during fetal development, in early postnatal life, and from puberty through to the adult (Shan and Hardy 1992; Zhang et al. 1994). Receptor expression is up-regulated by low levels of LH. Conversely, high concentrations of LH down-regulate the receptor in the adult, but not in the fetal or immature testis (Dufau 1988). Studies in the rat, confirmed in other species, have shown that the gene for the LH receptor is large (over 70 kb) and consists of 11 exons and 10 introns, although a single exon (exon 11) encodes the entire transmembrane and cytoplasmic domains of the receptor (Tsai-Morris et al. 1991). Multiple LH receptor messenger RNA transcripts which lack either the transmembrane domain or the transmembrane and cytoplasmic domains as the result of alternative splicing of the gene with partial deletions of exon 11 or complete deletions of exon 9, have been identified in several species, including the human (Shan and Hardy 1992; Wang et al. 1991b; Zhang et al. 1994). Additional transcript heterogeneity is provided by the presence of alternate polyadenylation domains in the 3′ region of the gene (Hu et al. 1994). The inherited condition of familial male precocious puberty (testotoxicosis), which is characterized by Leydig cell hyperplasia, the early onset of spermatogenesis, and low LH levels, results from activating mutations of the LH receptor within the transmembrane or cytoplasmic domains (Shenker et al. 1993). Conversely, mutations that result in an inactive receptor result in Leydig cell hypoplasia and male pseudohermaphroditism (Kremer et al. 1995).

4.2 Intracellular Signalling Events and cAMP

The intracellular regulation of Leydig cell steroidogenesis is complex, and many details remain to be fully understood. Primarily, LH exerts its effects through the LH-receptor/G-protein membrane complex via activation of adenyl cyclase to produce cAMP from ATP (Dufau 1988; Cooke et al. 1992; Payne and Youngblood 1995; Fig. 4). The elevated cAMP activates protein kinase A to induce protein phosphorylation within the cell, and the subsequent downstream events including new protein synthesis. Stimulation of steroidogenesis by LH in the Leydig cell is blocked by inhibitors of protein synthesis such as cycloheximide, indicating that new protein synthesis is essential (Irby and Hall 1971; Mendelson et al. 1975). The rapidity of the steroidogenic response of the Leydig cell to LH and cAMP, and the fact that it is uncertain whether new transcription is necessary for the immediate stimulation of steroidogenesis, suggest that the acute stimulation of testosterone synthesis may involve translation of pre-existing mRNA (Mendelson et al. 1975; Cooke et al. 1979).

4.3
Cholesterol Mobilization

4.3.1
Cholesterol Transport Proteins

Several proteins have been implicated in playing a regulatory role in acute steroidogenic regulation via their ability to influence the intracellular transport of cholesterol. Sterol carrier protein-2 (SCP_2), a steroid-carrying protein found in the liver and steroidogenic tissues has been suggested to promote the transport of cholesterol from the lipid droplets to the inner mitochondrial membrane (Chanderbhan et al. 1982). However, studies on its role in hormone-induced steroidogenesis have been equivocal: acute stimulation with LH causes a change in the intracellular distribution of SCP_2 but has no effect on the levels of this protein in the Leydig cell (van Noort et al. 1988; van Haren et al. 1992). Sterol activating peptide (SAP), a 30 amino acid peptide isolated from rat adrenal cells and Leydig tumor cells, has been found to stimulate production of pregnenolone in cholesterol-loaded adrenal mitochondria (Pedersen 1984, 1987). Although SAP is stimulated by tropic hormones in a number of steroidogenic cells, including the Leydig cell, its mechanism of action and functional role in steroidogenesis remains unclear.

The peripheral-type benzodiazepine receptor (PBR) complex is found at high levels in the outer mitochondrial membrane of steroidogenic tissues and is associated with regions of close contact between the inner and outer membranes of the mitochondria (Moran et al. 1990; Kinnally et al. 1993), suggesting an important role for this complex in the transfer of cholesterol across the mitochondrial intermembrane space (Papadopoulos 1993). In MA-10 mouse Leydig tumor cells, the endogenous ligand of this receptor, diazepam binding inhibitor (DBI) and its agonists stimulated steroidogenesis, whereas blockade of the action of DBI using antisense oligonucleotides prevented hormonally-stimulated steroidogenesis in these cells (Boujrad et al. 1993). However, DBI is not acutely regulated by tropic hormone stimulation in steroidogenic cells and has a comparatively long-half life (Brown et al. 1992), while the recovery of both DBI and PBR in response to hormonal stimulation of hypophysectomized rats is much slower than the recovery of steroidogenic activity (Cavallaro et al. 1993).

4.3.2
Steroidogenic Acute Regulatory Protein

In contrast to DBI/PBR, the mitochondrial steroidogenic acute regulatory protein (StAR) is rapidly up-regulated in response to LH and cAMP and this process is cycloheximide-sensitive (Clark et al. 1995). Moreover, transfection and expression of the StAR protein in MA-10 cells caused increased steroid production in the absence of any other steroidogenic stimulus (Clark et al. 1994). Most importantly, individuals who lack the functional StAR gene exhibit a condition called congenital lipoid adrenal hyperplasia, which is characterized by the inability to convert cholesterol to pregnenolone (Lin et al. 1995). Mechanistic studies indicate that StAR facilitates the transfer of the cholesterol across the mitochondrial membranes (Stocco and Clark 1997; Fig. 4), although many of the details of the mechanism of action remain to be elucidated. Altogether, these data strongly support the hypothesis that StAR mediates the acute stimulation of steroidogenesis by trophic hormones in steroidogenic cells, including LH-stimulated androgen synthesis by the Leydig cell.

4.3.3
The Role of the Cytoskeleton

In addition to the role of either soluble or membrane-bound proteins, there is evidence that the cytoskeleton is important in steroidogenesis. In the Leydig cell, inhibitors of actin microfilament formation have been shown to inhibit the response to LH by preventing cholesterol transport to the mitochondria (Hall et al. 1979). The further observation that both lipid droplets and mitochondria are tightly associated with the intermediate filaments of steroidogenic cells, including Leydig cells (Almahbobi et al. 1993), suggests that the cytoskeleton may be involved in the physical bringing together of lipid droplets and mitochondria during steroid synthesis.

4.4
Regulation of the Steroidogenic Enzymes

4.4.1
Chronic Regulation of the Steroidogenic Machinery

Acutely, LH and cAMP stimulate steroidogenesis by increasing the transfer of cholesterol to the mitochondria, which is the key rate-limiting step in Leydig cell steroidogenesis (Fig. 4). However, LH also regulates, in a chronic manner, the expression and activity of the steroidogenic enzymes (Nishihara et al. 1988; Payne and Sha 1991), and factors which reduce the levels or activity of these enzymes will limit the capacity of the Leydig cell to convert cholesterol to pregnenolone, and pregnenolone to androgens (Payne and Youngblood 1995).

All of the key steroidogenic enzymes of the Leydig cell are maintained or stimulated by LH and cAMP. The cholesterol side-chain cleavage enzyme (P450CSCC) gene has been isolated from adrenal tissue and is called *CYP11A1* (Morohashi et al. 1987). In the human, there is a single gene for the enzyme, located on chromosome 15 (Chung et al. 1986). The gene product encodes a 39 amino acid signal sequence which is required for translocation of the enzyme to the mitochondria (Kumamoto et al. 1989), and several cAMP-responsive elements (CRE) and CRE-like sequences have been identified in the 5′ DNA sequence, consistent with the observation that cAMP stimulates transcription of the gene (Morohashi et al. 1993; Watanabe et al. 1994). Both LH and cAMP stimulate microsomal P450C17 levels (Nishihara et al. 1988). The gene for this enzyme is called *CYP17*, has two upstream CRE-like sequences, and is found on human chromosome 10 (Matteson et al. 1986; Chung et al. 1987; Lund et al. 1990; Zhang et al. 1996).

There are multiple isoforms of 3βHSD, although in each species there appears to be a single testis-specific isoform (Rhéaume et al. 1991). Maximal expression of the 3βHSD enzyme in the Leydig cell is cAMP-dependent but, in contrast to the two P-450 steroidogenic enzymes, the level of expression of this enzyme is high even under basal conditions (Payne and Sha 1991). Three human genes have been identified which encode enzymes with 17βHSD activity (type I, II and III), but only type III appears to be specific to the testis (Geissler et al. 1994). At present, relatively little is known about the regulation of expression of 17βHSD in the Leydig cell.

4.4.2 *Transcriptional Regulation of Steroidogenesis*

Transcriptional regulation of the steroidogenic process in Leydig cells involves, as it does in other steroidogenic tissues, the transcription factor steroidogenic factor 1 (SF1). This transcription factor activates *CYP11A1* (P450CSCC) gene expression during development, and appears to be involved in mediating the response of both *CYP11A1* and *CYP17* to cAMP (Watanabe et al. 1994; Lund et al. 1997). The 5′-flanking region of the mouse StAR gene contains an SF1 binding site, and StAR expression is undetectable in mice which are deficient in the gene that encodes SF1 (Caron et al. 1997). The testicular 3βHSD gene also contains an SF1 binding site, which may be involved in the regulation of this enzyme (Clarke et al. 1996). Other transcription factors that may be involved include Sp1 and AP1: binding sites for these factors have been identified in the upstream region of both *CYP11A1* and the gene for ferredoxin, which is an essential component of the cholesterol side-chain cleavage complex (Chung et al. 1997).

4.5 Other Transducing Mechanisms

4.5.1 *Calcium*

While it is established that the cAMP pathway is the most important intracellular regulatory mechanism controlling Leydig cell steroidogenesis, other second messenger systems are clearly involved as well (Fig. 4). Increased testosterone secretion in response to low physiological doses of LH does not involve increases in intracellular cAMP, suggesting the existence of these alternative pathways (Themmen et al. 1985; Sullivan and Cooke 1986; Cooke 1990). Some factors other than LH which stimulate Leydig cell steroidogenesis in vitro, such as GnRH and epidermal growth factor (EGF), act through non-cAMP mediated mechanisms involving calcium (Sullivan and Cooke 1984,1986; Ascoli et al. 1987), and both LH/hCG and cAMP induce an increase in Ca^{2+} influx in Leydig cells (Sullivan and Cooke 1986; Kumar et al. 1994). Lowering extracellular Ca^{2+}, or the addition of calcium-channel inhibitors significantly reduces LH- or cAMP analog-stimulated Leydig cell steroid production (Janszen et al. 1976; Lin et al. 1979; Themmen et al. 1985), and the stimulation of secretion of inhibin by adult rat Leydig cells is completely abolished by these procedures (Simpson et al. 1991). Altogether, these data indicate a role for increased intracellular Ca^{2+} in both cAMP-mediated and cAMP-independent regulation of Leydig cell responses.

This calcium-mediated mechanism appears to involve the influx of extracellular Ca^{2+} via specific calcium channels in the Leydig cell plasma membrane, but release of intracellular stores might also be involved. Activation of phospholipase C (PLC) induces the release of intracellular Ca^{2+} via the generation of inositol 1,4,5 triphosphate (IP_3) in many cell types including ovarian steroidogenic cells, although a similar mechanism has yet to be demonstrated in Leydig cells (Ascoli et al. 1989). The exact mechanisms of Ca^{2+} action in Leydig cell steroidogenesis remain uncertain, but the calcium binding protein, calmodulin and a calmodulin-activated protein kinase mechanism have been implicated (Hall et al. 1981; Sullivan and Cooke 1985a).

4.5.2
Chloride

Chloride conductance is stimulated by LH and cAMP in rat Leydig cells (Noulin and Joffre 1993a, b), and basal and submaximal LH-stimulated and cAMP analog-stimulated steroidogenesis is increased by the absence of Cl^- in the culture medium (Choi and Cooke 1990; Ramnath et al. 1997). This enhancement involves increases in protein synthesis and StAR, but not of the activity of P450CSCC or 3βHSD (Ramnath et al. 1997). These data suggest a role for cAMP-mediated increase in Cl^- efflux from the Leydig cell during steroidogenesis, although the mechanism of action remains unclear.

4.5.3
Protein Kinase C

In contrast to protein kinase A, activation of protein kinase C can lead to either stimulation or inhibition of Leydig cell steroidogenesis (Nikula and Huhtaniemi 1989; Cooke et al. 1992). The main endogenous stimulators of protein kinase C are diacylglycerol (DAG) produced by the action of phospholipase C (PLC) and arachidonic acid which may be released by the calcium-mediated activation of phospholipase A_2 (PLA_2), or by metabolism of DAG by diacylglycerol lipase. There is indirect evidence that both PLC and PLA_2 may be stimulated by LH (Abeyasekara et al. 1990; Cooke et al. 1991; Fargin et al. 1991). Activation of protein kinase C appears to be involved in mediating the action of GnRH on the Leydig cell (Lin 1985) and has been implicated in Leydig cell desensitization following LH/hCG treatment (Inoue and Rebois 1989; West et al. 1991). It has been suggested that protein kinase C may also be involved in providing a constant negative regulatory control of Leydig cell steroidogenesis (Lopez-Ruiz et al. 1992).

4.5.4
Arachidonic Acid and Its Metabolites

Recent data have indicated that metabolites of arachidonic acid play a crucial role in Leydig cell regulation. Although the prostaglandins PGE_2 and $PGF_{2\alpha}$ inhibit Leydig cell function (Haour et al. 1979), blocking of the lipoxygenase, but not the cycloxygenase, pathway of arachidonic acid metabolism inhibits LH-stimulated steroidogenesis in vitro at a site distal to cAMP formation (Dix et al. 1984; Sullivan and Cooke 1985a). Conversely, while leukotriene B_4 (LTB_4) is produced by the Leydig cell, this leukotriene does not appear to affect steroidogenesis and it appears that intermediates within the lipoxygenase pathway may be responsible for stimulating Leydig cell steroidogenesis (Sullivan and Cooke 1985b; Reddy et al. 1993).

5
Extrinsic Regulation of the Leydig Cell by Factors Other than LH

5.1
Anterior Pituitary Hormones: FSH, Prolactin, and Growth Hormone

In addition to LH, other anterior pituitary hormones have direct effects on Leydig cell function. After hypophysectomy, FSH has the capacity to stimulate Leydig cell steroidogenesis and LH-receptor numbers (El Safoury and Bartke 1974; McNeilly et al. 1979), although these actions of FSH are assumed to be mediated via the Sertoli cell (Orth and Christensen 1977). In contrast to FSH, the Leydig cells have receptors for prolactin (Charreau et al. 1977), which appears to be involved in the regulation of LH-receptor numbers and in augmentation of the steroidogenic responsiveness of the Leydig cell to LH (Zipf et al. 1978). Paradoxically, high levels of prolactin have an inhibitory effect on androgen production by the Leydig cell (Purvis and Hansson 1978; Davis 1982). However, while hyperprolactinaemia is accompanied by testicular dysfunction, it is not clear whether these effects involve a direct action of prolactin on the testis, or effects on pituitary gonadotropin secretion. Finally, growth hormone (GH) is yet another anterior pituitary hormone, which has a stimulatory effect on Leydig cells especially during pubertal maturation, most probably by acting through stimulation of intratesticular IGF-1 production (Zipf et al. 1978; Chatelain et al. 1991).

5.2 Regulation by the Seminiferous Tubules

Evidence that the Leydig cell is under local control by the seminiferous tubule compartment came initially from experiments in which seminiferous tubule function was disrupted by chemical treatment, such as by anti-androgens and mitotic inhibitors, or physical means such as cryptorchidism and efferent duct ligation in rats. In these experimental models, seminiferous tubule damage was accompanied by a concomitant increase in Leydig cell steroidogenic capacity and functional morphology (Aoki and Fawcett 1978; Kerr et al. 1979; Rich et al. 1979; Risbridger et al. 1981). Further studies employing this approach, in conjunction with Leydig cell depletion and recovery after treatment with EDS, also indicated that tubule factors influenced Leydig cell growth and development (Kerr and Donachie 1986; O'Leary et al. 1986). In addition, the stimulatory effects of FSH on Leydig cell development indirectly implicate the Sertoli cells in Leydig cell regulation (Kerr and Sharpe 1985a, b; Teerds et al. 1989c).

At first it was believed that one or two discrete factors responsible for this communication might be identified and considerable effort was expended to identify the Leydig cell-regulatory activities in Sertoli cell and tubule cultures and in testicular interstitial fluid (Sharpe and Cooper 1984; Verhoeven and Cailleau 1987; Vihko and Huhtaniemi 1989; Hedger et al. 1990). The best characterized of these factors was a 70–80 kDa protein isolated from cultured rat and human Sertoli cells which stimulated basal MA-10 tumor Leydig cell steroidogenesis, and which was found to be a cathepsin L/metalloproteinase complex (Papadopoulos 1991; Boujrad et al. 1995). However, both Sertoli cell cultures and testicular extracts were found to contain multiple factors with both stimulatory and inhibitory actions on Leydig cell steroidogenesis (Syed et al. 1985; Benahmed et al. 1986; Hedger et al. 1994), which can be attributed to the fact that the Sertoli cells and germ cells produce a broad range of different growth factors. It is now apparent that communication between the two compartments is mediated by multiple mechanisms involving many factors, some of which are discussed below.

5.3 Cytokines and Growth factors

The list of cytokines and growth factors which have been shown to influence Leydig cell steroidogenesis in vitro is considerable and will not be reviewed in detail here. The best studied of these factors are IGF-1 and the TGF-β family members, which have also been implicated in controlling Leydig cell development. In general, IGF-1 increases LH/hCG- and cAMP-induced steroidogenesis (Lin et al. 1986; Perrard-Sapori et al. 1987; Chatelain et al. 1991; Gelber et al. 1992; Moore and Morris 1993), while TGFβ inhibits Leydig cell steroidogenesis (Avallet et al. 1987; Lin et al. 1987). The TGFβ-related proteins, inhibin and activin also have effects on steroidogenesis, particularly by immature Leydig cells, although the nature of these effects appear to be species-dependent (Hsueh et al. 1987; Lin et al. 1989; Mauduit et al. 1991a; Lejeune et al. 1997). Other growth factors for which there is clear evidence of both production within the testis and modulatory effects on Leydig cell steroidogenesis are stem cell factor (SCF; Yoshinaga et al. 1991), the platelet-derived growth factors (PDGF; Loveland et al. 1993) and fibroblast growth factors (FGF) 1 and 2 (Murono et al. 1993; Laslett et al. 1995).

In addition to the growth factors listed above, numerous immunoregulatory cytokines have effects on Leydig cell steroidogenesis. Some of these factors have already been discussed, and include both inflammatory cytokines, such as IL-1α, IL-1β, interleukin-2 (IL-2), tumor necrosis factor-α (TNFα), interferon-α (IFNα), and interferon-γ (IFNγ) and the anti-inflammatory cytokines, TGFβ and activin (Orava et al. 1985, 1989; Avallet et al. 1987; Hsueh et al. 1987; Lin et al. 1987, 1989; Calkins et al. 1990a, b; Guo et al. 1990; Mauduit et al. 1991b; Hales 1992; Mauduit et al. 1992; Meikle et al. 1992; Xiong and Hales 1993; Lejeune et al. 1997). In general, these cytokines inhibit Leydig cell steroidogenesis, although stimulatory effects of IL-1 particularly under basal conditions in vitro have been observed in several studies (Verhoeven et al. 1988; Warren et al. 1990; Moore and Moger 1991). Several of these cytokines, IL-1α, TNFα, TGFβ and activin, are produced normally within the testis and may be involved in testicular development and spermatogenesis (Gérard et al. 1991; De et al. 1993; Mullaney and Skinner 1993; Vliegen et al. 1993; Moore and Hutson 1994). However, their over-production by resident and infiltrating leukocytes (macrophages, lymphocytes and granulocytes) during testicular inflammation and immune responses is likely to have a considerable impact upon Leydig cell function and testicular function in general. Both systemic inflammation and local inflammation of the testis (orchitis), due to infection, cause a reduction in serum testosterone in men (Adamopoulos et al. 1978; Aiman et al. 1980; Cutolo et al. 1988), and similar results have been observed in experimental animals under inflammatory conditions (Bruot and Clemens 1992; Hales et al. 1992; Wallgren et al. 1993; Fountain et al. 1997).

5.4
Autocrine Regulation

5.4.1
Androgen-Mediated Autoregulation

As is the case for most cell types, Leydig cells are able to autoregulate their function, and this autoregulation involves both testosterone and non-androgenic mechanisms (Adashi and Hseuh 1981; Darney and Ewing 1981; Hedger and Eddy 1990). Leydig cells possess androgen receptors (Wilson and Smith 1975; Namiki et al. 1991), and androgens inhibit steroidogenesis in adult mouse Leydig cells via a receptor-mediated mechanism at the level of steady-state expression and cAMP-stimulated synthesis of P450C17 (Hales et al. 1987; Payne and Sha 1991; Burgos-Trinidad et al. 1997). Moreover, as already discussed, androgens stimulate the responsiveness of the Leydig cell to LH during development and in the regressed testis (Hardy et al. 1990; Shan and Hardy 1992; Misro et al. 1993; Duckett et al. 1997b).

5.4.2
Leydig Cell Desensitization

Leydig cell autoregulation has been implicated in the phenomenon of Leydig cell desensitization, whereby, after administration of a large dose of LH/hCG in vivo, adult Leydig cells lose the capacity to respond to a second stimulus for several days (Hsueh et al. 1977; Tsuruhara et al. 1977). In the human, this period of down-regulation or refractoriness to hCG is manifested by a biphasic response to a single injection of hCG, where the initial peak is followed by a plateau and a secondary rise in testosterone secretion 72–96 h later (Padron 1985). The secondary peak is seen only with hCG, and relates to the longer clearance rate relative to LH resulting in persisting hCG levels which stimulate the secondary rise in testosterone.

Studies in vitro have indicated that this Leydig cell desensitization is accompanied by a loss of surface LH-receptors, uncoupling of the G-protein-adenyl cyclase transducing mechanism, reduced transcription of mRNA for the LH-receptor mediated via a cAMP-mediated mechanism, and down-regulation of key steroidogenic enzymes (Hseuh et al. 1977; Tsuruhara et al. 1977; Dufau et al. 1989; Payne and Sha 1991; Wang et al. 1991c; West et al. 1991). The possibility that these responses are mediated by a discrete regulatory mechanism remains under investigation. Initially, considerable attention was given to the intermediary role of E_2, although roles for androgens and other Leydig cell products such as CRF and serotonin, also have been suggested (Kalla et al. 1980; Nozu et al. 1981; Hales et al. 1987; Dufau et al. 1989; Payne and Sha 1991). It should be recognized, however, that hyperstimulation of the Leydig cell also depletes intracellular cholesterol stores and generates oxygen free radicals which damage the P450 enzymes (Quinn et al. 1981; Quinn and Payne 1984; Georgiou et al. 1987; Hedger and Risbridger 1992) and the physiological significance of the Leydig cell down-regulation phenomenon, as opposed to its pathological and clinical importance, remains unresolved.

5.5 Glucocorticoids

Stress is a potent inhibitor of reproductive function, primarily because glucocorticoids, such as cortisol in the human and corticosterone in the rat, released during stress, inhibit the hypothalamo-pituitary-gonadal axis at several sites (see review by Hardy and Ganjam 1997). The Leydig cells themselves possess specific glucocorticoid receptors (Stalker et al. 1989), and glucocorticoids inhibit Leydig cell function by reducing expression of several steroidogenic enzymes (Bambino and Hsueh 1981; Welsh et al. 1982; Agular et al. 1992; Payne and Youngblood 1995; Gao et al. 1996), most likely at the promoter level of the enzyme genes, at least one of which, *CYP17* (P450C17), possesses a glucocorticoid response element (Nason et al. 1992).

Glucocorticoid activity is modulated by oxidative conversion to inactive steroids by the enzyme 11β-hydroxysteroid dehydrogenase (11βHSD) in mature Leydig cells (Phillips et al. 1989). Hence, the direct effect of glucocorticoids on the Leydig cell are determined by both the circulating levels of glucocorticoids and the level of testicular 11βHSD activity. The number of glucocorticoid receptors and 11βHSD activity in the Leydig cell are up-regulated during development: enzyme expression is lowest in the fetal testis and in Leydig cell precursors, with the highest levels in adult Leydig cells (Ge et al. 1997). As a result, the response of the testis to glucocorticoids varies throughout development.

5.6 Neuropeptides

Several locally produced neuropeptides have been implicated in controlling Leydig cell function. These include CRF, arginine vasopressin (AVP), the enkephalins, substance P, oxytocin, GnRH and GH-releasing factor (see review by Gnessi et al. 1997). The best studied of these peptides has been GnRH, which has both stimulatory and inhibitory effects on rat Leydig cell steroidogenesis, and has been widely used in studies on the intracellular regulation of the Leydig cell (Sullivan and Cooke 1984, 1985a, 1986; Lin 1985). Rat Leydig cells possess specific, high-affinity receptors for GnRH, although there is evidence that only low-affinity receptors are present in human Leydig cells (Bourne et al. 1980; Clayton and Huhtaniemi 1982; Bahk et al. 1995). Levels of GnRH in the normal testis are very low, and the functional significance of this peptide in the testis remains unclear (Hedger et al. 1985; Bahk et al. 1995).

5.7 Other Factors

Many intercellular regulatory molecules influence Leydig cell steroidogenesis. Nitric oxide, prostaglandins and serotonin inhibit Leydig cell function (Ellis 1972; Haour et al. 1979; del Punta et al. 1996). Catecholamines, vasoactive intestinal peptide and atrial natiuretic factor stimulate Leydig cell function (Anakwe and Moger 1986; Kasson et al. 1986b; Pandey et al. 1986). Many of these agents are routinely produced as part of the neural, immunological or metabolic activities of the testis, and presumably have a fundamental effect on Leydig cell homeostasis, although their production may increase during pathological processes with a concomitant increase in their functional significance.

6 Leydig Cell Function and Infertility

Alterations in Leydig cell function can be involved in the development of infertility, or may be the result of spermatogenic dysfunction. In the former case, the essential requirement of testosterone for spermatogenesis is the basis by which alterations of Leydig cell function can cause infertility. The lack of LH stimulation results in the subnormal production of testosterone, which in turn causes inadequate sperm production. Disorders of pubertal development such as idiopathic hypogonadotropic hypogonadism or Kallman's syndrome, through the production of an inadequate LH signal results in a failure of spermatogenic stimulation. Similarly, the onset of LH deficiency in adults from a variety of pathologies can result in infertility.

Inadequate Leydig cell function can result from LH receptor defects based on mutations in the gene encoding this protein. Conversely, mutations in the LH receptor in the sixth transmembrane domain results in constitutive activation of the receptor even in the absence of LH, and is responsible for familial precocious puberty and testotoxicosis (Shenker et al. 1993). Mutations in the androgen receptor can cause infertility by interfering with the action of testosterone on the seminiferous epithelium (McPhaul et al. 1993). These can vary from the complete failure of the receptor to bind testosterone, such as seen in testicular feminisation, to milder forms such as seen in Reifenstein's syndrome (Schweikert et al. 1987). In other studies, principally based on binding studies, a variable incidence of androgen receptor defects was proposed to explain male infertility, but the genetic defects in this group still require identification (Aiman and Griffin 1982; Bouchard et al. 1986; Morrow et al. 1987). Nevertheless partial deletions of the androgen receptor gene have been shown to result in azoospermia with a normal male phenotype (Akin et al. 1991). As well as point mutations in the androgen receptor, there is increasing data to suggest that the expression of CAG repeats in the amino terminal region of the androgen receptor may cause infertility (Tut et al. 1997). It has been accepted for some time that if the number of these repeats increases to 40 or more, then the males may develop Kennedy's disease, which is a neurodegenerative muscular disorder accompanied by infertility. More recently, a lesser expansion of these repeats has been linked to infertility without evidence of muscle dysfunction (Tut et al. 1997).

As discussed earlier in this review, there is data to suggest that the function of the Leydig cells can be influenced by the state of the seminiferous epithelium. These influences can be short-term, where a brief disruption of spermatogenesis, such as that induced by heating of the rat testis to 43 °C for 15 min., can alter Leydig cell activity for 1–2 weeks (Jegou et al. 1984). In studies where these changes are more prolonged, this results in increased LH and lower testosterone levels (Kerr et al. 1979; Rich et al. 1979). The latter experimental data parallel the results found in infertile men with severe damage to their seminiferous epithelium, where high LH concentrations are accompanied by low or low-normal testosterone levels (de Kretser et al. 1972). The presence of low-normal testosterone levels accompanied by higher LH concentrations represents a state of compensated Leydig cell failure. Further support for this view comes from the failure of such men to respond normally to an hCG challenge in terms of testosterone secretion (de Kretser et al. 1975). The most severe impairment of testosterone secretion is seen in patients with Klinefelter's syndrome, but may also be associated with the end-result of severe mumps orchitis, torsion of the testis or the seminiferous tubule damage found in men where testicular maldescent remained untreated until puberty.

References

Abayasekara DR, Band AM, Cooke BA (1990) Evidence for the involvement of phospholipase A2 in the regulation of luteinizing hormone-stimulated steroidogenesis in rat testis Leydig cells. Mol Cell Endocrinol 70:147–153

Adamopoulos DA, Lawrence DM, Vassilopoulos P, Contoyiannis PA, Swyer GIM (1978) Pituitary-testicular interrelationships in mumps orchitis and other viral infections. Br Med J 1:1177–1180

Adashi EY, Hseuh AJW (1981) Autoregulation of androgen production in a primary culture of rat testicular cells. Nature 293:737–738

Agular BM, Vinggaard AM, Vind C (1992) Regulation by dexamethasone of the 3β-hydroxysteroid dehydrogenase activity in adult rat Leydig cells. J Steroid Biochem Mol Biol 43:565–571

Aiman J, Griffin JE (1982) The frequency of androgen receptor deficiency in infertile men. J Clin Endocrinol Metab 54:725–732

Aiman J, Brenner PF, MacDonald PC (1980) Androgen and estrogen production in elderly men with gynecomastia and testicular atrophy after mumps orchitis. J Clin Endocrinol Metab 50:380–386

Akin JW, Behzadian A, Tho SPT, McDonough PG (1991) Evidence for a partial deletion in the androgen receptor in a phenotypic male with azoospermia. Am J Obstet Gynecol 165:1891–1894

Almahbobi G, Williams LJ, Han X-G, Hall PF (1993) Binding of lipid droplets and mitochondria to intermediate filaments in rat Leydig cells. J Reprod Fertil 98:209–217

Amat P, Paniagua R, Nistal M, Martin A (1986) Mitosis in adult human Leydig cells. Cell Tissue Res 243:219–221

Anakwe OO, Moger WH (1986) Catecholamine stimulation of androgen production by rat Leydig cells. Interaction with luteinizing hormone and luteinizing hormone-releasing hormone. Biol Reprod 35:806–814

Aoki A, Fawcett DW (1978) Is there a local feedback from the seminiferous tubules affecting activity of the Leydig cells? Biol Reprod 19:144–158

Ascoli M, Euffa J, Segaloff DL (1987) Epidermal growth factor activates steroid biosynthesis in cultured Leydig tumor cells without affecting the levels of cAMP and potentiates the activation of steroid biosynthesis by choriogonadotropin and cAMP. J Biol Chem 262:9196–9203

Ascoli M, Pignataro OP, Segaloff DL (1989) The inositol phosphate/diacylglycerol pathway in MA-10 Leydig tumor cells. J Biol Chem 264:6674–6681

Avallet O, Vigier M, Perrard-Sapori MH, Saez JM (1987) Transforming growth factor β inhibits Leydig cell functions. Biochem Biophys Res Commun 146:575–581

Bahk JY, Hyun JS, Chung SH, Lee H, Kim MO, Lee BH, Choi WS (1995) Stage-specific identification of the expression of GnRH mRNA and localization of the GnRH receptor in mature rat and adult human testis. J Urol 154:1958–1961

Bambino TH, Hseuh AJW (1981) Direct inhibitory effects of glucocorticoids upon testicular luteinizing hormone receptor and steroidogenesis in vivo and in vitro. Endocrinology 108: 2142–2148

Bardin CW, Catterall JF (1981) Testosterone, a major determinant of extragenital sexual dimorphism. Science 211:1285–1294

Bedrak E, Samuels LT (1969) Steroid biosynthesis by the equine testis. Endocrinology 85: 1186–1195

Belchetz PE, Plant TM, Nakai Y, Keogh EJ, Knobil E (1978) Hypophysial responses to continuous and intermittent delivery of hypotha-lamic gonadotropin-releasing hormone. Science 202:631–633

Benahmed M, Morera AM, Chauvin MA (1986) Evidence for a Sertoli cell, FSH-suppressible inhibiting factor(s) of testicular steroidogenic activity. Biochem Biophys Res Commun 139: 169–178

Benton L, Shan L-X, Hardy MP (1995) Differentiation of adult Leydig cells. J Steroid Biochem Mol Biol 53:61–68

Bergh A, Damber J-E (1993) Vascular controls in testicular physiology. In: de Kretser DM (ed) Molecular biology of the male reproductive tract. Academic Press, New York, pp 439–468

Boockfor FR, Wang D, Lin T, Nagpal ML, Spangelo BL (1994) Interleukin-6 secretion from rat Leydig cells in culture. Endocrinology 134:2150–2155

Bouchard P, Wright F, Portois MC, Couzinet B, Schaison G, Mowszowicz I (1986) Androgen insensitivity in oligospermic men: a reappraisal. J Clin Endocrinol Metab 63:1242–1246

Boujrad N, Hudson JR, Papadopoulos V (1993) Inhibition of hormone-stimulated steroidogenesis in cultured Leydig tumor cells by a cholesterol-linked phosphorothioate oligodeoxynucleotide antisense to diazepam-binding inhibitor. Proc Natl Acad Sci USA 90:5728–5731

Boujrad N, Ogwuegbu SO, Garnier M, Lee C-H, Martin BM, Papadopoulos V (1995) Identification of a stimulator of steroid hormone synthesis isolated from testis. Science 268:1609–1612

Bourne GA, Regiani S, Payne AH, Marshall JC (1980) Testicular GnRH receptors – characterization and localization on interstitial tissue. J Clin Endocrinol Metab 51:407–409

Boyar RM, Rosenfeld RS, Kapen S, Finkelstein JW, Roffwarg HP, Weitzman ED, Hellman L (1974) Human puberty. Simultaneous augmented secretion of luteinizing hormone and testosterone during sleep. J Clin Invest 54:609–618

Bremner WJ, Millar MR, Sharpe RM, Saunders PT (1994) Immunohistochemical localization of androgen receptors in the rat testis: evidence for stage-dependent expression and regulation by androgens. Endocrinology 135:1227–1234

Brosh N, Sternberg D, Honigwachs-Sha'anani J, Lee BC, Shav-Tal Y, Tzehoval E, Shulman LM, Toledo J, Hachman Y, Carmi P, Wen J, Sasse J, Horn F, Burstein Y, Zipori D (1995) The plasmacytoma growth inhibitor restrictin-P is an antagonist of interleukin 6 and interleukin 11. Identification as a stroma-derived activin A. J Biol Chem 270:29594–29600

Brown AS, Hall PF, Shoyab M, Papadopoulos V (1992) Endozepine/diazepam binding inhibitor in adrenocortical and Leydig cell lines: absence of hormonal regulation. Mol Cell Endocrinol 83:1–9

Bruot BC, Clemens JW (1992) Regulation of testosterone production in the adjuvant-induced arthritic rat. J Androl 13:87–92

Burger HG, Baker HWG (1984) Therapeutic considerations and results of gonadotropin treatment in male hypogonadotropic hypogonadism. Ann N Y Acad Sci 438:447–453

Burgos-Trinidad M, Youngblood GL, Maroto MR, Scheller A, Robins DM, Payne AH (1997) Repression of cAMP-induced expression of the mouse P450 17 α-hydroxylase/C17–20 lyase gene (*Cyp17*) by androgens. Mol Endocrinol 11:87–96

Calkins JH, Guo H, Sigel MM, Lin T (1990a) Tumor necrosis factor-α enhances inhibitory effects of interleukin-1β on Leydig cell steroidogenesis. Biochem Biophys Res Commun 166: 1313–1318

Calkins JH, Guo H, Sigel MM, Lin T (1990b) Differential effects of recombinant interleukin-1α and β on Leydig cell function. Biochem Biophys Res Commun 167:548–553

Caron KM, Ikeda Y, Soo SC, Stocco DM, Parker KL, Clark BJ (1997) Characterization of the promoter region of the mouse gene encoding the steroidogenic acute regulatory protein. Mol Endocrinol 11:138–147

Cavallaro S, Pani L, Guidotti A, Costa E (1993) ACTH-induced mitochondrial DBI receptor (MDR) and diazepam binding inhibitor (DBI) expression in adrenals of hypophysectomized rats is not cause-effect related to its immediate steroidogenic action. Life Sci 53: 1137–1147

Chanderbhan R, Noland BJ, Scallen TJ, Vahouny GV (1982) Sterol carrier protein 2. Delivery of cholesterol from adrenal lipid droplets to mitochondria for pregnenolone synthesis. J Biol Chem 257:8928–8934

Chandrashekar V, Bartke A (1992) The influence of beta-endorphin on testicular endocrine function in adult rats. Biol Reprod 47:1–5

Charreau EH, Attramadal A, Torjesen P, Purvis K, Calandra R, Hansson V (1977) Prolactin binding in rat testis: specific receptors in interstitial cells. Mol Cell Endocrinol 6:303–307

Chase DJ, Karle JA, Fogg RE (1992) Maintenance or stimulation of steroidogenic enzymes and testosterone production in rat Leydig cells by continuous and pulsatile infusions of luteinizing hormone during passive immunization against gonadotrophin-releasing hormone. J Reprod Fertil 95:657–667

Chatelain PG, Sanchez P, Saez JM (1991) Growth hormone and insulin-like growth factor I increase testicular luteinizing hormone receptors and steroidogenic responsiveness of growth hormone deficient dwarf mice. Endocrinology 128:1857–1862

Chemes HE, Gottlieb SE, Pasqualini T, Domenichini E, Rivarola MA, Bergada C (1985) Response to acute hCG stimulation and steroidogenic potential of Leydig cell fibroblastic precursors in humans. J Androl 6:102–112

Choi MS, Cooke BA (1990) Evidence for two independent pathways in the stimulation of steroidogenesis by luteinizing hormone involving chloride channels and cyclic AMP. FEBS Lett 261:402–404

Christensen AK, Gillim SW (1969) The correlation of fine structure and function in steroid-secreting cells, with emphasis on those of the gonad. In: McKearns KW (ed) The gonads. Appleton-Century-Croft, New York, pp 415–488

Christensen AK, Mason NR (1965) Comparative ability of seminiferous tubules and interstitial tissue of rat testes to synthesize androgens from progesterone-4-^{14}C *in vitro*. Endocrinology 76:646–656

Chung B, Matteson KJ, Voutilainen R, Mohandas TK, Miller WL (1986) Human cholesterol side chain cleavage enzyme, P450SCC: cDNA cloning, assignment of the gene to chromosome 15 and expression in the placenta. Proc Natl Acad Sci USA 83:8962–8966

Chung B, Picardo-Leonard J, Haniu M, Bienkowski M, Hall PF, Shively JE, Miller WL (1987) Cytochrome P450C17 cloning of human adrenal and testis cDNAs indicates the same gene is expressed in both tissues. Proc Natl Acad Sci USA 84:407–411

Chung B, Guo I-C, Chou S-J (1997) Transcriptional regulation of the *CYP11A1* and ferredoxin genes. Steroids 62:37–42

Cicero TJ, Adams ML, O'Connor LH, Nock B (1989) In vivo evidence for a direct effect of naloxone on steroidogenesis in the male rat. Endocrinology 125:957–963

Clark BJ, Wells J, King SR, Stocco DM (1994) The purification, cloning, and expression of a novel luteinizing hormone-induced mitochondrial protein in MA-10 mouse Leydig tumor cells. Characterization of the steroidogenic acute regulatory protein (StAR). J Biol Chem 269: 28314–28322

Clark BJ, Soo SC, Caron KM, Ikeda Y, Parker KL, Stocco DM (1995) Hormonal and developmental regulation of the steroidogenic acute regulatory (StAR) protein. Mol Endocrinol 9: 1346–1355

Clark SJ, Ellis N, Styne DM, Gluckman PD, Kaplan SL, Grumbach MM (1984) Hormone ontogeny in the ovine fetus. XVII. Demonstration of pulsatile luteinizing hormone secretion by the fetal pituitary gland. Endocrinology 115:1774–1779

Clarke TR, Bain PA, Burmeister M, Payne AH (1996) Isolation and characterization of several members of the murine Hsd3β gene family. DNA Cell Biol 15:387–399

Clayton RN (1996) Gonadotrophin receptors. Baillieres Clin Endocrinol Metab 10:1–8

Clayton RN, Huhtaniemi IT (1982) Absence of gonadotropin-releasing hormone receptors in human gonadal tissue. Nature 299:56–59

Clements JA, Reyes FI, Winter JSD, Faiman C (1976) Studies on human sexual development: III. Fetal pituitary, serum and amniotic fluid concentrations of LH, CG and FSH. J Clin Endocrinol Metab 42:9–19

Cochran RC, Schuetz AW, Ewing LL (1979) Age-related changes in conversion of 5α-androstan-17β-ol-3-one to 5α-androstane-3β,17β-diol and 5α-androstane-3α,17β-diol by rat testicular cells in vitro. J Reprod Fertil 57:143–147

Collin O, Bergh A (1996) Leydig cells secrete factors which increase vascular permeability and endothelial cell proliferation. Int J Androl 19:221–228

Collin O, Bergh A, Damber J-E, Widmark A (1993) Control of testicular vasomotion by testosterone and tubular factors in rats. J Reprod Fertil 97:115–121

Cooke BA (1990) Is cyclic AMP an obligatory second messenger for luteinizing hormone? Mol Cell Endocrinol 69:C11–C15

Cooke BA, Janszen FHA, van Driel MJA, van der Molen HJ (1979) Evidence for the involvement of lutropin-independent RNA synthesis in Leydig cell steroidogenesis. Mol Cell Endocrinol 14:181–189

Cooke BA, Dirami G, Chaudry L, Choi MSK, Abayasekara DRE, Phipp L (1991) Release of arachidonic acid and the effects of corticosteroids on steroidogenesis in rat testis Leydig cells. J Steroid Biochem Mol Biol 40:465–471

Cooke BA, Choi MCK, Dirami G, Lopez-Ruiz MP, West AP (1992) Control of steroidogenesis in Leydig cells. J Steroid Biochem Mol Biol 43:445–449

Cooke PS, Meisami E (1991) Early postnatal hypothyroidism increases adult size of testis and other reproductive organs, but does not increase testosterone levels. Endocrinology 129: 237–243

Cudicini C, Lejeune H, Gomez E, Bosmans E, Ballet F, Saez J, Jegou B (1997) Human Leydig cells and Sertoli cells are producers of interleukins-1 and -6. J Clin Endocrinol Metab 82: 1426–1433

Cunningham GR, Huckins C (1979) Persistence of complete spermatogenesis in the presence of low intratesticular concentrations of testosterone. Endocrinology 105:177–186

Cutolo M, Balleari E, Giusti M, Monachesi M, Accardo S (1988) Sex hormone status of patients with rheumatoid arthritis: evidence of low serum concentrations at baseline and after human chorionic gonadotrophin stimulation. Arthritis Rheum 31:1314–1317

Damber JE, Bergh A, Widmark A (1987) Testicular blood flow and microcirculation in rats after treatment with ethane dimethane sulfonate. Biol Reprod 37:191–1296

Darney KJ, Ewing L (1981) Autoregulation of testosterone secretion in perfused rat testis. Endocrinology 109:993–995

Davis JL (1982) Lowering prolactin levels in a hyperprolactinaemic man. Responses of luteinizing hormone, follicle-stimulating hormone, and testosterone. Arch Intern Med 142: 146–148

De SK, Chen HL, Pace JL, Hunt JS, Terranova PF, Enders GC (1993) Expression of tumor necrosis factor-α in mouse spermatogenic cells. Endocrinology 133:389–396

de Kretser DM (1967) The fine structure of the testicular interstitial cells in men of normal androgenic status. Z Zellforsch 80:594–609

de Kretser DM, Catt KJ, Burger HG, Smith GC (1969) Radioautographic studies on the localization of ^{125}I-labeled human luteinizing and growth hormone in immature male rats. J Endocrinol 43:105–111

de Kretser DM, Burger HG, Fortune D, Hudson B, Long AR, Paulsen CA Taft HP (1972) Hormonal, histological and chromosomal studies in adult males with testicular disorders. J Clin Endocrinol Metab 35:392–401

de Kretser DM, Burger HG, Hudson B, Keogh EJ (1975) The hCG stimulation test in men with testicular disorders. Clin Endocrinol 4:591–596

Del Punta K, Charreau EH, Pignataro OP (1996) Nitric oxide inhibits Leydig cell steroidogenesis. Endocrinology 137:5337–5343

Desjardins C (1996) Fluid exchange and transport of endocrine and paracrine solutes supporting the Leydig cell. In: Payne AH, Hardy MP, Russell LD (eds) The Leydig cell. Cache River Press, Vienna, pp 507–521

de Winter JP, Timmerman MA, Vanderstichele HMJ, Klaij IA, Grootenhuis AJ, Rommerts FFG, de Jong FH (1992) Testicular Leydig cells in vitro secrete only inhibin α-subunits, whereas Leydig cell tumors can secrete bioactive inhibin. Mol Cell Endocrinol 83:105–115

Dix CJ, Habberfield AD, Sullivan MHF, Cooke BA (1984) Inhibition of steroid production in Leydig cells by non-steroidal anti-inflammatory and related compounds: evidence for the involvement of lipoxygenase products in steroidogenesis. Biochem J 219:529–537

Duckett RJ, Hedger MP, McLachlan RI, Wreford NG (1997a) The effects of gonadotropin-releasing hormone-immunisation and recombinant follicle-stimulating hormone on the Leydig cell and macrophage populations of the adult rat testis. J Androl 18:417–423

Duckett RJ, Wreford NG, Meachem SJ, McLachlan RI, Hedger MP (1997b) The effect of chorionic gonadotropin and flutamide on Leydig cell and macrophage populations in the testosterone-estradiol implanted rat. J Androl 18:656–662

Dufau ML (1988) Endocrine regulation and communicating functions of the Leydig cell. Annu Rev Physiol 50:483–508

Dufau ML, Ulisse S, Khanum A, Buczko E, Kitamura M, Fabbri A, Namiki M (1989) LH action in the Leydig cell: modulation by angiotensin II and corticotropin releasing hormone, and regulation of $P450_{17\alpha}$ mRNA. J Steroid Biochem 34:205–217

Dufau ML, Miyagawa Y, Takada S, Khanum A, Miyagawa H, Buczko E (1997) Regulation of androgen synthesis: the late steroidogenic pathway. Steroids 62:128–132

Eddy EM, Washburn TF, Bunch DO, Goulding EH, Gladen BC, Lubahn DB, Korach KS (1996) Targeted disruption of the estrogen receptor gene in male mice causes alteration of spermatogenesis and infertility. Endocrinology 137:4796–4805

Eisenhauer KM, McCue PM, Nayden DK, Osawa Y, Roser JF (1994) Localization of aromatase in equine Leydig cells. Domest Anim Endocrinol 11:291–298

Ellis LC (1972) Inhibition of rat testicular androgen synthesis in vitro by melatonin and serotonin. Endocrinology 90:17–28

El Safoury S, Bartke A (1974) Effects of follicle-stimulating hormone and luteinizing hormone on plasma testosterone levels in hypophysectomized and intact immature and adult male rats. J Endocrinol 61:193–198

Faiman C, Winter JSD (1971) Sex differences in gonadotrophin concentrations in infancy. Nature 232:130–131

Fargin A, Yamamoto K, Cotecchia S, Goldsmith PK, Spiegel AM, Lapetina EG, Caron MG, Lefkowitz RJ (1991) Dual coupling of the cloned 5-HT1A-receptor to both adenylyl cyclase and phospholipase-C is mediated via the same gi-protein. Cell Signal 3:547–557

Fawcett DW, Neaves WR, Flores MN (1973) Comparative observations on intertubular lymphatics and the organization of the interstitial tissue of the mammalian testis. Biol Reprod 9:500–532

Fisher CR, Graves KH, Parlow AF, Simpson ER (1998) Characterization of mice deficient in aromatase (ArKO) because of targeted disruption of the *cyp19* gene. Prac Natl Acad Sci USA 95:6995–6970

Fountain S, Holland MK, Hinds LA, Janssens PA, Kerr PJ (1997) Interstitial orchitis with impaired steroidogenesis and spermatogenesis in the testes of rabbits infected with an attenuated strain of myxoma virus. J Reprod Fertil 110:161–169

Gao HB, Shan LX, Monder C, Hardy MP (1996) Suppression of endogenous corticosterone levels in vivo increases the steroidogenic capacity of purified rat Leydig cells in vitro. Endocrinology 137:1714–1718

Gautier C, Levacher C, Avallet O, Vigier M, Rouiller-Fabre V, Lecerf L, Saez J, Habert R (1994) Immunohistochemical localization of transforming growth factor-β1 in the fetal and neonatal rat testis. Mol Cell Endocrinol 99:55–61

Gaytan F, Bellido C, Morales C, Reymundo C, Aguilar E, van Rooijen N (1994a) Effects of macrophage depletion at different times after treatment with ethylene dimethane sulfonate (EDS) on the regeneration of Leydig cells in the adult rat. J Androl 15:558–564

Gaytan F, Bellido C, Morales C, Reymundo C, Aguilar E, van Rooijen N (1994b) Selective depletion of testicular macrophages and prevention of Leydig cell repopulation after treatment with ethylene dimethane sulfonate in rats. J Reprod Fertil 101:175–182

Gaytan F, Pinilla L, Romero JL, Aguilar E (1994c) Differential effects of the administration of human chorionic gonadotropin to postnatal rats. J Endocrinol 142:527–534

Gaytan F, Bellido C, Aguilar E, van Rooijen N (1994d) Requirement for testicular macrophages in Leydig cell proliferation and differentiation during prepubertal development in rats. J Reprod Fertil 102:393–399

Gaytan F, Bellido C, Morales C, van Rooijen N, Aguilar E (1995) Role of testicular macrophages in the response of Leydig cells to gonadotropins in young hypophysectomized rats. J Endocrinol 147:463–471

Ge R-S, Hardy DO, Catterall JF, Hardy MP (1997) Developmental changes in glucocorticoid receptor and 11β-hydroxysteroid dehydrogenase oxidative and reductive activities in rat Leydig cells. Endocrinology 138:5089–5095

Geissler WM, Davis DL, Wu L, Bradshaw KD, Patel S, Mendonca BB, Elliston KO, Wilson JD, Russell DW, Andersson S (1994) Male pseudohermaphroditism caused by mutations of testicular 17β-hydroxysteroid dehydrogenase 3. Nat Genet 7:34–39

Gelber SJ, Hardy MP, Mendis-Handagama SMLC, Casella SJ (1992) Effects of insulin-like growth factor-1 on androgen production by highly purified pubertal and adult rat Leydig cells. J Androl 13:125–130

Georgiou MG, Perkins LM, Payne AH (1987) Steroid synthesis-dependent, oxygen-mediated damage of mitochondrial and microsomal cytochrome P450 enzymes in rat Leydig cell cultures. Endocrinology 121:1390–1399

Gérard N, Syed V, Bardin CW, Genetet N, Jégou B (1991) Sertoli cells are the site of interleukin-1α synthesis in rat testis. Mol Cell Endocrinol 82:R13–R16

Gnessi L, Fabbri A, Spera G (1997) Gonadal peptides as mediators of development and functional control of the testis: an integrated system with hormones and local environment. Endocr Rev 18:541–609

Guo H, Calkins JH, Sigel MM, Lin T (1990) Interleukin-2 is a potent inhibitor of Leydig cell steroidogenesis. Endocrinology 127:1234–1239
Habert R, Picon R (1984) Testosterone, dihydrotestosterone and estradiol-17 beta levels in maternal and fetal plasma and in fetal testes in the rat. J Steroid Biochem 21:193–198
Hagen C, McNeilly AS (1975) Identification of human luteinizing hormone, follicle-stimulating hormone, luteinizing hormone β-subunit and gonadotropin α-subunit in foetal and adult pituitary glands. J Endocrinol 67:49–57
Hales DB (1992) Interleukin-1 inhibits Leydig cell steroidogenesis primarily by decreasing 17α-hydroxylase/C17-20 lyase cytochrome P450 expression. Endocrinology 131:2165–2172
Hales DB, Sha L, Payne AH (1987) Testosterone inhibits cAMP-induced de novo synthesis of Leydig cell cytochrome P-$450_{17\alpha}$ by an androgen receptor-mediated mechanism. J Biol Chem 262:11200–11206
Hales DB, Xiong Y, Tur-Kaspa I (1992) The role of cytokines in the regulation of Leydig cell P450c17 gene expression. J Steroid Biochem Mol Biol 43:907–914
Hall PF (1994) Testicular steroid synthesis: organization and regulation. In: Knobil E, Neill JD (eds) The physiology of reproduction, 2nd edn. Raven Press, New York, pp 1335–1362
Hall PF, Young DG (1968) Site of action of trophic hormones upon the biosynthetic pathways to steroid hormones. Endocrinology 82:559–565
Hall PF, Irby DC, de Kretser DM (1969) Conversion of cholesterol to androgens by rat testes: comparison of interstitial cells and seminiferous tubules. Endocrinology 84:488–496
Hall PF, Charponnier C, Nakamura M, Gabbiani G (1979) The role of microfilaments in the response of Leydig cells to luteinizing hormone. J Steroid Biochem 11:1361–1369
Hall PF, Osawa S, Mrotek J (1981) The influence of calmodulin on steroid synthesis in Leydig cells from rat testis. Endocrinology 109:1677–1682
Haour F, Kouznetzova B, Dray F, Saez JM (1979) hCG-induced prostaglandin E_2 and $F_{2\alpha}$ release in adult rat testis: role in Leydig cell desensitization to hCG. Life Sci 24:2151–2158
Hardy MP, Ganjam VK (1997) Stress, 11β-HSD, and Leydig cell function. J Androl 18: 475–479
Hardy MP, Zirkin BR, Ewing LL (1989) Kinetic studies on the development of the adult population of Leydig cells in testes of the pubertal rat. Endocrinology 124:762–770
Hardy MP, Kelce WR, Klinefelter GR, Ewing LL (1990) Differentiation of Leydig cell precursors in vitro: a role for androgen. Endocrinology 127:488–490
Hardy MP, Kirby JD, Hess RA, Cooke PS (1993) Leydig cells increase their numbers but decline in steroidogenic function in the adult rat after neonatal hypothyroidism. Endocrinology 132:2417–2420
Hardy MP, Sharma RS, Arambepola NK, Sottas CM, Russell LD, Bunick D, Hess RA, Cooke PS (1996) Increased proliferation of Leydig cells induced by neonatal hypothyroidism in the rat. J Androl 17:231–238
Harley VR (1993) Genetic control of testis determination. In: de Kretser DM (ed) Molecular biology of the male reproductive system. Academic Press, New York, pp 1–20
Hayes R, Chalmers SJ, Nikolic-Paterson DP, Atkins RC, Hedger MP (1996) Secretion of bioactive interleukin-1 by rat testicular macrophages in vitro. J Androl 17:41–49
He L, Hedger MP, Clements JA, Risbridger GP (1991) Localization of immunoreactive β-endorphin and adrenocorticotropic hormone, and pro-opiomelanocortin mRNA to testicular interstitial tissue macrophages. Biol Reprod 45:282–289
Hedger MP (1997) Testicular leukocytes: what are they doing? Rev Reprod 2:38–47
Hedger MP, Clarke L (1993) Isolation of rat blood lymphocytes using a two-step Percoll density gradient: effect of activin (erythroid differentiation factor) on peripheral T lymphocyte proliferation in vitro. J Immunol Meth 163:133–136
Hedger MP, Eddy EM (1990) Leydig cell cooperation in vitro: evidence for communication between adult rat Leydig cells. J Androl 11:9–16
Hedger MP, Risbridger GP (1992) Effect of serum and serum lipoproteins on testosterone production by adult rat Leydig cell in vitro. J Steroid Biochem Mol Biol 43:581–589

Hedger MP, Robertson DM, Browne CA, de Kretser DM (1985) The isolation and measurement of luteinizing hormone-releasing hormone (LHRH) from the rat testis. Mol Cell Endocrinol 42:163–174

Hedger MP, Robertson DM, de Kretser DM, Risbridger GP (1990) The quantification of steroidogenesis-stimulating activity in testicular interstitial fluid by an in vitro bioassay employing adult rat Leydig cells. Endocrinology 127:1967–1977

Hedger MP, McFarlane JR, de Kretser DM, Risbridger GP (1994) Multiple factors with steroidogenesis-regulating activity in testicular intertubular fluid from normal and experimentally cryptorchid adult rats. Steroids 59:676–685

Holash JA, Harik SI, Perry G, Stewart PA (1993) Barrier properties of testis microvessels. Proc. Natl Acad Sci USA 90:11 069–11 073

Hsueh AJW, Dufau ML, Catt KJ (1977) Gonadotropin-induced regulation of luteinizing hormone receptors and desensitization of testicular 3′,5′-cyclic AMP and testosterone responses. Proc Natl Acad Sci USA 74:592–595

Hsueh AJW, Dahl KD, Vaughan J, Tucker E, Rivier J, Bardin CW, Vale W (1987) Heterodimers and homodimers of inhibin subunits have different paracrine action in the modulation of luteinizing hormone-stimulated androgen biosynthesis. Proc Natl Acad Sci USA 84:5082–5086

Hu Z-Z, Buczko E, Zhuang L, Dufau ML (1994) Sequence of the 3′-noncoding region of the luteinizing hormone receptor gene and identification of two polyadenylation domains that generate the major mRNA forms. Biochim Biophys Acta 1220:330–337

Huhtaniemi I (1977) Studies on steroidogenesis and its regulation in human fetal adrenal and testis. J Steroid Biochem 8:491–497

Hutson JC (1992) Development of cytoplasmic digitations between Leydig cells and testicular macrophages of the rat. Cell Tissue Res 267:385–389

Hutson JC (1994) Testicular macrophages. Int Rev Cytol 149:99–143

Inoue Y, Rebois RV (1989) Protein kinase C can desensitize the gonadotropin-responsive adenylate cyclase in Leydig tumor cells. J Biol Chem 264:8504–8908

Irby DC, Hall PF (1971) Stimulation by ICSH of protein biosynthesis in isolated Leydig cells from hypophysectomized rats. Endocrinology 89:1367–1374

Jackson AE, O'Leary PC, Ayers MM, de Kretser DM (1986) The effects of ethylene dimethane sulphonate (EDS) on rat Leydig cells: evidence to support a connective tissue origin of Leydig cells. Biol Reprod 35:425–437

Janszen FHA, Cooke BA, van Driel MJA, van der Molen HJ (1976) The effect of calcium ions on testosterone production in Leydig cells from rat testis. Biochem J 160:433–437

Jégou B, Laws AO, de Kretser DM (1984) Changes in testicular function induced by short-term exposure of the rat testis to heat: further evidence for interaction of germ cells, Sertoli cells and Leydig cells. Int J Androl 7:244–257

Kaler LW, Neaves WB (1978) Attrition of the human Leydig cell population with advancing age. Anat Rec 192:513–518

Kalla NR, Nisula BC, Menard R, Loriaux DL (1980) The effect of estradiol on testicular testosterone biosynthesis. Endocrinology 106:35–39

Kaplan SL, Grumbach MM, Aubert ML (1976) The ontogenesis of pituitary hormones and hypothalamic factors in the human fetus. Maturation of central nervous system regulation of anterior pituitary function. Recent Prog Horm Res 32:161–243

Kasson BG, Adashi EY, Hsueh AJW (1986a) Arginine vasopressin in the testis: an intragonadal peptide control system. Endocr Rev 7:156–168

Kasson BG, Lim P, Hsueh AJW (1986b) Vasoactive intestinal peptide stimulates androgen biosynthesis by cultured neonatal testicular cells. Mol Cell Endocrinol 48:21–29

Keeney DS, Ewing LL (1990) Effects of hypophysectomy and alterations in spermatogenic function on Leydig cell volume, number, and proliferation in adult rats. J Androl 11:367–378

Keeney DS, Mendis-Handagama SMLC, Zirkin BR, Ewing LL (1988) Effect of long-term deprivation of luteinizing hormone on Leydig cell volume, Leydig cell number, and steroidogenic capacity of the rat testis. Endocrinology 123:2906–2915

Keeney DS, Sprando RL, Robaire B, Zirkin BR, Ewing LL (1990) Reversal of long-term LH deprivation on testosterone secretion and Leydig cell volume, number and proliferation in adult rats. J Endocrinol 127:47–58

Kern S, Robertson SA, Mau VJ, Maddocks S (1995) Cytokine secretion by macrophages in the rat testis. Biol Reprod 53:1407–1416

Kerr JB, Donachie K (1986) Regeneration of Leydig cells in unilaterally cryptorchid rats: evidence for stimulation by local testicular factors. Cell Tissue Res 245:649–655

Kerr JB, Knell CM (1988) The fate of fetal Leydig cells during the development of the fetal and postnatal rat testis. Development 103:535–544

Kerr JB, Sharpe RM (1985a) Follicle-stimulating hormone induction of Leydig cell maturation. Endocrinology 116:2592–2604

Kerr JB, Sharpe RM (1985b) Stimulatory effect of follicle-stimulating hormone on rat Leydig cells: a morphometric and ultrastructural study. Cell Tissue Res 239:405–415

Kerr JB, Rich KA, de Kretser DM (1979) Alterations of fine structure and androgen secretion of the interstitial cells in the experimentally cryptorchid rat testis. Biol Reprod 20:409–422

Kerr JB, Bartlett JMS, Donachie K, Sharpe RM (1987) Origin of regenerating Leydig cells in the testis of the adult rat. Cell Tissue Res 249:367–377

Khan S, Teerds K, Dorrington J (1992a) Growth factor requirements for DNA synthesis by Leydig cells from the immature rat. Biol Reprod 46:335–341

Khan S, Khan SJ, Dorrington JH (1992b) Interleukin-1 stimulates deoxyribonucleic acid synthesis in immature rat Leydig cells in vitro. Endocrinology 131:1853–1857

Kinnally KW, Zorov DB, Antonenko YN, Snyder SH, McEnery MW, Tedesschi H (1993) Mitochondrial benzodiazepine receptor linked to inner membrane ion channels by nanomolar actions of ligands. Proc Natl Acad Sci USA 90:1374–1378

Kremer H, Kraaij R, Toledo SP, Post M, Fridman JB, Hayashida CY, van Reen M, Milgrom E, Ropers HH, Mariman E et al. (1995) Male pseudohermaphroditism due to a homozygous missense mutation of the luteinizing hormone receptor gene. Nat Genet 9:160–164

Kumamoto T, Ito A, Omura T (1989) Critical region in the extension peptide for the import of cytochrome P450(SCC) precursor into mitochondria. J Biochem 105:72–78

Kumar S, Blumberg DL, Canas JA, Maddaiah VT (1994) Human chorionic gonadotropin (hCG) increases cytosolic free calcium in adult rat Leydig cells. Cell Calcium 15:349–355

Landy H, Boepple PA, Mansfield MJ, Whitcomb RW, Schneyer AL, Crawford JD, Crigler JF Jr, Crowley WF Jr (1991) Altered patterns of pituitary secretion and renal secretion of free alpha subunit during gonadotropin-releasing hormone agonist-induced pituitary desensitization. J Clin Endocrinol Metab 72:711–717

Laslett AL, McFarlane JR, Hearn MTW, Risbridger GP (1995) Requirement for heparin sulphate proteoglycans to mediate basic fibroblast growth factor (FGF-2)-induced stimulation of Leydig cell steroidogenesis. J Steroid Biochem Mol Biol 54:245–250

Lee W, Mason AJ, Schwall R, Szonyi E, Mather JP (1989) Secretion of activin by interstitial cells in the testis. Science 243:396–398

Lejeune H, Chuzel F, Sanchez P, Durand P, Mather JP, Saez JM (1997) Stimulating effect of both human recombinant inhibin A and activin A on immature porcine Leydig cell functions in vitro. Endocrinology 138:4783–4791

Leydig F (1850) Zur anatomie der männlichen geschlechtsorgane und analdrüsen der säugetiere [On the anatomy of the male sex organs and anal glands of mammals]. Z Wiss Zool 2:1–57

Lin D, Sugawara T, Strauss JF, Clark BJ, Stocco DM, Saenger P, Rogol A, Miller WL (1995) Indispensable role of steroidogenic acute regulatory protein in adrenal and gonadal steroidogenesis. Science 267:1828–1831

Lin T (1985) Mechanism of action of gonadotropin releasing hormone stimulated Leydig cell steroidogenesis III. The role of arachidonic acid and calcium/phospholipid dependent protein kinase. Life Sci 36:1255–1264

Lin T, Murono E, Osterman J, Troen P, Nankin HR (1979) The effects of verapamil on interstitial cell steroidogenesis. Int J Androl 2:549–558

Lin T, Haskell J, Vinson N, Terracio L (1986) Direct stimulatory effects of insulin-like growth factor-I on Leydig cell steroidogenesis. Biochem Biophys Res Commun 137:950–956
Lin T, Blaisdell J, Haskell JF (1987) Transforming growth factor-β inhibits Leydig cell steroidogenesis in primary culture. Biochem Biophys Res Commun 146:387–394
Lin T, Blaisdell J, Haskell JF (1988) Hormonal regulation of type I insulin-like growth factor receptors of Leydig cells in hypophysectomized rats. Endocrinology 123:134–139
Lin T, Calkins JH Morris PL, Vale W, Bardin CW (1989) Regulation of Leydig cell function in primary culture by inhibin and activin. Endocrinology 125:2134–2140
Lin T, Wang D, Nagpal ML, Chang W, Calkins JH (1992) Down-regulation of Leydig cell insulin-like growth factor-I gene expression by interleukin-1. Endocrinology 130:1217–1224
Lin T, Wang D, Nagpal ML (1993) Human chorionic gonadotropin induces interleukin-1 gene expression in rat Leydig cells in vivo. Mol Cell Endocrinol 95:139–145
Lindzey J, Wetsel WC, Couse JF, Stoker T, Cooper R, Korach KS (1998) Effects of castration and chronic steroid treatments on hypothalamic gonadotropin-releasing hormone content and pituitary gonadotropins in male wild-type and estrogen receptor-α knockout mice. Endocrinology 139:4092–4101
Lopez-Ruiz MP, Choi MS, Rose MP, West AP, Cooke BA (1992) Direct effect of arachidonic acid on protein kinase C and LH-stimulated steroidogenesis in rat Leydig cells; evidence for tonic inhibitory control of steroidogenesis by protein kinase C. Endocrinology 130:1122–1130
Lording DW, de Kretser DM (1972) Comparative ultrastructural and histochemical studies of the interstitial cells of the rat testis during fetal and postnatal development. J Reprod Fertil 29: 261–269
Lorence MC, Naville D, Graham-Lorence SE, Mack SO, Murry BA, Trant JM, Mason JI (1991) 3β-hydroxysteroid dehydrogenase/$\Delta^{5\rightarrow4}$-isomerase expression in rat and characterization of the testis isoform. Mol Cell Endocrinol 80:21–31
Loveland KL, Hedger MP, Risbridger GP, Herszfeld D, de Kretser DM (1993) Identification of receptor tyrosine kinases in the rat testis. Mol Reprod Dev 36:440–447
Lubahn DB, Brown TR, Simental JA, Higgs HN, Migeon CJ, Wilson EM, French FS (1989) Sequence of the intron/exon junctions of the coding region of the human androgen receptor gene and identification of a point mutation in a family with complete androgen insensitivity. Proc Natl Acad Sci USA 86:9534–9538
Lund J, Ahlgren R, Wu DH, Kagimoto M, Simpson ER, Waterman MR (1990) Transcriptional regulation of the bovine CYP17 ($P\text{-}450_{17\alpha}$) gene. Identification of two cAMP regulatory regions lacking the consensus cAMP-responsive element (CRE). J Biol Chem 265:3304–3312
Lund J, Bakke M, Mellgren G, Morohashi K, Døskeland S-O (1997) Transcriptional regulation of the bovine CYP17 gene by cAMP. Steroids 62:43–45
Maddocks S, Sharpe RM (1989) Interstitial fluid volume in the rat testis: androgen-dependent regulation by the seminiferous tubules. J Endocrinol 120:215–222
Mancini RE, Vilar O, Lavieri JC, Andrada JA, Heinrich JJ (1963) Development of Leydig cells in the normal human testis. A cytological, cytochemical and quantitative study. Am J Anat 112:203–210
Matteson KL, Picado-Leonard J, Chung B, Mohandas TK, Miller WL (1986) Assignment of the gene for adrenal P-450C17 (steroid 17α-hydroxylase/17,20-lyase) to human chromosome 10. J Clin Endocrinol Metab 63:798–791
Mauduit C, Chauvin MA, de Peretti E, Morera AM, Benahmed M (1991a) Effect of activin A on dehydroepiandrosterone and testosterone secretion by primary immature porcine Leydig cells. Biol Reprod 45:101–109
Mauduit C, Hartmann DJ, Chauvin MA, Revol A, Morera AM, Benahmed M (1991b) Tumor necrosis factor α inhibits gonadotropin action in cultured porcine Leydig cells: site(s) of action. Endocrinology 129:2933–2940
Mauduit C, Chauvin MA, Hartmann DJ, Revol A, Morera AM, Benahmed M (1992) Interleukin-1α as a potent inhibitor of gonadotropin action in porcine Leydig cells: site(s) of action. Biol Reprod 46:1119–1126

Mayerhofer A, Seidl K, Lahr G, Bitter-Suermann D, Christoph A, Barthels D, Wille W, Gratzl M (1992) Leydig cells express neural cell adhesion molecules in vivo and in vitro. Biol Reprod 47:656–664

McFarland KC, Sprengel R, Phillips HS, Kohler M, Rosemblit N, Nikolics K, Segaloff DL, Seeberg PH (1989) Lutropin-choriogonadotropin receptor; an unusual member of the G protein coupled receptor family. Science 245:494–499

McFarlane JR, Laslett A, de Kretser DM, Risbridger GP (1996) Evidence that heparin binding autocrine factors modulate testosterone production by the adult rat Leydig cell. Mol Cell Endocrinol 118:57–63

McNeilly AS, de Kretser DM, Sharpe RM (1979) Modulation of prolactin, luteinizing hormone (LH) and follicle stimulating hormone (FSH) secretion by LHRH and bromocriptine (CB154) in the hypophysectomized pituitary-grafted male rat and its effects on testicular LH receptors and testosterone output. Biol Reprod 21:141–147

McPhaul MJ, Marcelli M, Zoppi S, Griffin JE, Wilson JD (1993) Genetic basis of endocrine disease. 4. The spectrum of mutations in the androgen receptor gene that causes androgen resistance. J Clin Endocrinol Metab 76:17–23

Meikle AW, Cardoso de Sousa JC, Dacosta N, Bishop DK, Samlowski WE (1992) Direct and indirect effects of murine interleukin-2, gamma interferon, and tumor necrosis factor on testosterone synthesis in mouse Leydig cells. J Androl 13:437–443

Meinhardt A, Bacher M, McFarlane JR, Metz CN, Seitz J, Hedger MP, de Kretser DM, Bucala R (1996) Macrophage migration inhibitory factor (MIF) production by rat Leydig cells: evidence for a role in local regulation of testicular function. Endocrinology 137:5090–5095

Mendelson C, Dufau ML, Catt KJ (1975) Dependence of gonadotropin-induced steroidogenesis on RNA and protein synthesis in the interstitial cells of the rat testis. Biochim Biophys Acta 411:222–230

Mendis-Handagama SMLC, Risbridger GP, de Kretser DM (1987) Morphometric analysis of the components of the neonatal and the adult rat testis interstitium. Int J Androl 10:525–534

Miller SC, Bowman BM, Rowland HG (1983) Structure, cytochemistry, endocytic activity, and immunoglobulin (Fc) receptors of rat testicular interstitial-tissue macrophages. Am J Anat 168:1–13

Misro MM, Ganguly A, Das RP (1993) Is testosterone essential for the maintenance of normal morphology in immature rat Leydig cells? Int J Androl 16:221–226

Mock EJ, Norton HW, Frankel AI (1978) Daily rhythmicity of serum testosterone concentration in the male laboratory rat. Endocrinology 103:1111–1121

Molenaar R, de Rooij DG, Rommerts FFG, van der Molen (1986) Repopulation of Leydig cells in mature rats after selective destruction of the existent Leydig cells with ethylene dimethane sulfonate is dependent on luteinizing hormone and not follicle-stimulating hormone. Endocrinology 118:2546–2554

Moore A, Morris ID (1993) The involvement of insulin-like growth factor I in local control of steroidogenesis and DNA synthesis of Leydig and non-Leydig cells in the rat testicular interstitium. J Endocrinol 138:107–114

Moore C, Hutson JC (1994) Physiological relevance of tumor necrosis factor in mediating macrophage-Leydig cell interactions. Endocrinology 134:63–69

Moore C, Moger WH (1991) Interleukin-1 alpha-induced changes in androgen and cyclic adenosine 3′,5′-monophosphate release in adult rat Leydig cells in culture. J Endocrinol 129: 381–390

Moran O, Sandri G, Panfili E, Stuhmer W, Sorgato MC (1990) Electrophysiological characterization of contact sites in brain mitochondria. J Biol Chem 265:908–913

Morohashi K, Sogawa K, Omura T, Fujii-Kuriyama Y (1987) Gene structure of human cytochrome P-450 (SCC), cholesterol desmolase. J Biochem 101:879–887

Morohashi K, Zanger UM, Honda S, Hara M, Waterman MR, Omura T (1993) Activation of CYP11A and CYP11B gene promoters by the steroidogenic cell-specific transcription factor, Ad4BP. Mol Endocrinol 7:1196–1204

Morris AJ, Taylor MF, Morris ID (1997) Leydig cell apoptosis in response to ethane sulphonate after both in vivo and in vitro treatment. J Androl 18:274–280

Morris ID, Phillips DM, Bardin CW (1986) Ethylene dimethanesulfonate destroys Leydig cells in the rat testis. Endocrinology 118:709–719

Morris MD, Chaikoff IL (1959) The origin of cholesterol in liver, small intestines, adrenal gland and testis of the rat: dietary versus endogenous contributions. J Biol Chem 234:1095–1097

Morrow AF, Gyorki S, Warne GL, Burger HG, Bangah ML, Outch KH, Mirovics A, Baker HWG (1987) Variable androgen receptor levels in infertile men. J Clin Endocrinol Metab 64: 1115–1121

Mulder E, Peters MJ, de Vries J, van der Molen HJ (1975) Characterization of a nuclear receptor for testosterone in seminiferous tubules of mature rat testes. Mol Cell Endocrinol 2:171–182

Mullaney BP, Skinner MK (1993) Transforming growth factor-β (β1, β2, and β3) gene expression and action during pubertal development of the seminiferous tubule: potential role at the onset of spermatogenesis. Mol Endocrinol 7:67–76

Murono EP, Washburn AL, Goforth DP, Wu N (1993) Fibroblast growth factor-induced increase in ^{125}I-human chorionic gonadotropin binding to luteinizing hormone receptors in cultured immature Leydig cells is mediated by binding to heparan sulfate proteoglycans. Mol Cell Endocrinol 97:109–114

Naftolin F, Feder HH (1973) Suppression of luteinizing hormone secretion in male rats by 5α-androstan-17β-ol-3-one (dihydrotestosterone) propionate. J Endocrinol 56:155–156

Nakajin S, Shively J, Yuan PM, Hall PF (1981) Microsomal cytochrome P-450 from neonatal pig testes: two enzymatic activities (17α-hydroxylase and C_{17-20}-lyase) associated with one protein. Biochemistry 20:4037–4045

Namiki M, Yokokawa K, Okuyama A, Koh E, Kiyohara H, Nakao M, Sakoda S, Matsumoto K, Sonoda T (1991) Evidence for the presence of androgen receptors in human Leydig cells. J Steroid Biochem Mol Biol 38:79–82

Nason TF, Han XG, Hall PF (1992) Cyclic AMP regulates expression of the rat gene for steroid 17α-hydroxylase/C_{17-20}-lyase P-450 (CYP17) in rat Leydig cells. Biochim Biophys Acta 1171: 73–80

Nikula H, Huhtaniemi I (1989) Effects of protein kinase C activation on cyclic AMP and testosterone production of rat Leydig cells in vitro. Acta Endocrinol 121:327–333

Nishihara M, Winters CA, Buczko E, Waterman MR, Dufau ML (1988) Hormonal regulation of rat Leydig cell cytochrome $P450_{17\alpha}$ mRNA levels and characterization of a partial length rat $P450_{17\alpha}$ cDNA. Biochem Biophys Res Commun 154:151–158

Nistal M, Paniagua R, Regardera J, Santamaria L, Amat P (1986) A quantitative morphological study of human Leydig cells from birth to adulthood. Cell Tissue Res 246:229–236

Noulin JF, Joffre M (1993a) Cyclic AMP- and calcium-activated chloride currents in Leydig cells isolated from mature rat testis. Arch Int Physiol Biochim Biophys 101:35–41

Noulin JF, Joffre M (1993b) Characterization and cyclic AMP-dependence of a hyperpolarization-activated chloride conductance in Leydig cells from mature rat testis. J Membr Biol 133:1–15

Nozu K, Dufau ML, Catt KJ (1981) Estradiol receptor-mediated regulation of steroidogenesis in gonadotropin-desensitized Leydig cells. J Biol Chem 256:1915–1922

O'Donnell L, Stanton PG, Wreford NG, Robertson DM, McLachlan RI (1996) Inhibition of 5α-reductase activity impairs the testosterone-dependent restoration of spermiogenesis in adult rats. Endocrinology 137:2703–2710

Okuda Y, Bardin CW, Hodgskin LR, Morris PL (1995) Interleukins-1α and -1β regulate interleukin-6 expression in Leydig and Sertoli cells. Rec Prog Horm Res 50:367–372

O'Leary P, Jackson AE, Averill S, de Kretser DM (1986) The effects of ethane dimethane sulphonate (EDS) on bilaterally cryptorchid rat testes. Mol Cell Endocrinol 45:183–190

Orava M, Cantell K, Vihko R (1985) Human leukocyte interferon inhibits human chorionic gonadotropin stimulated testosterone production by porcine Leydig cells in culture. Biochem Biophys Res Commun 127:809–815

Orava M, Voutilainen R, Vihko R (1989) Interferon-γ inhibits steroidogenesis and accumulation of mRNA of the steroidogenic enzymes, $P450_{SCC}$ and $P450_{C}17$ in cultured porcine Leydig cells. Mol Endocrinol 3:887–894

Orth J, Christensen AK (1977) Localization of ^{125}I-labeled FSH in the testes of hypophysectomized rats by autoradiography at the light and electron microscope levels. Endocrinology 101:262–278

Padron Duran RS (1985) Respuesta testicular a diferentes dosis de gonadotropina corionica humana en hombres normales [Testicular response to various doses of human chorionic gonadotropin in normal men]. Rev Invest Clin 37:17–19

Pandey KN, Pavlou SN, Kovacs WJ, Inagami T (1986) Atrial natriuretic factor regulates steroidogenic responsiveness and cyclic nucleotide levels in mouse Leydig cells *in vitro*. Biochem Biophys Res Commun 138:399–404

Paniagua R, Amat P, Nistal M, Martin A (1986) Ultrastructure of Leydig cells in human ageing testis. J Anat 146:173–183

Papadopoulos V (1991) Identification and purification of a human Sertoli cell-secreted protein (hSCSP-80) stimulating Leydig cell steroid biosynthesis. J Clin Endocrinol Metab 72:1332–1339

Papadopoulos V (1993) Peripheral-type benzo-diazepine/diazepam binding inhibitor receptor: biological role in steroidogenic cell function. Endocr Rev 14:222–240

Payne AH, Sha L (1991) Multiple mechanisms for regulation of 3β-hydroxysteroid dehydrogenase/Δ^5-Δ^4-isomerase, 17α-hydroxylase/C_{17-20} lyase cytochrome P450, and cholesterol side-chain cleavage P450 messenger ribonucleic acid levels in primary cultures of mouse Leydig cells. Endocrinology 129:1429–1435

Payne AH, Youngblood GL (1995) Regulation of expression of steroidogenic enzymes in Leydig cells. Biol Reprod 52:217–225

Payne AH, Kelch RP, Musich SS, Halpern ME (1976) Intratesticular site of aromatization in the human. J Clin Endocrinol Metab 42:1081–1087

Pedersen RC (1984) Polypeptide activators of cholesterol side-chain cleavage. Endocr Res 10: 533–561

Pedersen RC (1987) Steroidogenesis activator polypeptide (SAP) in the rat ovary and testis. J Steroid Biochem 27:731–735

Pelliniemi LJ, Niemi M (1969) Fine structure of the human foetal testis. I. The interstitial tissue. Z Zellforsch 99:507–522

Perrard-Sapori MH, Chatelain PC, Rogemond N, Saez JM (1987) Modulation of Leydig cell functions by culture with Sertoli cells or with Sertoli cell-conditioned medium: effect of insulin, somatomedin-C and FSH. Mol Cell Endocrinol 50:193–201

Phillips DM, Lakshmi V, Monder C (1989) Corticosteroid 11β-dehydrogenase in rat testis. Endocrinology 125:209–216

Purvis K, Hansson V (1978) Hormonal regulation of Leydig cell function. Mol Cell Endocrinol 12:123–128

Quinn PG, Payne AH (1984) Oxygen-mediated damage of microsomal cytochrome P450 enzymes in cultured Leydig cells. Role in steroidogenic desensitization. J Biol Chem 259: 4130–4135

Quinn PG, Dombrausky LJ, Chen Y-DI, Payne AH (1981) Serum lipoproteins increase testosterone production in hCG-desensitized Leydig cells. Endocrinology 109:1790–1792

Raeside JI, Renaud RL (1983) Estrogen and androgen production by purified Leydig cells of mature boars. Biol Reprod 28:727–733

Ramnath HI, Peterson S, Michael AE, Stocco DM, Cooke BA (1997) Modulation of steroidogenesis by chloride ions in MA-10 mouse tumor Leydig cells: roles of calcium, protein synthesis, and the steroidogenic acute regulatory protein. Endocrinology 138:2308–2314

Reddy GP, Prasad M, Sailesh S, Kumar YVK, Reddanna P (1993) Arachidonic acid metabolites as intratesticular factors controlling androgen production. Int J Androl 16:227–233

Resko JA, Eik-Nes KB (1966) Diurnal testosterone levels in peripheral plasma of human male subjects. J Clin Endocrinol Metab 26:573–576

Reyes FI, Boroditsky RS, Winter JSD, Faiman C (1974) Studies on human sexual development. II. Fetal and maternal serum gonadotropin and sex steroid concentrations. J Clin Endocrinol Metab 38:612–617

Rhéaume E, Lachance Y, Zhao Z-F, Breton N, Dumont M, de Launoit Y, Trudel C, Luu-The V, Simard J, Labrie F (1991) Structure and expression of a new complementary DNA encoding the almost exclusive 3β-hydroxysteroid dehydrogenase/$\Delta^5 \rightarrow \Delta^4$-isomerase in human adrenals and gonads. Mol Endocrinol 5:1147–1157

Rich KA, Kerr JB, de Kretser DM (1979) Evidence for Leydig cell dysfunction in rats with seminiferous tubule damage. Mol Cell Endocrinol 13:123–135

Risbridger GP, Kerr JB, Peake RA, de Kretser DM (1981) An assessment of Leydig cell function after bilateral or unilateral efferent duct ligation: further evidence for local control of Leydig cell function. Endocrinology 109:1234–1241

Risbridger GP, Clements J, Robertson DM, Drummond AE, Muir J, Burger HG, de Kretser DM (1989) Immuno- and bioactive inhibin and inhibin α-subunit expression in rat Leydig cell cultures. Mol Cell Endocrinol 66:119–122

Russell DW, Wilson JD (1994) Steroid 5α-reductase: two genes/two enzymes. Annu Rev Biochem 63:25–61

Russell LD, Amlani SR, Vogl AW, Weber JE (1987) Characterization of filaments within Leydig cells of the rat testis. Am J Anat 178:231–240

Russell LD, Corbin TJ, Ren HP, Amador A, Bartke A, Ghosh S (1992) Structural changes in rat Leydig cells posthypophysectomy: a morphometric and endocrine study. Endocrinology 131:498–508

Ryan KJ, Naftolin F, Reddy V, Flores F, Petro Z (1972) Estrogen formation in the brain. Am J Obstet Gynecol 114:454–460

Santen RJ (1975) Is aromatization of testosterone to estradiol required for inhibition of luteinizing hormone secretion in men? J Clin Invest 56:1555–1563

Santen RJ (1977) Independent effects of testosterone and estradiol on the secretion of gonadotropins in men. In: Troen P, Nankin HR (eds) The testis in normal and infertile men. Raven Press, New York, pp 197–211

Santen RJ, Bardin CW (1973) Episodic luteinizing hormone secretion in man. Pulse analysis, clinical interpretation, physiologic mechanisms. J Clin Invest 52:2617–2628

Shenker A, Laue L, Kosugl S, Merendino JJ Jr, Minegishi T, Cutler GB Jr (1993) A constitutively activating mutation of the luteinizing hormone receptor in familial male precocious puberty. Nature 365:652–654

Schulze W, Davidoff MS, Holstein A-F (1987) Are Leydig cells of neural origin? Substance P-like immunoactivity in human testicular tissue. Acta Endocrinol 115:373–377

Schweikert HU, Knauf W, Romalo G, Holler W, Bidlingmaier F, Knorr D (1987) Androgen binding in cultured human fibroblasts from patients with idiopathic hypopadias. Horm Metab Res 19:497–501

Setchell BP, Sharpe RM (1981) The effects of human chorionic gonadotrophin on capillary permeability, extracellular fluid volume and flow of lymph and blood in the testis of rats. J Endocrinol 91:245–254

Setchell BP, Maddocks S, Brooks DE (1994) Anatomy, vasculature, innervation, and fluids of the male reproductive tract. In: Knobil E, Neill JD (eds) The physiology of reproduction, 2nd edn. Raven Press, New York, pp 1063–1173

Shan LX, Hardy MP (1992) Developmental changes in levels of luteinizing hormone receptor and androgen receptor in rat Leydig cells. Endocrinology 131:1107–1114

Shan L, Hardy DO, Catterall JF, Hardy MP (1995) Effects of luteinizing hormone (LH) and androgen on steady-state levels of messenger ribonucleic acid for LH receptors, androgen receptors, and steroidogenic enzymes in rat Leydig cell progenitors in vivo. Endocrinology 136:1686–1693

Sharpe RM, Cooper I (1984) Intratesticular secretion of a factor(s) with major stimulatory effects on Leydig cell testosterone secretion in vitro. Mol Cell Endocrinol 37:159–168

Sharpe RM, Bartlett JMS, Allenby G (1991) Evidence for the control of testicular interstitial fluid volume in the rat by specific germ cell types. J Endocrinol 128:359–367

Shenker A, Laue L, Kosugi S, Meredino JJ, Minegishi T, Cutler GB (1993) A constitutively activating mutation of the luteinizing hormone receptor in familial male precocious puberty. Nature 365:652–654

Simpson BJB, Risbridger GP, Hedger MP, de Kretser DM (1991) The role of calcium in luteinizing hormone/human chorionic gonadotrophin stimulation of Leydig cell immunoactive inhibin secretion in vitro. Mol Cell Endocrinol 75:49–56

Skinner MK, Fritz IB (1985) Testicular peritubular cells secrete a protein under androgen control that modulates Sertoli cell functions. Proc Natl Acad Sci USA 82:114–118

Stalker A, Hermo L, Antakly T (1989) Covalent affinity labeling, autoradiography, and immunocytochemistry localize the glucocorticoid receptor in rat testicular cells. Am J Anat 186: 369–377

Stocco DM, Clark BJ (1997) The role of the steroidogenic acute regulatory protein in steroidogenesis. Steroids 62:29–36

Sullivan MHF, Cooke BA (1984) Role of calcium in luteinizing hormone releasing hormone agonist (ICI 118630) stimulated steroidogenesis in rat Leydig cells. Biochem J 218:621–624

Sullivan MHF, Cooke BA (1985a) Effects of calmodulin and lipoxygenase inhibitors on LH- and LHRH agonist-stimulated steroidogenesis in rat Leydig cells. Biochem J 232:55–59

Sullivan MHF, Cooke BA (1985b) Control and production of leukotriene B4 in rat tumor and testicular Leydig cells. Biochem J 230:821–824

Sullivan MHF, Cooke BA (1986) The role of Ca^{2+} in steroidogenesis in Leydig cells: stimulation of intracellular Ca^{2+} levels by luteinizing hormone, luteinizing hormone releasing hormone agonist and cyclic AMP. Biochem J 236:45–51

Sun YT, Wreford NG, Robertson DM, de Kretser DM (1990) Quantitative cytological studies of spermatogenesis in intact and hypophysectomized rats: identification of androgen-dependent stages. Endocrinology 127:1215–1223

Syed V, Khan SA, Ritzen EM (1985) Stage-specific inhibition of interstitial cell testosterone secretion by rat seminiferous tubules in vitro. Mol Cell Endocrinol 40:257–264

Talbot JA, Rodger RS, Shalet SM, Littley MD, Robertson WR (1990) The pulsatile secretion of bioactive luteinising hormone in normal adult men. Acta Endocrinol 122:643–650

Tan J, Joseph DR, Quarmby VE, Lubahn DB, Sar M, French FS, Wilson EM (1988) The rat androgen receptor primary structure, autoregulation of its messenger ribonucleic acid and immunocytochemical localization of the receptor protein. Mol Endocrinol 2:1276–1285

Teerds KJ, Dorrington JH (1993) Localization of transforming growth factor beta1 and beta2 during testicular development in the rat. Biol Reprod 48:40–45

Teerds KJ, Closset J, Rommerts FFG, de Rooij DG, Stocco DM, Colenbrander B, Wensing CJG, Hennen G (1989a) Effects of pure FSH and LH preparations on the number and function of Leydig cells in immature hypophysectomized rats. J Endocrinol 120:97–106

Teerds KJ, de Rooij DG, Rommerts FFG, van den Hurk R, Wensing CJG (1989b) Proliferation and differentiation of possible Leydig cell precursors after destruction of the existing Leydig cells with ethane dimethane sulphonate: the role of LH/human chorionic gonadotrophin. J Endocrinol 122:689–696

Teerds KJ, de Rooij DG, Rommerts FFG, van den Hurk R, Wensing CJG (1989c) Stimulation of the proliferation and differentiation of Leydig cell precursors after the destruction of existing Leydig cells with ethane sulphonate (EDS) can take place in the absence of LH. J Androl 10: 472–477

Teerds KJ, Rommerts FF, Dorrington JH (1990) Immunohistochemical detection of transforming growth factor-α in Leydig cells during the development of the rat testis. Mol Cell Endocrinol 69:R1–6

Themmen APN, Hoogerbrugge JW, Rommerts FFG, van der Molen HJ (1985) Is cAMP the obligatory second messenger in the action of lutropin on Leydig cell steroidogenesis? Biochem Biophys Res Commun 128:1164–1172

Tilley WD, Marchelli M, Wilson JD, McPhaul MJ (1989) Characterization and expression of a cDNA encoding the human androgen receptor. Proc Natl Acad Sci USA 86:327–331

Tinajero JC, Fabbri A, Ciocca DR, Dufau ML (1993) Serotonin secretion from rat Leydig cells. Endocrinology 113:3026–3029

Tsai-Morris CH, Buczko E, Wang W, Xie X-Z, Dufau ML (1991) Structural organization of the rat luteinizing hormone (LH) receptor gene. J Biol Chem 266:11355–11359

Tsuruhara T, Dufau ML, Cigorraga S, Catt KJ (1977) Hormonal regulation of testicular luteinizing hormone receptors. Effects on cyclic AMP and testosterone responses in isolated Leydig cells. J Biol Chem 252:9002–9009

Turner TT, Caplis L, Miller DW (1996) Testicular microvascular blood flow: alteration after Leydig cell eradication and ischemia but not experimental varicocele. J Androl 17:239–248

Tut TG, Ghadessy FJ, Trifiro MA, Pinsky L, Yong EL (1997) Long polyglutamine tracts in the androgen receptor are associated with reduced *trans*-activation, impaired sperm production, and male infertility. J Clin Endocrinol Metab 82:3777–3782

van Haren L, Teerds KJ, Ossendorp BC, van Heusden GPH, Orly J, Stocco DM, Wirtz KWA, Rommerts FFG (1992) Sterol carrier protein 2 (non-specific lipid transfer protein) is localized in membranous fractions of Leydig cells and Sertoli cells, but not in germ cells. Biochim Biophys Acta 1124:288–296

van Noort M, Rommerts FFG, van Amerongen A, Wirtz KWA (1988) Intracellular redistribution of SCP2 in Leydig cells after hormonal stimulation may contribute to increased pregnenolone production. Biochem Biophys Res Commun 154:60–65

van Vorstenbosch CJAHV, Colenbrander B, Wensing CJG, Ramaekers FCS, Vooijs GP (1984) Cytoplasmic filaments in fetal and neonatal pig testis. Eur J Cell Biol 34:292–299

Verhoeven G, Cailleau J (1987) A Leydig cell stimulatory factor produced by human testicular tubules. Mol Cell Endocrinol 49:137–147

Verhoeven G, Cailleau J, Damme J, Billiau A (1988) Interleukin-1 stimulates steroidogenesis in cultured rat Leydig cells. Mol Cell Endocrinol 57:51–60

Vihko KK, Huhtaniemi I (1989) A rat seminiferous epithelial factor that inhibits Leydig cell cAMP and testosterone production: mechanism of action, stage-specific secretion and partial characterization. Mol Cell Endocrinol 65:119–127

Vliegen MK, Schlatt S, Weinbauer GF, Bergmann M, Groome NP, Nieschlag E (1993) Localization of inhibin/activin subunits in the testis of adult non-human primates and men. Cell Tissue Res 273:261–268

Wallgren M, Kindahl H, Rodriguez-Martinez H (1993) Alterations in testicular function after endotoxin injection in the boar. Int J Androl 16:235–243

Wang DL, Nagpal ML, Calkins JH, Chang WW, Sigel MM, Lin T (1991a) Interleukin-1β induces interleukin-1α messenger ribonucleic acid expression in primary cultures of Leydig cells. Endocrinology 129:2862–2866

Wang H, Ascoli M, Segaloff DL (1991b) Multiple luteinizing hormone/chorionic gonadotropin receptor messenger ribonucleic acid transcripts. Endocrinology 129:133–138

Wang H, Segaloff DL, Ascoli M (1991c) Lutropin/choriogonadotropin down-regulates its receptor by both receptor-mediated endocytosis and a cAMP-dependent reduction in receptor mRNA. J Biol Chem 266:780–785

Warren DW, Pasupuleti V, Lu Y, Platler BW, Horton R (1990) Tumor necrosis factor and interleukin-1 stimulate testosterone secretion in adult male rat Leydig cells in vitro. J Androl 11:353–360

Watanabe N, Inoue H, Fujii-Kuriyama Y (1994) Regulatory mechanisms of cAMP-dependent and cell-specific expression of human steroidogenic cytochrome P450SCC (CYP11A1) gene. Eur J Biochem 222:825–834

Waters BL, Trainer TD (1996) Development of the human fetal testis. Ped Pathol Lab Med 16:9–23

Weinbauer GF, Göckeler E, Nieschlag E (1988) Testosterone prevents complete suppression of spermatogenesis in the gonadotropin-releasing hormone (GnRH) antagonist-treated nonhuman primate (*Macaca fascicularis*). J Clin Endocrinol Metab 67:284–290

Welsh TH, Bambino TH, Hsueh AJ (1982) Mechanism of glucocorticoid-induced suppression of testicular androgen biosynthesis in vitro. Biol Reprod 27:1138–1146

West AP, Lopez-Ruiz MP, Cooke BA (1991) Differences in LH receptor down-regulation between rat and mouse Leydig cells: effects of 3′,5′-cyclic AMP and phorbol esters. Mol Cell Endocrinol 77:R7–R11

Wilson E, Smith AA (1975) Localization of androgen receptors in rat testis: biochemical studies. In: French FS, Hansson V, Ritzén EM, Nayfeh SN (eds) Hormonal regulation of spermatogenesis. Plenum Press, New York, pp 281–286

Wing T-Y, Ewing LL, Zegeye B, Zirkin BR (1985) Restoration effects of exogenous luteinizing hormone on the testicular steroidogenesis and Leydig cell ultrastructure. Endocrinology 117:1779–1787

Xiong Y, Hales DB (1993) The role of tumor necrosis factor-α in the regulation of mouse Leydig cell steroidogenesis. Endocrinology 132:2438–2444

Yoshinaga K, Nishikawa S, Ogawa M, Hayashi S-I, Kunisada T, Fujimoto T, Nishikawa S-I (1991) Role of c-kit in mouse spermatogenesis: identification of spermatogonia as a specific site of c-kit expression and function. Development 113:689–699

Yu J, Dolter KE (1997) Production of activin A and its roles in inflamation and hematopoiesis. Cyto Cell Mol Ther 3:169–177

Zhang F-P, Hämäläinen T, Kaipia A, Pakarinen P, Huhtaniemi I (1994) Ontogeny of luteinizing hormone receptor gene expression in the rat testis. Endocrinology 134:2206–2213

Zhang P, Han XG, Mellon SH, Hall PF (1996) Expression of the gene for cytochrome P-450 17α-hydroxylase/C17–20 lyase (CYP17) in porcine Leydig cells: identification of a DNA sequence that mediates cAMP response. Biochim Biophys Acta 1307:73–82

Zipf WB, Payne AH, Kelch RP (1978) Prolactin, growth hormone and luteinizing hormone in the maintenance of testicular luteinizing hormone receptors. Endocrinology 103:595–600

Post-transcriptional Control and Male Infertility

Robert E. Braun

1 Introduction

The importance of RNA binding proteins in male fertility is gaining great momentum. There are now several examples in humans and mice where mutations in genes encoding RNA binding proteins cause infertility. These include male-specific fertility factors that map to the Y chromosome, and autosomal genes whose functions are not restricted to the testis. In most cases the exact function of the RNA binding proteins is unknown, as are their in vivo RNA substrates. The need for post-transcriptional control during germ cell differentiation has also been demonstrated by studying specific mRNAs, principally the protamine mRNAs, whose temporal translational regulation in spermatids is essential for normal spermatogenesis. The search for message-specific RNA binding proteins has led to the discovery of several novel proteins that localize to different compartments in differentiating spermatogenic cells. Emerging genetic evidence suggests that these proteins have essential functions in developing germ cells.

RNA binding proteins mediate diverse aspects of RNA behavior. Post-transcriptional control can involve alternative splicing and 3′ processing of nascent mRNAs, regulated transport or spatial localization of mRNAs, repression or activation of translation, and control of message stability. Evidence for some of these phenomena exist in mammalian male germ cells. This chapter focuses on two primary topics: (1) the mechanism and importance of translational regulation during spermiogenesis, and (2) the evidence that RNA binding proteins provide essential functions in germ cell differentiation.

Translational control is a frequently used and essential mechanism of gene regulation during spermatogenesis. Because mRNA synthesis is terminated prior to the completion of spermatogenesis, many genes are transcribed in meiotic spermatocytes and early post-meiotic spermatids, and their products stored for varying lengths of time until they are recruited for translation in

Department of Genetics, Box 357360, 1959 NE Pacific, University of Washington, Seattle, Washington 98195, USA

Results and Problems in Cell Differentiation, Vol. 28
McElreavey (Ed.): The Genetic Basis of Male Infertility

elongating and elongated spermatids. The best documented cases include the murine and rat mRNAs encoding the basic nuclear proteins protamine 1 (Prm1) and 2 (Prm2; Balhorn et al. 1984; Kleene et al. 1984; Kleene 1989), and the transition proteins 1 (Tnp1) and 2 (Tnp2; Heidaran and Kistler 1987; Heidaran et al. 1988; Yelick et al. 1989; Schlicker et al. 1997). However, not all mRNAs fall into this stereotype and it is clear that many mRNAs are translationally repressed and activated prior to any changes in transcriptional silencing. The central questions for all cases of translational control include: (1) What is the mechanism of translational repression? (2) What is the mechanism of translational activation? (3) How is translational control coupled to morphogenesis? (4) What is the significance and importance of translational regulation for spermatogenesis? (5) Does loss of translational control lead to infertility?

To answer these questions it is important to determine the cis-acting sequences in the mRNA that are important for translational regulation, and to identify the factors that interact with them. In general, sequences in the 3′ untranslated region (3′ UTR) of mRNAs are important for a wide variety of mRNA behavior including translational repression, mRNA stability, message localization, and mRNA transport (Curtis et al. 1995; Wickens et al. 1997). However, regulatory elements can also lie in the 5′ untranslated region (5′ UTR), and one should not assume that the 3′ UTR contains the requisite regulatory elements. Sequences that control translation during spermatogenesis have been shown to map in the 5′ and 3′ UTRs. Repression and activation of translation presumably requires proteins or antisense RNAs that bind to regulatory elements in the RNA and regulate the translatability of the messages. Discovery of the factors that interact with the cis-acting regulatory elements has lagged behind the discovery of the cis-acting elements, but in many instances candidate RNA binding proteins are now in hand.

The best characterized examples of translational control in mammalian spermatogenesis are the protamine mRNAs. However, many other mRNAs are also under translational control, and their temporal regulation is different from that of the protamines. It seems unlikely that these different classes of translationally repressed mRNAs will share the same cis-acting sequences and trans-acting proteins for their translational repression, and that the underlying mechanisms mediating the repression and activation of such mRNAs will be different from that of the protamines. Because nearly all of our knowledge of the mechanism of translational control during murine spermatogenesis stems from the study of the protamine mRNAs, this review is primarily devoted to their translational regulation. However, if we are to gain a complete understanding of the importance of translational regulation, and the mechanisms controlling it during spermatogenesis, it will be important to study the translational regulation of other mRNAs as well. For a comprehensive listing of mRNAs that are under translational control during spermatogenesis the reader is advised to see a recent review by Kleene (1996).

Loss of post-transcriptional control may be responsible for a significant percentage of male infertility in humans. Approximately 13% of human azoospermia can be attributed to deletion of a region of the Y chromosome referred to as the *azoospermic factor* (*AZF;* Reijo et al. 1995; Vereb et al. 1997). Within this region are two families of genes that encode RNA binding proteins of unknown function. These orphan RNA binding proteins presumably function in some aspect of post-transcriptional regulation, and in some cases it is likely that they participate directly in translational control. The genetic causes of the remaining cases of idiopathic male infertility, both azoospermia and oligospermia, are largely unknown. The discovery of novel spermatogenic RNA binding proteins in mice provide new candidate genes for additional male fertility factors. Discovering the RNA targets of orphan proteins will complement the search for proteins that bind to known RNAs. It would not be surprising if these two paths eventually converge.

2 The Need for Translational Control

Murine spermatogenesis takes approximately 35 days and consists of the mitotic proliferation of spermatogonial cells, the recombination and segregation of chromosomes during meiosis, and the differentiation of haploid spermatids into spermatozoa, spermiogenesis (Fig. 1). Transcription is ongoing throughout spermatogonial proliferation, meiosis, and early spermiogenesis,

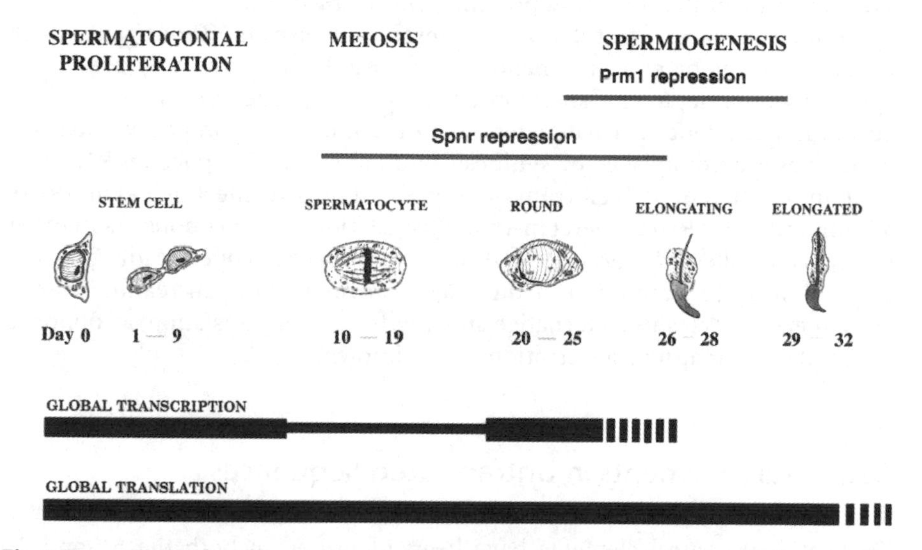

Fig. 1. Transcriptional and translational patterns during murine spermatogenesis. Shown at the *bottom* are the periods of global transcription and translation during murine spermatogenesis. At the *top* are the windows of translational repression for the *Spnr* and *Prm1* genes

but then ceases several days prior to the completion of spermiogenesis (Monesi 1964; Kierszenbaum and Tres 1975). The available data do not allow us to distinguish between transcriptional silencing at the round spermatid to elongating spermatid transition (~ step 9), or the elongating spermatid to elongated spermatid transition (~ step 13). Distinguishing between these two possibilities is important if one is to understand the reason for transcriptional silencing. Cessation of transcription at step 9 would likely involve developmental changes in the transcriptional apparatus, whereas transcriptional arrest at step 13 could be attributed to changes in chromatin structure that occur as chromosome condensation commences.

In addition to those mRNAs that are translationally repressed until after transcription ceases in elongating spermatids (e.g. *Prm1*, *Prm2*, *Tnp1*, *Tnp2*), many other mRNAs are translationally repressed during meiosis and in early round spermatids, and are translated in late round or elongating spermatids. RNAs that are repressed during meiosis and translated during spermiogenesis include *Spnr* (Schumacher et al. 1995b), *Tenr* (Schumacher et al. 1995a) and *Pgk2* (Robinson and Simon 1991; McCarrey et al. 1992; Gold et al. 1983). Spnr is a microtubule-RNA binding protein that associates with the manchette in elongating and elongated spermatids and may function in translational activation of stored mRNAs (Schumacher et al. 1995b; Schumacher et al. 1998). The Tenr protein is a testis-specific RNA binding protein that is expressed post-meiotically and is present in a lattice-like structure within the round spermatid nucleus. The reason for translational regulation during meiosis is unknown, nor is it necessary that the translational control mechanisms be the same as those of the transition proteins and the protamines.

An unusual case of translational repression involves the *Dnmt* gene which encodes the major cytosine methyltransferase. Dnmt activity is present in spermatogonia, leptotene spermatocytes, and zygotene spermatocytes, but absent in pachytene spermatocytes. Loss of Dnmt activity in pachytene spermatocytes is accomplished by synthesis of an alternatively spliced mRNA that contains a different 5′ UTR from the expressed mRNA (Mertineit et al. 1998). Within the 5′ UTR are several in-frame translational start codons. As in other systems (Geballe and Morris 1994), these short upstream open reading frames are inhibitory to translation of the major downstream open reading frame. Why spermatocytes use alternative splicing to silence translation, as opposed to simply terminating transcription, is not known.

3 Regulatory Elements in Untranslated Sequences

Translational control elements have been identified in both the 5′ and 3′ untranslated regions of mRNAs under translational control (Curtis et al. 1995; Wickens et al. 1997). It is likely that the mechanisms of translational repression differ for 5′ UTR versus 3′ UTR-mediated control, although it is possible that some aspects of the mechanisms will be the same. In general, regulatory elements located in the 5′ UTR are more likely to function by sterically hindering the assembly or translocation of the 43 S initiation complex (Hershey 1991). Control elements in the 3′ UTR may also inhibit translational initiation, but are more likely to do so by nucleating the assembly of a translationally inert mRNP (ribonucleoprotein particle), by sequestering the mRNA into a subcellular location within the cell, or by interfering with poly(A)-mediated initiation and eIF-4F (cap-binding complex) binding to the 5′ end of the message.

Translational repression of *Prm1* mRNA is controlled by sequences in its 3′ UTR. Comparison of the nucleotide sequence of the mouse, human, and bovine protamine 1 genes reveals that all three genes share at least 75% similarity in their 3′ UTRs (Krawetz et al. 1987; Lee et al. 1987; Peschon et al. 1987). Transgenic chimeric mRNAs containing the human growth hormone coding sequences fused to the *Prm1* 3′ UTR are translationally repressed in round and elongating spermatids, and translationally active in elongated spermatids, in the same manner as the endogenous *Prm1* mRNA (Braun et al. 1989b). Deletion of sequences in the *Prm1* 5′ UTR has no effect on translational control of a heterologous message. Thus, sequences in the 3′ UTR mediate both proper translational repression in round spermatids, and translational activation in elongated spermatids. In the case of *Pgk2* mRNA, transgenic analysis has shown that the 5′ UTR is not sufficient for translational repression, implying that sequences elsewhere in the mRNA must contain the regulatory elements (Robinson and Simon 1991).

The *Prm1* mRNA is first detected in step 7 round spermatids (Braun et al. 1989b; Mali et al. 1989) and its mRNA first translated in step 12 elongating spermatids (Balhorn et al. 1984). The mRNA for *Prm1* is initially found in round spermatids as a 560 nucleotide transcript and later in elongating spermatids as a heterogeneous size class of molecules down to as small as 400 nucleotides (Kleene et al. 1984). This decrease in mRNA size is due to a shortening of the poly(A) tail that occurs at the same time as synthesis of Prm1 commences. The presence of the fully adenylated form of *Prm1* mRNA on polysomes suggests that deadenylation is not required for translational activation, and that deadenylation probably reflects the initial stages of decay of a translated mRNA. A decrease in poly(A) tail length coincident with translation initiation has also been observed for *Prm2* (Hecht 1989), as well as for the mRNAs for the transition proteins (Yelick et al. 1989).

At least two separate regions of the 156 nt *Prm1* 3′ UTR are capable of repressing the translation of a reporter mRNA in transgenic mice. The regions are nonoverlapping, suggesting that there are redundant elements capable of translational repression. The first of these regions maps to the 5′-most 37 nucleotides of the *Prm1* 3′ UTR (Fajardo et al. 1997; Fig. 2). Within this region is a 9 nucleotide stretch that is conserved at seven of nine positions in the

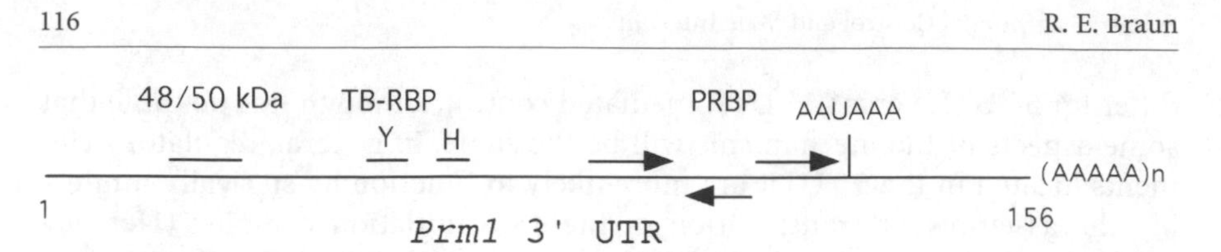

Fig. 2. Regulatory elements in the *Prm1* 3′ UTR. The 48/50 kDa protein binds to a conserved sequence between nucleotides 17–25. TB-RBP binds to the Y element between nucleotides 116–129. Prbp binds to either of two putative stem-loops formed between sequences marked by *arrows*

Prm2 3′UTR. The presence of a 48/50 kDa germ cell-specific protein that binds this sequence (Fajardo et al. 1994) suggests that the site is responsible for translational repression, although formal proof of this awaits mutation of the sequence in transgenic mice. The second region that is sufficient for translational repression is contained in the 3′-most 62 nucleotides of the *Prm1* 3′ UTR (Braun 1990; Fig. 2). Within this region is a 17 nucleotide sequence which is also present in the *Prm2* 3′ UTR and shares identity at 14 out of 17 positions. A second copy of this conserved sequence is also present in the *Prm1* 3′ UTR and maps between nucleotides 80–95. It is likely that this conserved sequence mediates translational repression but this too awaits mutational analysis in vivo. Although independent transgenic analysis of both the 5′-most 37 nucleotides and the 3′-most 62 nucleotides of the *Prm1* 3′ UTR has shown that each is capable of translational repression in vivo, neither region appears to repress translation as well as the full-length *Prm1* 3′ UTR. Completely wild-type levels of translational repression appear to require the presence of both halves of the *Prm1* 3′ UTR. This added stringency may be beneficial to the process of spermatogenesis given the detrimental consequences of premature expression of Prm1 protein on the developing spermatid (Lee et al. 1995). Additionally, these elements may participate in establishing an mRNP conformation that is recognized by the factors responsible for proper temporal translational activation.

Translational activation of the *Prm1* mRNA may require separate elements from those involved in translational repression. While analyzing the expression of the reporter transgene containing the translational silencer in the first 37 nucleotides of the *Prm1* 3′ UTR, but lacking a portion of the 3′-most region of the 3′ UTR, Fajardo et al. (1997) noticed that the human growth hormone reporter mRNA was poorly activated for translation in elongated spermatids. Failure to activate translation could be due to the absence of specific sequences required for activation. Transgenes that contain the 3′ most 62 nucleotides of the *Prm1* 3′ UTR activate translation normally, suggesting that sequences required for activation map to that region (Braun 1990). It is not known if the same sequences that mediate repression through the 3′-most 62 nucleotides, possibly the conserved 17 nucleotide sequence, also mediate translational activation.

Using RNA electrophoretic mobility shift assays (EMSAs), Kwon and Hecht (1991) have identified two regions in the *Prm1* and *Prm2* 3′ UTRs, referred to as the Y and H elements, that bind a protein present in testis extracts (Fig. 2). The Y and H elements of *Prm2* have been shown to mediate translational repression of a reporter mRNA in a rabbit reticulocyte lysate supplemented with a testis extract enriched for the binding factor, suggesting that the site is capable of mediating translational repression (Kwon and Hecht 1993). However, the ability of a complex to inhibit translation in vitro does not mean that it necessarily does so in vivo. It will be important to test the Y and H elements in transgenic animals.

4
The Protamine mRNA Is Stored in a Ribonucleoprotein Particle

Translational repression mediated by sequences located in the 3′ UTR of the protamine mRNA necessitates communication between the two ends of the message. Several models can be considered. Sequences in the 3′ UTR could function to localize the mRNA to a subcellular compartment within the cell that is devoid of ribosomes. In the case of *Prm1* mRNA there is no evidence of compartmentalization of the mRNA in the cytoplasm of round and elongating spermatids (Morales et al. 1991). Alternatively, proteins or antisense RNAs could bind to the *Prm1* 3′ UTR and to a region in the 5′ end of the message, for example the 5′ cap (^{7}M-GpppN), and interfere with the assembly of the 43 S ribosomal subunit on the end of the message. Factors bound to the 3′ UTR could also initiate a process of complete mRNA masking, by the same protein, or other proteins, thus resulting in a translationally compromised ribonucleoprotein particle. Lastly, proteins bound to the 3′ UTR could interfere with the binding of poly(A) binding protein to the poly(A) tail. Recent evidence in yeast suggest that poly(A) binding protein can physically couple the two ends of the message, and that this coupling is important for translation initiation (Tarun and Sachs 1995, 1996).

Elucidating the components of the *Prm1* mRNA ribonucleoprotein particle (mRNP) may facilitate our understanding of the mechanism of translational control. For example, the presence of the cap-binding complex eIF-4F in the translationally inert mRNP would suggest that the 3′ UTR regulatory elements do not block this initial step in translation initiation. Unfortunately, without a specific component of the *Prm1* mRNP in hand, it is difficult to purify it from the other mRNPs present in round spermatids and to characterize its complexity. Nonetheless, it is known that the *Prm1* mRNA sediments at less than 80 S in Nycodenz gradients (Tafuri et al. 1993), and that it may contain poly(A) binding protein, Pabp (Gu et al. 1995), as well as several other proteins. The presence of Pabp in the protamine mRNP suggests that the 3′ UTR does not inhibit translation initiation by preventing binding of Pabp to the poly(A) tail. However, the stoichiometry of Pabp bound to the *Prm1* message is unknown, and may be less than optimal for mediating translation initiation. In addition, regulatory proteins bound to the 3′ UTR may prevent Pabp from interacting with the eIF-4G initiation factor.

Also contained in the *Prm1* mRNP is at least one Y box protein. Y box proteins are believed to be sequence-specific DNA binding proteins, and non-specific RNA binding proteins (Wolffe 1994). Y box proteins of 54 and 56 kDa were originally identified as protein components of stored maternal mRNPs in cytoplasmic extracts prepared from *Xenopus laevis* oocytes (Richter and Evers 1984; Tafuri and Wolffe 1990; Murray et al. 1991, 1992; Tafuri et al. 1993). Proteins sharing antigenicity with the *Xenopus* p54/p56 proteins have been described in mouse testicular extracts and shown to have apparent molecular weights of 48/52 kDa (Kwon et al. 1993), and the gene for the 52 kDa protein has been cloned and is referred to as mouse Y box protein, Msy1, (Tafuri et al. 1993). Msy1 sediments in the 60–80 S mRNP fraction of testis extracts suggesting it may be associated with untranslated RNAs (Tafuri et al. 1993). The mouse p48 and p52 Y box proteins are highly enriched in the testis and have been shown to bind non-specifically to various RNAs in vitro, including *Prm1*, *Prm2* , *Tnp1*, *hGH*, and pGem-2 RNA (Kwon et al. 1993). Within the testis, p48 and p52 are first detected in the cytoplasm of mid-pachytene cells, and continue to persist through the elongating spermatid stage, consistent with a role in translational repression (Oko et al. 1996).

5 Sequence-Specific RNA Binding Proteins

Given the non-specific RNA binding properties of the Msy1 Y box protein, and the evidence that the protamine mRNPS are assembled at a time when other mRNAs are translationally active, one would also expect there to be sequence-specific protamine RNA binding proteins. Such proteins might initiate or direct the assembly of the protamine mRNP, and thereby confer the necessary specificity required for selective translational inhibition. Several candidate protamine RNA binding proteins have been described.

RNA EMSAs with radiolabeled *Prm1* 3′ UTR and testis protein extracts detect a closely spaced doublet on nondenaturing polyacrylamide gels (Fajardo et al. 1994). Crosslinking studies indicate that the protein components of the complex are approximately 48 and 50 kDa. The binding activity is present in the cytoplasm of pachytene spermatocytes and round spermatids, and is reduced in elongated spermatids. Competition assays reveal that the complex is specific for *Prm1* and *Prm2* mRNAs. Mutational analysis and RNAse footprinting studies have been used to map the binding site to a specific region in the 5′-end of the 3′ UTRs (Fig. 2). Comparison of the binding sites in *Prm1* and *Prm2* show that there is a nine nucleotide sequence that is identical at seven out of the nine nucleotides, and that single nucleotide mutations abolish protein binding (F. Giorgini and R.E. Braun, unpubl.). The binding specificity of the 48/50 kDa proteins, coupled with their spatial and temporal localization, are properties one would expect for a translational repressor of *Prm1* mRNA. Transgenic analysis of a reporter RNA that contains the binding site for the 48/50 kDa proteins is translationally repressed in round spermatids, although repression may be incomplete (Fajardo et al. 1997). A candidate gene encoding the 48/50 kDa binding activity has been cloned from a mouse testis cDNA library using the yeast three-hybrid system (H.G. Davies and R.E. Braun, unpubl.). Surprisingly, the RNA binding protein is a new member of the Y box family of RNA binding proteins. As described above, initial studies of the Y box family of RNA binding proteins suggested that they are components of the translational masking machinery and that they bind RNA nonspecifically (Wolffe 1994). However, using a method of in vitro selection and amplification, Bouvet et al. (Bouvet et al. 1995) have recently shown that at least some members of this family of proteins can recognize specific target sequences. The 48/50 kDa Y box protein that binds to the *Prm1* 3′ UTR clearly binds RNA in a sequence-specific manner, as point mutations in the binding site eliminate binding in the three-hybrid assay as well as in an RNA EMSA (H.G. Davies, F. Giorgini and R.E. Braun, unpubl.). It may be that other Y box proteins, previously considered to be nonspecific RNA binding proteins, are also sequence-specific RNA binding proteins, but whose in vivo target RNAs are unknown.

The second region of the *Prm1* 3′ UTR capable of translational repression in transgenic mice maps within the 3′-most 62 nucleotides of the 3′ UTR (Fig. 2; Braun 1990). The 48/50 kDa protein does not bind to the 62 nucleotide region, thus there must be other translational control factors that mediate the repression through this region. One possibility is the protamine RNA binding protein, Prbp. The *Prbp* gene was cloned in an expression screen for cDNAs that encode *Prm1* 3′ UTR RNA binding proteins (Lee et al, 1996). Prbp is a 40 kDa protein that contains two copies of a well-conserved double-stranded RNA binding motif (Green and Mathews 1992; St-Johnston et al, 1992). Prbp is localized to the cytoplasm of pachytene spermatocytes and round spermatids, consistent with it mediating translational control. In an RNA EMSA, bacterially expressed Prbp binds to a region of secondary structure in the 3′-most region of the *Prm1* 3′ UTR. Prbp binding to the *Prm1* 3′ UTR can be competed with poly(I)/poly(C) RNA, suggesting that, in vitro, Prbp binds regions of dsRNA. Prbp is capable of inhibiting translation in vitro, but it is not specific to mRNAs containing the *Prm1* 3′ UTR. Despite the fact that Prbp binds to a region of the 3′ UTR shown to be capable of inhibiting translation in vivo, its classification as a general double stranded RNA binding protein weakens its involvement as a specific mediator of *Prm1* translational inhibition. Disruption of the *Prbp* gene by homologous recombination in embryonic stem cells causes severe oligospermia and male sterility (Zhong et al., 1999).

Kwon and Hecht (1991, 1993) have described two regions in the 3′ UTRs of *Prm1* and *Prm2* mRNA, the Y and H boxes, that interact with proteins present in testis extracts. They have shown that the Y box in *Prm2* can be UV cross-linked to a testis-brain RNA binding protein, TB-RBP, and that the H and Y boxes of *Prm2* are able to confer translational repression on a heterologous mRNA in rabbit reticulocyte extracts supplemented with partially fractionated testis protein extracts enriched for TB-RBP (Kwon and Hecht 1993). The gene encoding TB-RBP has been cloned and is the mouse homologue of the human *TRANSLIN* gene. TRANSLIN binds single-stranded DNA in vitro and has been proposed to be associated with regions of breakpoint junctions of chromosomal translocations in some human malignant lymphoid cells (Aoki et al. 1995). The potential involvement of TB-RBP in DNA-associated events in the nucleus, and RNA-associated functions in the cytoplasm, may be regulated by post-translational modifications (Wu et al. 1997). The sequence-specific binding of TB-RBP to the H and Y boxes, and the ability of TB-RBP-enriched fractions to selectively inhibit translation in vitro, suggest that it could be a regulator of *Prm1* and *Prm2* translation in vivo. Further support for such a function would be the demonstration that the H and Y boxes can repress the translation of a heterologous mRNA in transgenic mice, and that mutation of TB-RBP leads to premature translation in vivo.

6
Premature Translation of Prm1 mRNA

Translational inhibition of *Prm1* mRNA is essential for the completion of spermiogenesis (Lee et al. 1995). Premature translation of a *Prm1* transgene results in precocious chromosome condensation and dominant male sterility. Phenotypes vary depending on the levels of premature translation and range from a complete arrest in spermatid differentiation at the round spermatid stage, to late-stage spermatid defects that include abnormal head morphogenesis and incomplete processing of the Prm2 protein. The dominant male sterility is due to the sharing of postmeiotic gene products between spermatids (Braun et al. 1989a) made possible by the intercellular bridges which connect cells within developing clones of spermatids (Burgos and Fawcett 1955).

Failure to repress translation of the protamine or transition protein mRNAs, or constitutive repression of their translation due to a failure to activate translation in elongated spermatids, could be responsible for some cases of idiopathic infertility in men. Several studies have shown that sperm nuclear condensation is dramatically altered in some infertile men (Dadoune et al. 1988; Engh et al. 1992; Foresta et al. 1992). Instances of both hypocondensation and hypercondensation have been found, and in some studies it has been shown that there are alterations in nucleoprotamine ratios in epididymal spermatozoa (Chevaillier et al. 1987; Balhorn et al. 1988; de Yebra et al. 1993;

Schlicker et al. 1994). Mutations in the protamine genes themselves, for instance regulatory mutations in the 3′ UTR, could lead to precocious translation. Alternatively, transcriptional regulatory mutations, or structural gene mutations, could lead to the absence of protamine synthesis. Thus far, no mutations in the human protamine genes have been shown to be responsible for infertility. One would also expect loss-of-function mutations in genes that normally repress or activate protamine translation to also cause male sterility. This is a class of mutations that would be worth screening for among infertile males once candidate genes have been identified in the mouse. As discussed earlier, disruption of the murine *Prbp* gene causes male sterility in mice, suggesting it may be an infertility factor in humans.

7
Activation of Translationally Repressed mRNAs

Translational activation requires that the mRNA be mobilized from repressed mRNPs onto polysomes in elongated spermatids. By some unknown mechanism the repression apparatus must be modified to release the mRNA for translation. Synthesis of a translational activator like a protein kinase or phosphatase that modifies the mRNP is a possibility. The RNA binding properties of TB-RBP appear to be coupled to its phosphorylation status (Kwon and Hecht 1993), and are consistent with such a model. Of course, synthesis of a translational activator of the protamine mRNAs only tells us how activation of protamine translation is achieved. It does not tell us how the synthesis of the activator is regulated in the absence of new transcription. The ultimate question is what is the initial trigger that initiates synthesis of the activator? Coupling translational activation with a morphological event occurring in the elongated spermatid is an exciting possibility to consider. The finding that newly synthesized Prm1 protein is found only in the nuclear fraction of sonication-resistant nuclei suggests that the protein is immediately deposited into the nucleus upon synthesis (Green et al. 1994; Lee et al. 1995). One interpretation of this observation is that Prm1 protein synthesis occurs at or near the nuclear pore. To facilitate this restricted subcellular synthesis the *Prm1* mRNA may be localized to the nuclear periphery perhaps through interaction with the cytoskeleton. Developmentally regulated protein production near the site of action has been established for a number of mRNAs such as those involved in pattern formation during embryogenesis of *Drosophila,* and in addition there is correlative evidence for the involvement of the cytoskeleton in translational control processes (reviewed by Hesketh 1994). The recent evidence that mRNPs and polysomes can associate with cytoskeletal elements supports the notion that this is a mechanism for targeted protein synthesis. In elongating spermatids a specialized microtubule array called the manchette forms and may function in nuclear shaping (Fawcett et al. 1971; Russell et al. 1991).

It has also been suggested this structure plays a role in overall sperm differentiation by serving as a "track" that is utilized for the movement of organelles, vesicles, and mRNPs (Fawcett et al. 1971; MacKinnon and Abraham 1972; Schumacher et al. 1995).

In the molecular screen that yielded *Prbp*, *Spnr* (spermatid perinuclear RNA-binding protein) was also cloned and characterized (Schumacher et al. 1995b). *Spnr* encodes an RNA-binding protein that is highly expressed in elongating haploid germ cells, is localized to the manchette structure, and appears to be a microtubule-associated protein (MAP; Schumacher et al. 1998). Given that *Spnr* was cloned based on its ability to bind the *Prm1* 3′ UTR, it may function as a MAP that links the *Prm1* mRNA to the manchette and thereby plays a role in the putative subcellular localization of protamine mRNA molecules that are destined to be activated for translation at the nuclear periphery. Since the elongating spermatid is a highly polarized cell where many mRNAs are translationally regulated, it is plausible that certain mRNAs get localized to specific subcellular regions where translation ensues and the protein product is immediately used. The protamine mRNAs might be targeted by such a regulatory mechanism.

8 Orphan RNA Binding Proteins

Several genes that encode putative RNA binding proteins expressed during spermatogenesis have been cloned (Table 1). In most cases the RNA targets of these proteins have not yet been identified. Furthermore, the biochemical

Table 1. RNA binding proteins expressed in germ cells

Protein	Expression	Subcellular localization	Mutant phenotype
Dazla	Sg, ls, zs, ps Fetal oogonia	Cytoplasmic Cytoplasmic	Fetal germ cell loss
DAZ	Fetal spermatogonia	Nuclear	
RBM	Fetal, prepubertal, adult germ cells	Nuclear	
Prbp	Ps, rs	Cytoplasmic	Oligospermia
Tiar	Primordial germ cells	Nuclear	Primordial germ cell loss
Spnr	Es	Manchette	
Tenr	Rs	Nuclear (lattice)	
Msy1; p48/p52; 48/50 kDa	Ps, rs	Cytoplasmic	

Sg, spermatogonia; ls, leptotene spermatocyte; zs, zygotene spermatocyte; ps, pachytene spermatocyte; rs, round spermatid; es, elongated spermatid function of these RNA binding proteins is unknown. However, the possible involvement of some of these RNA binding proteins in regulating post-transcriptional gene expression during spermatogenesis seems likely, and in one case has been elucidated through the study of men that fail to produce normal levels of sperm.

The presence of Y chromosome deletions in several men with azoospermia or oligospermia has led to the hypothesis of a spermatogenic locus on the euchromatic portion of Yq, referred to as the *azoospermia factor* (*AZF;* Tiepolo and Zuffardi 1976; Johnson et al, 1989; Skare et al. 1990; Ma et al. 1993; Reijo et al. 1995, 1996; Vogt et al. 1996). The analysis of infertile men with microdeletions within Yq has led to the subdivision of *AZF* into three regions referred to as *AZFa*, *AZFb*, and *AZFc* (Eberhart and Wasserman 1995; Vogt et al. 1996; Elliott and Cooke 1997). Within this region are at least two gene families that encode RNA binding proteins that may be responsible for the spermatogenic defects observed in men deleted for *AZF*. The first of these genes, referred to as *RBM*, is present in dozens of copies on both arms of the human Y chromosome. RBM contains an RNA recognition motif, is highly similar to the autosomally encoded hnRNP G protein, and may be its germ cell homologue. The hnRNPs are RNA binding proteins found associated with nascent nuclear polyadenylated RNAs (Weighardt et al. 1996). hnRNPs probably play some function in the processing of nascent mRNAs, perhaps affecting splice site selection or alternative 3′ end processing. Some hnRNPs shuttle between the nucleus and the cytoplasm and may function in the nuclear to cytoplasmic transport of mRNAs. It has been demonstrated that RBM is a nuclear protein and that it is present in human fetal, prepubertal, and adult germ cells (Elliott et al. 1997). Men that carry microdeletions encompassing the *AZFb* region have reduced levels of RBM and an interruption of spermatogenesis at the meiotic to postmeiotic transition (Elliott and Cooke 1997). Mice contain an *RBM* gene family that is also located on the Y chromosome (Elliott et al. 1996). The mouse *Rbm* is expressed as early as 4 days after birth, peaks at its highest level 14 days after birth, and is expressed at low levels in adult testis. As is true for most hnRNPs, the actual RNA target(s) of RBM are unknown. RBM may interact with a general class of pre-mRNAs, or alternatively, may have a few selected mRNAs for which it performs some important function.

A second candidate gene family, referred to as *deleted in azoospermia* (*DAZ*), has also been proposed to be the *azoospermia factor*. The *DAZ* gene encodes a putative RNA binding protein that also contains an RNA binding motif and that has an unknown function (Reijo et al. 1995). Interestingly, the *DAZ* gene is the first Y chromosome gene identified that appears to have arisen from transposition and then amplification of an autosomal gene (Saxena et al. 1996). *DAZ* is present on the Y chromosome and on an autosome in humans, but in mice is restricted to a single copy gene on chromosome 17 (Reijo et al. 1996b; Saxena et al. 1996; Ruggiu et al. 1997;). The mouse protein,

Dazla, has been detected at low levels in the cytoplasm of B spermatogonia and preleptotene and zygotene spermatoctyes, and at high levels in pachytene cells (Ruggiu et al. 1997). Interestingly, the human DAZ protein is localized to the nucleus in fetal gonocytes (R. Reijo, pers. comm.). A *Dazla* homologue, referred to as *boule*, has been identified in *Drosophila* and mutations in *boule* result in pachytene arrest and male sterility (Eberhart et al. 1996). The *Dazla* gene has been deleted by homologous recombination in the mouse and both male and female homozygotes are sterile (Ruggiu et al. 1997). A depletion of germ cells is first detected in the fetal gonad, and in adults there is complete germ cell loss beyond the spermatogonial stage. Interestingly, heterozygous males, although fertile, have a reduced sperm count and an increase in the number of abnormal spermatozoa. It may be that Dazla has several roles in male, and perhaps female, germ cell development. One function would be in the fetal meiotic cells in females, and in the fetal postmitotic germ cells in males; and a second function in differentiating germ cells in adult males. The localization of Dazla to the cytoplasm of germ cells suggest that it functions in cytoplasmic post-transcriptional events, perhaps in translational control or in mRNA stabilization.

RNA binding proteins have also been shown to be required for primordial germ cell (PGC) development in male and female embryos. Disruption of the gene encoding Tiar, an RNA recognition motif/ribonucleoprotein-type RNA binding protein that is highly expressed in primordial germ cells, results in failure to populate the genital ridges of E13.5 embryos (Beck et al. 1998). Tiar appears to affect the survival of PGCs during their migratory period from the base of the allantois to the genital ridge. Like RBM and DAZ, the RNA targets of Tiar are unknown.

9 Perspectives

There are now several examples in humans and mice where mutations in genes encoding RNA binding proteins cause infertility. These include male-specific fertility factors that map to the Y chromosome, and autosomal genes whose functions are not restricted to the testis. In most cases the exact function of the RNA binding proteins is unknown, as are their in vivo RNA substrates. The discovery of orphan RNA binding proteins that are uniquely required for spermatogenesis tells us that control of RNA metabolism is important for spermatogenesis. Identifying the biochemical and molecular function of these orphan RNA binding proteins, as well as their RNA targets, are the obvious and important tasks that lie ahead. Premature translation of the *Prm1* message also causes male sterility, suggesting the possibility that mutations in genes that regulate translational repression and activation may be infertility factors in humans.

Despite the prevalence of translational control during spermatogenesis, relatively little is known about the mechanisms that regulate the translation of individual messages. Part of the reason for this stems from the lack of an in vitro system to study mammalian spermatogenesis. Development of such a system would greatly benefit the field and lead to significant advancements. Nonetheless, the availability of transgenesis and gene knockout technologies in the mouse allows one to convincingly demonstrate function, and the relative ease with which these can now be performed, should permit the unambiguous assignment of function to individual genes.

Numerous mRNAs have been shown to be under translational control during spermatogenesis, yet in only one case, *Prm1*, has it been demonstrated that sequences in the 3′ UTR repress its translation. The *Prm1* 3′ UTR appears to contain redundant translational repression elements and perhaps separate sequences required for translational activation. There is a need to study the regulation of other messages to determine if they contain similar or different cis-acting regulatory elements. Despite significant progress in identifying *Prm1* RNA binding proteins, none of the proteins thus far described, Prbp, TB-RBP, or the 48/50 kDa protein, have been shown to mediate the translational repression of the protamine mRNAs in vivo. Proof that any of these genes are involved in translational control during spermatogenesis awaits genetic analysis of their function in the mouse. The possibility of redundancy may complicate the genetic dissection and require the construction of multiply mutant animals.

References

Aoki K, Suzuki K, Sugano T, Tasaka T, Nakahara K, Kuge O, Omori A, Kasai, M (1995) A novel gene, Translin, encodes a recombination hotspot binding protein associated with chromosomal translocations. Nat Genet 10:167–174

Balhorn R, Weston S, Thomas C, Wyrobek AJ (1984) DNA packaging in mouse spermatids. Synthesis of protamine variants and four transition proteins. Exp Cell Res 150:298–308

Balhorn R, Reed S, Tanphaichitr N (1988) Aberrant protamine 1/protamine 2 ratios in sperm of infertile human males. Experientia 44:52–55

Beck ARP, Miller IJ, Anderson P, Streuli M (1998) RNA-binding protein TIAR is essential for primordial germ cell development. Proc Natl Acad Sci USA 95:2331–2336

Bouvet P, Matsumoto K, Wolffe AP (1995) Sequence-specific RNA recognition by the Xenopus Y-box proteins. An essential role for the cold shock domain. J Biol Chem 270: 28297–28303

Braun RE (1990) Temporal translational regulation of the protamine 1 gene during mouse spermatogenesis. Enzyme 44:120–128

Braun RE, Behringer RR, Peschon JJ, Brinster RL, Palmiter RD (1989a) Genetically haploid spermatids are phenotypically diploid. Nature 337:373–376

Braun RE, Peschon JJ, Behringer RR, Brinster RL, Palmiter RD (1989b) Protamine 3′-untranslated sequences regulate temporal translational control and subcellular localization of growth hormone in spermatids of transgenic mice. Genes Dev 3:793–802

Burgos MH, Fawcett DW (1955) Studies on the fine structure of the mammalian testis. I. Differentiation of the spermatids in the cat *(Felis domestica)*. J Biophys Biochem Cytol 1: 287–300

Chevaillier P, Mauro N, Feneux D, Jouannet P, David G (1987) Anomalous protein complement of sperm nuclei in some infertile men [letter]. Lancet 2:806–807

Curtis D, Lehmann R, Zamore PD (1995) Translational regulation in development. Cell 81:171–178

Dadoune JP, Mayaux MJ, Guihard Moscato ML (1988) Correlation between defects in chromatin condensation of human spermatozoa stained by aniline blue and semen characteristics. Andrologia 20:211–217

de Yebra L, Ballesca JL, Vanrell JA, Bassas L, Oliva R (1993) Complete selective absence of protamine P2 in humans. J Biol Chem 268:10553–10557

Eberhart CG, Wasserman SA (1995) The *pelota* locus encodes a protein required for meiotic cell division: an analysis of G2/M arrest in *Drosophila* spermatogenesis. Development 121: 3477–3486

Eberhart CG, Maines JZ, Wasserman SA (1996) Meiotic cell cycle requirement for a fly homologue of human Deleted in Azoospermia. Nature 381:783–785

Elliott DJ, Cooke HJ (1997) The molecular genetics of male infertility. Bioessays 19:801–809

Elliott DJ, Ma K, Kerr SM, Thakrar R, Speed R, Chandley AC, Cooke H (1996) An RBM homologue maps to the mouse Y chromosome and is expressed in germ cells. Hum Mol Genet 5: 869–874

Elliott DJ, Millar MR, Oghene K, Ross A, Kiesewetter F, Pryor J, McIntyre M, Hargreave TB, Saunders PT, Vogt PH, Chandley AC, Cooke H (1997) Expression of RBM in the nuclei of human germ cells is dependent on a critical region of the Y chromosome long arm. Proc Natl Acad Sci USA 94:3848–3853

Engh E, Clausen OP, Scholberg A, Tollefsrud A, Purvis K (1992) Relationship between sperm quality and chromatin condensation measured by sperm DNA fluorescence using flow cytometry. Int J Androl 15:407–415

Fajardo MA, Butner KA, Lee K, Braun RE (1994) Germ cell-specific proteins interact with the 3′ untranslated regions of *Prm-1* and *Prm-2* mRNA. Dev Biol 166:643–653

Fajardo MA, Haugen HS, Clegg CH, Braun RE (1997) Separate elements in the 3′ untranslated region of the mouse protamine 1 mRNA regulate translational repression and activation during murine spermatogenesis. Dev Biol 191:42–52

Fawcett DW, Anderson WA, Phillips DM (1971) Morphogenetic factors influencing the shape of the sperm head. Dev Biol 26:220–251

Foresta C, Zorzi M, Rossato M, Varotto A (1992) Sperm nuclear instability and staining with aniline blue: abnormal persistence of histones in spermatozoa in infertile men. Int J Androl 15: 330–337

Geballe AP, Morris DR (1994) Initiation codons within 5′-leaders of mRNAs as regulators of translation. Trends Biochem Sci 19:159–164

Gold B, Fujimoto H, Kramer JM, Erickson RP, Hecht NB (1983) Haploid accumulation and translational control of phosphoglycerate kinase-2 messenger RNA during mouse spermatogenesis. Dev Biol 98:392–399

Green GR, Balhorn R, Poccia DL, Hecht NB (1994) Synthesis and processing of mammalian protamines and transition proteins. Mol Reprod Dev 37:255–263

Green SR, Mathews MB (1992) Two RNA-binding motifs in the double-stranded RNA-activated protein kinase, DAI. Genes Dev 6:2478–2490

Gu,W, Kwon Y, Oko R, Hermo L, Hecht NB (1995) Poly (A) binding protein is bound to both stored and polysomal mRNAs in the mammalian testis. Mol Reprod Dev 40:273–285

Hecht NB (1989) Mammalian protamines and their expression. In: Hnilico G, Stein G, Stein J (eds) Histones and other basic nuclear proteins. CRC Press, Boca Raton

Heidaran MA, Kistler WS (1987) Transcriptional and translational control of the message for transition protein 1, a major chromosomal protein of mammalian spermatids. J Biol Chem 262:13309–13315

Heidaran MA, Showman RM, Kistler WS (1988) A cytochemical study of the transcriptional and translational regulation of nuclear transition protein 1 (TP1), a major chromosomal protein of mammalian spermatids. J Cell Biol 106:1427–1433
Hershey JW (1991) Translational control in mammalian cells. Annu Rev Biochem 60:717–755
Hesketh J (1994) Translation and the cytoskeleton: a mechanism for targeted protein synthesis. Mol Biol Rep 19:233–243
Johnson MD, Tho SP, Behzadian A, McDonough PG (1989) Molecular scanning of Yq11 (interval 6) in men with Sertoli-cell-only syndrome. Am J Obstet Gynecol 161:1732–1737
Kierszenbaum AL, Tres IL (1975) Structural and transcriptional features of the mouse spermatid genome. J Cell Biol 65:258–270
Kleene KC (1989) Poly(A) shortening accompanies the activation of translation of five mRNAs during spermiogenesis in the mouse. Development 106:367–373
Kleene KC (1996) Patterns of translational regulation in the mammalian testis. Mol Reprod Dev 43:268–281
Kleene KC, Distel RJ, Hecht NB (1984) Translational regulation and deadenylation of a protamine mRNA during spermiogenesis in the mouse. Dev Biol 105:71–79
Krawetz SA, Connor W, Dixon GH (1987) Cloning of bovine P1 protamine cDNA and the evolution of vertebrate P1 protamines. DNA 6:47–57
Kwon YK, Hecht NB (1991) Cytoplasmic protein binding to highly conserved sequences in the 3′ untranslated region of mouse protamine 2 mRNA, a translationally regulated transcript of male germ cells. Proc Natl Acad Sci USA 88:3584–3588
Kwon YK, Hecht NB (1993) Binding of a phosphoprotein to the 3′ untranslated region of the mouse protamine 2 mRNA temporally represses its translation. Mol Cell Biol 13:6547–6557
Kwon YK, Murray MT, Hecht NB (1993) Proteins homologous to the Xenopus germ cell-specific RNA-binding proteins p54/p56 are temporally expressed in mouse male germ cells. Dev Biol 158:99–100
Lee CH, Hoyer FS, Engel W (1987) The nucleotide sequence of a human protamine 1 cDNA. Nucleic Acids Res 15:7639
Lee K, Haugen HS, Clegg CH, Braun RE (1995) Premature translation of protamine 1 mRNA causes precocious nuclear condensation and arrests spermatid differentiation in mice. Proc Natl Acad Sci USA 92:12451–12455
Lee K, Fajardo MA, Braun RE (1996) A testis cytoplasmic RNA-binding protein that has the properties of a translational repressor. Mol Cell Biol 16:3023–3034
Ma K, Inglis JD, Sharkey A, Bickmore WA, Hill RE, Prosser EJ, Speed RM, Thomson EJ, Jobling M, Taylor K, et al (1993) A Y chromosome gene family with RNA-binding protein homology: candidates for the azoospermia factor AZF controlling human spermatogenesis. Cell 75: 1287–1295
MacKinnon EA, Abraham JP (1972) The manchette in stage 14 rat spermatids: a possible structural relationship with the redundant nuclear envelope. Z Zellforsch Mikrosk Anat 124: 1–11
Mali P, Kaipia A, Kangasniemi M, Toppari J, Sandberg M, Hecht NB, Parvinen M (1989) Stage-specific expression of nucleoprotein mRNAs during rat and mouse spermiogenesis. Reprod Fertil Dev 1:369–382
McCarrey JR, Berg WM, Paragioudakis SJ, Zhang PL, Dilworth DD, Arnold BL, Rossi JJ (1992) Differential transcription of Pgk genes during spermatogenesis in the mouse. Dev Biol 154: 160–168
Mertineit C, Yoder JA, Taketo T, Laird DW, Trasler JM, Bestor TH (1998) Sex-specific exons control DNA methyltransferase in mammalian germ cells. Development 125:889–897
Monesi V (1964) Ribonucleic acid synthesis during mitosis and meiosis in the mouse testis. J Cell Biol 22:521–532
Morales CR, Kwon YK, Hecht NB (1991) Cytoplasmic localization during storage and translation of the mRNAs of transition protein 1 and protamine 1, two translationally regulated transcripts of the mammalian testis. J Cell Sci 100:119–131

Murray MT, Krohne G, Franke WW (1991) Different forms of soluble cytoplasmic mRNA binding proteins and particles in Xenopus laevis oocytes and embryos. J Cell Biol 112:1–11

Murray MT, Schiller DL, Franke WW (1992) Sequence analysis of cytoplasmic mRNA-binding proteins of Xenopus oocytes identifies a family of RNA-binding proteins. Proc Natl Acad Sci USA 89:11–15

Oko R, Korley R, Murray MT, Hecht NB, Hermo L (1996) Germ cell-specific DNA and RNA binding proteins p48/52 are expressed at specific stages of male germ cell development and are present in the chromatoid body. Mol Reprod Dev 44:1–13

Peschon JJ, Behringer RR, Brinster RL, Palmiter RD (1987) Spermatid-specific expression of protamine 1 in transgenic mice. Proc Natl Acad Sci USA 84:5316–5319

Reijo R, Lee TY, Salo P, Alagappan R, Brown LG, Rosenberg M, Rozen S, Jaffe T, Straus D, Hovatta O, et al (1995) Diverse spermatogenic defects in humans caused by Y chromosome deletions encompassing a novel RNA-binding protein gene. Nat Genet 10:383–393

Reijo R, Alagappan RK, Patrizio P, Page DC (1996a) Severe oligozoospermia resulting from deletions of azoospermia factor gene on the Y chromosome [see comments]. Lancet 347: 1290–1293

Reijo R, Seligman J, Dinulos MB, Jaffe T, Brown LG, Disteche CM, Page DC (1996b) Mouse autosomal homolog of DAZ, a candidate male sterility gene in humans, is expressed in male germ cells before and after puberty. Genomics 35:346–352

Richter JD, Evers DC (1984) A monoclonal antibody to an oocyte-specific poly(A) RNA-binding protein. J Biol Chem 259:2190–2194

Robinson MO, Simon MI (1991) Determining transcript number using the polymerase chain reaction: Pgk-2, mP2, and Pgk-2 transgene mRNA levels during spermatogenesis. Nucleic Acids Res 19:1557–1562

Ruggiu M, Speed R, Taggart M, McKay SJ, Kilanowski F, Saunders P, Dorin J, Cooke HJ (1997) The mouse Dazla gene encodes a cytoplasmic protein essential for gametogenesis. Nature 389:73–77

Russell L D, Russell JA, MacGregor GR, Meistrich ML (1991) Linkage of manchette microtubules to the nuclear envelope and observations of the role of the manchette in nuclear shaping during spermiogenesis in rodents. Am J Anat 192:97–120

Saxena R, Brown LG, Hawkins T, Alagappan RK, Skaletsky H, Reeve MP, Reijo R, Rozen S, Dinulos MB, Disteche CM, Page DC (1996) The DAZ gene cluster on the human Y chromosome arose from an autosomal gene that was transposed, repeatedly amplified and pruned. Nat Genet 14:292–299

Schlicker M, Schnulle V, Schneppel L, Vorob'ev VI, Engel W (1994) Disturbances of nuclear condensation in human spermatozoa: search for mutations in the genes for protamine 1, protamine 2 and transition protein 1. Hum Reprod 9:2313–2317

Schlicker M, Reim K, Schluter G, Engel W (1997) Specific binding of a 47-kilodalton protein to the 3′ untranslated region of rat transition protein 2 messenger ribonucleic acid. Biol Reprod 56:697–706

Schumacher JM, Lee K, Edelhoff S, Braun RE (1995a) Distribution of Tenr, an RNA Binding Protein, in a Lattice-Like Network within the Spermatid Nucleus in the Mouse. Biol of Reprod 52:1274–1283

Schumacher JM, Lee K, Edelhoff S, Braun RE (1995b) Spnr, a Murine RNA-Binding Protein that is Localized to Cytoplasmic Microtubules. J Cell Biol 129:1023–1032

Schumacher JM, Artzt K, Braun RE (1998) Spermatid perinuclear ribonucleic acid-binding protein binds microtubules in vitro and associates with abnormal manchettes in vivo in mice. Biol Reprod 59:69–76

Skare J, Drwinga H, Wyandt H, vanderSpek J, Troxler R, Milunsky A (1990) Interstitial deletion involving most of Yq. Am J Med Genet 36:394–397

St-Johnston D, Brown NH, Gall JG, Janesch M (1992) A conserved double-stranded RNA-binding domain. Proc Natl Acad Sci USA 89:10979–10983

Tafuri SR, Wolffe AP (1990) *Xenopus* Y-box transcription factors: Molecular cloning, functional analysis, and developmental regulation. Proc Natl Acad Sci USA 87:9028–9032

Tafuri SR, Familari M, Wolffe AP (1993) A mouse Y box protein, MSY1, is associated with paternal mRNA in spermatocytes. J Biol Chem 268:12213–12220

Tarun SZ, Jr, Sachs AB (1995) A common function for mRNA 5′ and 3′ ends in translation initiation in yeast. Genes Dev 9:2997–3007

Tarun SZ, Jr, Sachs AB (1996) Association of the yeast poly(A) tail binding protein with translation initiation factor eIF-4G. Embo J 15:7168–7177

Tiepolo L, Zuffardi O (1976) Localization of factors controlling spermatogenesis in the nonfluorescent portion of the human Y chromosome long arm. Hum Genet 34:119–124

Vereb M, Agulnik AI, Houston JT, Lipschultz LI, Lamb DJ, Bishop CE (1997) Absence of DAZ gene mutations in cases of non-obstructed azoospermia. Mol Hum Reprod 3:55–59

Vogt PH, Edelmann A, Kirsch S, Henegariu O, Hirschmann P, Kiesewetter F, Kohn FM, Schill WB, Farah S, Ramos C, Hartmann M, Hartschuh W, Meschede D, Behre HM, Castel A, Nieschlag E, Weidner W, Grone HJ, Jung A, Engel W, Haidl G (1996) Human Y chromosome azoospermia factors (AZF) mapped to different subregions in Yq11. Hum Mol Genet 5: 933–943

Weighardt F, Biamonti G, Riva S (1996) The roles of heterogeneous nuclear ribonucleoproteins (hnRNP) in RNA metabolism. Bioessays 18:747–756

Wickens M, Anderson P, Jackson RJ (1997) Life and death in the cytoplasm: messages from the 3′ end. Curr Opin Genet Dev 7:220–232

Wolffe AP (1994) Structural and functional properties of the evolutionarily ancient Y-box family of nucleic acid binding proteins. Bioessays 16:245–251

Wu XQ, Gu W, Meng X, Hecht NB (1997) The RNA-binding protein, TB-RBP, is the mouse homologue of translin, a recombination protein associated with chromosomal translocations. Proc Natl Acad Sci USA 94:5640–5645

Yelick PC, Kwon YH, Flynn JF, Borzorgzadeh A, Kleene KC, Hecht NB (1989) Mouse transition protein 1 is translationally regulated during the postmeiotic stages of spermatogenesis. Mol Reprod Dev 1:193–200

Zhong J, Peters AHFM, Lee K, Braun RE (1999) A double-stranded RNA binding protein required for activation of repressed messages in mammalian germ cells. Nat Genet 22: 175–178

An Integration of Old and New Perspectives of Mammalian Meiotic Sterility

Terry Ashley

1 Introduction

Male sterility arises when an individual produces an insufficient number of functional sperm to ensure fertilization. Obviously, mutations in structural proteins can lead to misshapen or immobile sperm that result in infertility. This review, however, will focus on disruptions in the meiotic process that lead to cell death during meiosis or early spermatogenesis. In other words, it will focus on events that occur prior to the differentiation and maturation of the haploid gametes into sperm.

Meiosis is the unique dual cell division that produces eggs or sperm. The first division reduces the chromosome number from diploid to haploid, the second, without an intervening interphase, segregates sister chromatids. It is a complex and protracted process that requires nearly two weeks in mouse spermatocytes, and even longer in human spermatocytes. Meiosis consists of many potentially hazardous steps that are subject to errors, any one of which might result in disruption of meiotic progression. Unfortunately, studying the source of these errors in humans requires a testicular biopsy to examine the affected spermatocytes. Furthermore, by definition the proband is sterile and selected matings of other family members is obviously ethically impossible. Consequently there is a relatively small literature on human meiotic errors and sterility (for review, see de Boer and de Jong 1989; Speed 1989); therefore much of the information discussed here has come from studies on mice and other model organisms. As we shall see, most meiotic-associated cell death occurs between pachynema and spermiogenesis, with high attrition rates in pachynema and metaphase I.

Before beginning an examination of defects in meiosis that have been linked to sterility, we will review the meiotic process and the meiotic-specific structures associated with different stages of meiotic prophase. Next we will survey the list of currently identified meiotic proteins and review their pre-

Department of Genetics, Yale University School of Medicine, 333 Cedar Street, New Haven, Connecticut 06510, USA

Results and Problems in Cell Differentiation, Vol. 28
McElreavey (Ed.): The Genetic Basis of Male Infertility

sumed roles in meiosis. With that background we will then examine meiotic sterility from a series of different perspectives. We will begin by viewing meiotic sterility from a "process-oriented" perspective focusing on cytologically observable defects. In this context we will review theories that have been advanced that attempt to link the cytological observations with the sterility phenotype. We will then shift to a "cell cycle control" perspective and compare the unique requirements of cell cycle controls in meiosis with the better defined controls of the mitotic cell cycle before interpreting some of the meiotic defects we have been discussing from this view point. At each level we will attempt to integrate the information previously discussed, while adding new information. Hopefully this review will provide a viewpoint for thinking about and approaching the study of meiotic sterility.

1.1 The Meiotic Cell Division

The function of mammalian meiosis is to deliver a haploid set of chromosomes to each gamete, be it egg or sperm. This is accomplished by two successive divisions without an intervening replication. The first (reductional) division reduces the chromosome number from diploid to haploid by segregating homologous chromosomes without a centromere division, while the second (equational) division segregates sister chromatids and in that regard is considered mitosis-like. Prophase of the reductional division in particular involves a complex series of events that are unique in the life cycle of the organism. It is divided into substages: leptonema (during which chromosomes begin to condense), zygonema (during which homologous chromosomes synapse) pachynema (during which homologous chromosomes recombine), diplonema (during which homologues begin to repel one another except at sites where recombination has occurred) and diakinesis (during which chromosome condensation continues). One of the first of the unique meiotic events appears to be progressive alignment of homologous pairs of chromosomes (Scherthan et al. 1996). The mechanism for this alignment remains elusive, although it has been suggested that it involves paranemic "side-by-side" interactions (Kleckner et al. 1991; Kleckner and Weiner 1993). Once homologous chromosomes come into close contact, they begin to synapse. Synapsis is preceded or accompanied by a further sequence-specific check for homology facilitated by the subnuclear organelles described below. In mammals, the molecular mechanism by which this check is accomplished remains unresolved.

Pairs of homologues are called bivalents and normally behave as units from the time they synapse in prophase through their eventual movement onto the metaphase plate. Once synapsed, homologues recombine, or crossover, a step believed to be necessary (but insufficient) to hold them together until anaphase I. Meiotic recombination is thought to involve many of the same enzymes that are involved in DNA repair/recombination in mitotically dividing somatic cells. In mammalian meiosis, the sites of recombination can be visualized in late prophase and metaphase I as chiasmata. In addition to recombination, a "chiasma binder" is thought to be necessary to assure adhesion of homologues distal to the site of the crossover. A convincing argument has been presented which links sister chromatid cohesion to chiasma maintenance (Maguire 1978; 1993), see also (Moore and Orr-Weaver 1998). Bivalents then move to the metaphase plate, where homologues orient to opposite poles in preparation for their segregation at anaphase I. In spermatocytes, once the anaphase I division is completed, the haploid set of chromosomes moves to the metaphase II plate without an intervening interphase. There the centromeres of sister chromatids orient to opposite poles in preparation for their segregation at anaphase II. Since segregation of sister chromatids also occurs in mitosis, it can be expected that most of the problems unique to meiosis arise during the first division.

1.2 Meiotic-Specific Structures

Meiotic prophase events are facilitated by meiosis-specific subnuclear organelles. To facilitate the discussion below, a diagram of these structures and a time line of their appearance in meiotic prophase is shown in Fig. 1. The first of these structures to be described was the synaptonemal complex or SC (Fawcett 1956; Moses 1956). The SC is an electron dense, tripartite proteinaceous structure located between homologous chromosomes during meiotic

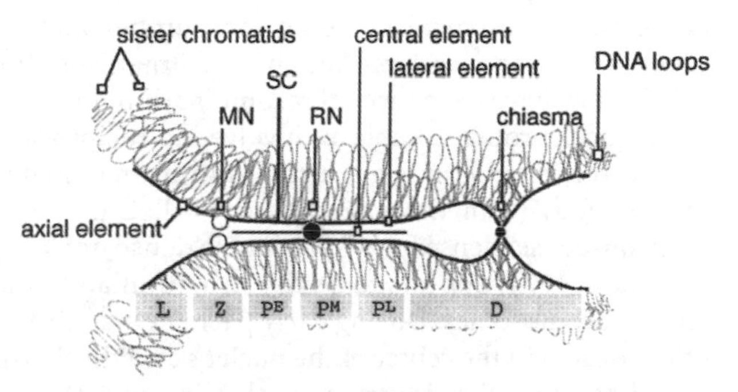

Fig. 1. Diagram of chromosome organization during meiotic prophase in mouse spermatocytes. Meiotic prophase stages are shown at bottom. The diagram does not reflect the relative duration of these stages. During leptonema (*L*) the chromatin begins to condense and the axial element begins to form between sister chromatids of each homologue. Homologous chromosomes synapse during zygonema (*Z*) as the axial elements of the two homologues come into progressively closer alignment. As homologues synapse, a complete synaptonemal complex (*SC*) is formed between them. Meiotic nodules *(MNs)* are present on both axial elements and SCs during zygonema and early pachynema *(p^E)* Recombination nodules *(RNs)* are present on SCs during mid-pachynema *(p^M). As* homologues begin to separate during diplonema (*D*), they remain held together at chiasmata sites

prophase. It consists of two axial/lateral elements, a central element, and numerous transverse filaments. The earliest observable event in meiotic prophase is the development of an axial element between the sister chromatids of each homologue. The chromatin itself is organized into loops with only the chromatin at the base of the loops in contact with the axial element (Weith and Traut 1980). As the axial elements of homologous chromosomes come into closer parallel alignment, a central element begins to form between them, and transverse filaments span the region between the homologous axes. Once the SC is formed the axial elements are called lateral elements.

As axial elements form in plants, electron-dense "nodular" material can be seen associated along their length and (briefly) along the newly formed SC. These variably sized (30–200 nm) structures have been called zygotene nodules (Albini and Jones 1987), meiotic nodules (Stack et al. 1993) or early recombination nodules (Carpenter 1987). Since they are present from leptonema through early pachynema, and since their function remains unknown, they will be called meiotic nodules or MNs in the following discussion. MNs are postulated to be nucleoprotein complexes involved in the check for homology that precedes, or accompanies synapsis (Albini and Jones 1987; Carpenter 1987; Anderson and Stack 1988). Direct interaction between homologues is probably confined to these DNA-nucleoprotein complexes. Although MNs are not obvious in electron micrographs of early mammalian prophase nuclei, similar nucleoprotein complexes do exist in mammalian spermatocytes (Plug et al. 1998). During early pachynema the number of MNs decreases, but another nodular structure can now be seen along the SC. The number and distribution of these more uniform (100 nm in their longest dimension) electron-dense structures corresponds to the number and distribution of crossover events observed in a large number of organisms (for review see Carpenter 1984). They have been termed recombination nodules (RNs) and are assumed to be nucleoprotein complexes involved in reciprocal recombination.

Two other structures are also found in mammalian spermatocyte nuclei: the sex body (sometimes erroneously called the sex vesicle) and a spherical electron-dense "dense body" or "double dense body". The sex body (for review see Solari, 1974) appears to be the structural and functional domain of the X and Y chromosomes. During early pachynema the X and Y axial elements are often located in the center of the nucleus and the density of the chromatin surrounding these axes is similar to that in the rest of the nucleus. However, as pachynema progresses, the X and Y move to the periphery of nucleus and the surrounding chromatin becomes more compact and stains more darkly with both conventional light and electron microscopic staining techniques. This change in position and condensation of the X and Y chromatin corresponds to the development of the sex body. In electron micrographs, the (electron) dense, or double dense body (whether single, or double is species specific) is usually found in the vicinity of the sex chromosomes and exhibits a staining pattern similar to nucleoli (Dresser and Moses 1980). Although nucleoli are evident in mitotic interphase nuclei, they are also present and active in meiotic prophase nuclei (Moses 1980).

As a caution before ending this section, it is important to keep in mind that nuclear DNA is organized into chromatin by the coiling of DNA around octamers of histones to form nucleosomes which are then twisted into higher order structures and stabilized by the "linker" protein, histone H1 (McGhee and Felsenfeld 1980). Processes such as replication, transcription or DNA repair/recombination require the relaxation of these structures. This relaxation can be detected by increased sensitivity to endonuclease cleavage, altered nucleosome spacing, and regions of nucleosome-free DNA (for review see Elgin 1988; Kornberg and Lorch 1995; Krude and Elgin 1996). Relaxation of the chromatin structure is also important in meiosis as evidenced by the detection and mapping of nuclease sensitivity sites along the lengths of mammalian meiotic chromosomes (Chandley and McBeath 1987). Studies in yeast have shown that recombination is not random, but occurs at nucleosome-free sites with a more open chromatin configuration (Ohta et al. 1994; Wu and Lichten 1994, 1995; Fan and Petes 1996). As we will see in the discussion below, attempts have been made to link meiotic sterility to defects in replication, transcription and DNA repair/recombination during meiotic prophase. Each of these processes occurs during meiotic prophase and requires an open-chromatin configuration. The fact that these processes share common enzymatic pathways further complicates the task of determining which process, or processes is being critically disrupted. For example, transcription and repair/recombination (White et al. 1993; Svejstrup et al. 1995), transcription/replication (Masai and Arai 1997) and replication/recombination (Coverley et al. 1991; He et al. 1995; Wold and Kelly 1988) require some of the same proteins. Replication and transcription of the same sequences have been noted in meiotic prophase in plants (Hotta et al. 1985). Moreover, both errors in replication and errors in DNA repair/recombination trigger similar cell cycle checkpoints and lead to cell cycle arrest or apoptosis (Murray and Hunt 1993). Consequently, tracing the links between gross disruptions of meiotic progression and a specific molecular pathway is proving to be a formidable job.

1.3 Identified Protein Components of Meiotic-Specific Structures

Before examining the meiotic sterility phenotypes from a more molecular perspective, it will be helpful to consider the meiotic proteins that have been identified and review some of the roles that have been attributed to them. The following discussion focuses on proteins that have been identified in mammalian meiotic prophase. However, reference will be made, where appropriate, to proteins described in plants or yeast. Although much of the mammalian information is derived from antibody localization studies, several additional strategies are supplementing the information obtained from these efforts. One approach examines the meiotic phenotype of mice homozygous for the targeted disruption of individual genes coding some of the above-mentioned proteins. Another examines differences in antibody distribution on the axes of mice heterozygous for various chromosome aberrations that disrupt the synaptic, recombinational, or disjunctional process. It can be anticipated that clarification of the roles of the rapidly growing list of meiotic proteins will lead to a better understanding of the molecular mechanism(s) involved in the check for homology, synapsis, recombination and segregation of meiotic chromosomes.

1.3.1 *Components of the Synaptonemal Complex*

Several components of SCs have been identified.

SCP3/COR1. SCP3 from rat (Lammers et al. 1994), or COR1, the same protein from Syrian hamster (Dobson et al. 1994) is clearly a component of the axial/lateral elements. COR1/SCP3 is a component of the axial elements from the time that these structures begin to form early in leptonema until they break down in the transition between diplonema and diakinesis. One of the key differences between meiosis and mitosis is the restriction of interactions between sister chromatids during meiotic prophase and the simultaneous promotion of interaction between homologues. COR1 appears between sister chromatids before homologues begin to synapse in zygonema, and is present during pachynema, when reciprocal recombination is thought to occur. However, COR1 does not entirely disappear as the SCs disassemble during diplonema. It is also a component of meiotic kinetochores and remains associated with them throughout the first division, before finally disappearing from these locations around anaphase II (Dobson et al. 1994). The continued presence of the COR1 at sites where sister chromatids remain associated (Dobson et al. 1994) suggests that this protein may play an essential role in sister chromatid cohesion throughout meiosis. Since COR1/SCP3 is continually present through out meiotic prophase, antibodies against this protein are invaluable reference markers for identifying sites (MNs or RNs) and times of activities of other proteins.

Red1/Hop1. In yeast, axial/lateral element formation normally occurs concurrent with synapsis. Although not a COR1/SCP3 primary structural homolog, Red1 is a component of the lateral elements in *S. cerevisiae* (Rockmill and Roeder 1990; Smith and Roeder 1997). *zip1* null mutants of yeast do not form a complete SC, nonetheless Red1 antibodies localize to the asynapsed axes (Smith and Roeder 1997). *red1* mutants show an increased incidence of nondisjunction of *recombinant* chromosomes (Rockmill and Roeder 1990) supporting a role for this axial/lateral element protein in sister chromatid cohesion and chiasma maintenance. Hop1 is also a component of axial elements in

S. cerevisiae (Hollingsworth and Byers 1989; Hollingsworth et al. 1990). Although it is essential for synapsis, it is not required for axial element formation (Hollingsworth and Byers 1989).

SCP1/SYN1 and Zip1. As mentioned above, homologous synapsis is accomplished by the formation of the SC, a process which requires the assembly of a central element and transverse filaments between aligned lateral elements. SCP1 in rat (Meuwissen et al. 1992, 1997) or SYN1 in hamster (Dobson et al. 1994) is a component of the central region of the SC. Although not homologous to SCP1/SYN1, Zip1 is a component of the central region of the SC in *S. cerevisiae* (Sym et al. 1993). All three proteins share a coiled-coil motif (Meuwissen et al. 1992; Sym et al. 1993; Dobson et al. 1994; Liu et al. 1996b). Synapsis (complete SC formation) of autosomal homologues is required for meiotic progression in mammals, but not yeast. Another function of the SC seems to be regulation of the position of reciprocal recombination events along the length of pachytene bivalents, a role shared by all organisms that have been studied (Sym and Roeder 1994; Kleckner 1996; Roeder 1997). However, without data on the result of targeted disruption of SYN1/SCP1, it is impossible to definitively conclude whether or not this particular protein is essential for either of these events.

HSP70-2. Mammalian HSP70-2 is a heat shock protein and a member of the family of molecular chaperonins that mediate protein folding and assembly of other proteins. HSP70-2 has elevated expression during meiotic prophase in males, but not females (Zakeri et al. 1988) and antibody localization studies show that it is a component of the SCs of spermatocytes, but not oocytes (Allen et al. 1996). Disruption of the *Hsp70-2* gene results in sterility in male, but not in female mice (Dix et al. 1996a). It is the only meiotic protein identified to date that exhibits sex-specific expression.

UBC9. UBC9, a ubiquitin conjugating enzyme that is homologous to the *S. cerevisiae* Ubc9 protein (Seufert et al. 1995) and *S. pombe* gene *hus5* (Al-Khodairy et al. 1995) displays an irregular localization pattern along SCs (Kovalenko et al. 1996). More recently UBC9 has been pulled from a two hybrid screen with both COR1 and SYN1, suggesting a potential direct interaction between these proteins (Tarsounas et al. 1997). Many cellular processes, including selective protein degradation, apoptosis, cell cycle control, ribosomal biogenesis, and DNA repair are regulated by ubiquitination (Finley and Chau 1991; Jentsch 1992). In addition, some ubiquitin conjugating enzymes are also involved in chromatin reconfiguration (Jentsch 1992). Since ubiquitin mediated proteolysis is a key component of cell cycle control, the meiotic role of this protein is particularly interesting in the discussion developed below. However, which of these functions UBC9 actually performs in meiosis remains to be determined.

1.3.2 *Meiotic nodules*

Although standard electron microscopic techniques revealed the existence of MNs in plants, similar structures have been more difficult to identify in mammalian prophase nuclei. However, antibody localization studies of mammalian meiotic material not only suggest that these nucleoprotein complexes exist in mammals, but these studies are producing a rapidly growing list of individual protein components. MNs are postulated to consist of proteins involved in the check for homology that precedes or accompanies synapsis. Therefore, before beginning an examination of the potential role of some of the components of MNs, a brief synopsis of two popular models of synaptic initiation will provide the background necessary for interpreting some of the results discussed below. These models are the "gene conversion" model and the "delayed replication" model. The gene conversion model suggests that the sequence-specific check for homology as homologous chromosomes synapse is accomplished by gene conversion events facilitated by one or more homologues of the *E. coli* protein RecA. In *E. coli,* the initiating event in either gene conversion or reciprocal recombination is thought to be a double-strand break (DSB), followed by resection of the 3′ end to form a 5′ single strand (ss) tail and polymerization of the RecA protein onto the ssDNA. The resulting presynaptic filament then facilitates recognition of homologous sequences in the duplex DNA of the homologue and promotes invasion of the single strand into the homologous duplex (West 1992; Kowalczykowski et al. 1994). Subsequent branch migration results in the formation of double Holliday junction intermediates which, depending on the direction of resolution of the Holliday junctions, can result in either gene conversion or reciprocal recombination. Extrapolated to eukaryotic meiosis, this model implies that DSBs are initiating events in both homologous synapsis and crossing over.

The second model, delayed replication, is based on work on meiosis in plants showing that replication of a small (0.1–0.3%) GC-rich subfraction of the genome of lily is not completed in S phase, but is delayed until zygonema, when homologous chromosomes synapse (Hotta et al. 1966; Hotta and Stern 1971; Stern and Hotta 1985). Inhibition of replication of this DNA, termed zygDNA (zygotene DNA), blocks synapsis and degeneration of the meiocytes follows (Ito et al. 1967; Roth and Ito 1967). Since zygDNA is delayed in replication in premeiotic S phase, the model based on this data is here termed the "delayed replication" model. Using this model it is logical to expect that complementary sequences on each homologue are delayed in replication and that each unreplicated strand can be unwound and the single strands used in a paranemic search for homology prior to their replication (Plug et al. 1996).

RecA homologues. As mentioned above, eukaryotic homologues of RecA are expected to be critical early participants in gene conversion events and are predicted to be linked to homologous synapsis. In *S. cerevisiae*, this prediction seems to be borne out by the discovery that Dmc1, a meiotic-specific homologue of RecA (Bishop et al. 1992) and Rad51, another RecA homolog, appear at about the same time as DSBs and colocalize with one another (Bishop 1994; Rockmill et al. 1995; Fan and Petes 1996; Dresser et al. 1997). The maximum number of Dmc1/Rad51 foci (~50) observed in yeast prophase nuclei is always significantly less than the total number of recombination events (~260) when both gene conversion and crossovers are counted as recombination events. This discrepancy has been attributed to the presumed quick turnover of Rad51/Dmc1 complexes linked to a rapid processing of recombination intermediates (Bishop 1994).

A RAD51 antibody that cross reacts with the DMC1 protein also localizes to early meiotic prophase nuclei in mouse spermatocytes and oocytes (Ashley et al. 1995; Plug et al. 1996). As the axial elements form (evidenced by the advent of COR1), RAD51 localizes as foci to sites along the asynapsed axes and, as homologues synapse during zygonema, RAD51 foci are also found along the newly synapsed SCs. The number of RAD51 foci then begins to drop rapidly and most RAD51 foci disappear by the transition from early to mid pachynema (Plug et al. 1998). However, two aspects of mammalian RAD51 localization are difficult to reconcile with the gene conversion model, if DSBs are the initiating event: (1) RAD51 foci are already present on asynapsed axial elements in leptonema, and (2) RAD51 foci are often present at corresponding locations along homologous axes. The first of these observations is surprising because of the duration (around 36 h) and the number of RAD51 foci (>200) present during the leptotene-zygotene interval in mouse (Plug et al. 1996; Moens et al. 1997). Under the tenets of the gene conversion model, during much of this time double-strand breaks (DSBs) would remain open, leaving the genome vulnerable to nucleolic attack and extensive genome rearrangements. The gene conversion model also predicts polymerization of RAD51 on ssDNA followed by a search for homology on the intact double-stranded DNA homologue. The presence of RAD51 foci at corresponding sites on both homologues is contrary to this expectation. However, if a subfraction of the DNA is not replicated until zygonema, the same sequences might be expected to be delayed on each homologue (Plug et al. 1996). The RAD51 localization data therefore appears to favor the delayed replication model.

Differences in RAD51 behavior in yeast and mammals are also evident in mitotically dividing cells. *rad51* mutations in vegetative cells of *S. cerevisiae* only affect the cell's response to DNA damage, not their viability, as would be expected for a gene whose presumed role involves repair/recombination. In contrast, mice homozygous for a disrupted *Rad51* gene die in utero during early embryonic development (Lim and Hasty 1996; Teruhisa et al. 1996), suggesting a critical role for RAD51 in normally dividing somatic cells. In *S. pombe*, *rhp51* strains that lack the *Rad51* homologue are slow growing and appear to have replication defects while *rhp51, rad2* double mutants (which lack the frap endonuclease-*fen1* involved in DNA replication) is not viable (Lehmann et al. 1995). Hence, in *S. pombe* and in mammalian cells, *RAD51* is important for replication, as well as repair. Consistent with this assumption, antibodies against RAD51 localize to S and G2 phase nuclei in human cells in culture (Haaf et al. 1995; Scully et al. 1997). This colocalization raises the possibility that mammalian RAD51 may also be involved in DNA replication, or some additional, as yet unidentified role. In *S. cerevisiae*, mutations in *dmc1* or *rad51* result in meiotic arrest and cell death in pachynema (Bishop et al. 1992; Bishop 1994; Rockmill et al. 1995). Although the prenatal death of *Rad51*$^{-/-}$ mice precludes an assessment of the meiotic effect of disruption of the gene, the cytological link between asynapsis and progression of meiosis (see below), suggests that the effect would be catastrophic.

BRCA1. The involvement of mutations in the *BRCA1* gene in between 5 and 10% of familial breast and ovarian cancers has lead to the speculation that the protein may be involved in a DNA repair/recombination pathway (Scully et al. 1997). Antibodies to BRCA1 colocalize with RAD51 during leptonema and early pachynema at sites along the asynapsed axes of human spermatocytes (Scully et al. 1997), but unlike RAD51, BRCA1 foci disappear as soon as homologues synapse. This localization pattern suggests that BRCA1 is a component of presynaptic MNs. However, if a chromosome has no pairing partner, as is the case of the X chromosome in spermatocytes, or is delayed in synapsis, as sometimes occurs with autosomal homologues, BRCA1 appears to accumulate on the asynapsed axial elements.

Since *BRCA1* has no homologue in yeast, no clues to its function are forthcoming from that direction. However, a mutation of *BRCA1* has been created by targeted disruption of the gene sequence and mice homozygous for the disrupted gene also die in utero at about the same time as do the *Rad51*$^{-/-}$ mice (Gowen et al. 1996; Hakem et al. 1996; Liu et al. 1996a). BRCA1 colocalization with RAD51 is not confined to meiotic prophase; the two proteins also colocalize in undamaged human S-phase cells in culture (Scully et al. 1997). Since *Brca1*$^{-/-}$ mice die before the meiotic effect of the gene disruption can be evaluated, the same cautions should be taken in assigning a role to BRCA1 in repair/recombination vs. DNA replication, or some other roles. Again, considering the placement of BRCA1 on meiotic chromatin prior to synapsis in early meiotic prophase, it appears likely that synapsis and meiotic progression would not precede in the absence of BRCA1 and that the resulting asynapsis would lead to sterility.

ATR. ATR is a mammalian phosphatidylinositol 3-kinase (PIK)-like kinase (Cimprich et al. 1996; Keegan et al. 1996). It is closely related to ATM, another PIK-kinase that is mutated in the human hereditary disease ataxia-telangiectasia (Savitsky et al. 1995). Both proteins share primary sequence homology with Mec1 and Tel1 of *S. cerevisiae* (Greenwell et al. 1995), rad3 of *S. pombe*

(Seaton et al. 1992), and Mei41 of *Drosophila melanogaster* (Hari et al. 1995). With the possible exception of Tel1, these proteins have all been implicated in cell cycle checkpoint control (Carr 1996; Hoekstra 1997). ATR shows its highest level of expression in the testis, and antibody studies place it at sites along the asynapsed axial elements of spermatocytes (Keegan et al. 1996) where it colocalizes with RAD51 (Plug et al. 1998). Like BRCA1, ATR foci disappear as soon as homologues synapse and like BRCA1, ATR accumulates on the asynapsed axial element of the X chromosome and on autosomal axial elements that exhibit delayed synapsis (Keegan et al. 1996). As discussed below, this meiotic localization pattern is complementary to that observed for the related PIK-kinase, ATM. Information is not yet available on the meiotic effect of the targeted disruption of *Atr.*

RPA. Mammalian RPA is a trimeric protein, consisting of a 70 kDa, a 34 kDa, and a 14 kDa subunit. The 70 kDa subunit binds to single-stranded (ss)DNA (Wold and Kelly 1988; Heyer et al. 1990). RPA is a critical component of the replication complex (Wold and Kelly 1988) and is also involved in DNA excision repair (Coverley et al. 1991, 1992). Since it is essential for replication, disruption of the gene is lethal. Meiotic localization studies show that RPA foci are found only on synapsed SCs (not asynapsed axes). At these sites it colocalizes with RAD51 (Plug et al. 1998). This localization pattern suggests RPA is a component of postsynaptic MNs (Plug et al. 1998). Interestingly, in vitro experiments show that in the absence of RPA, or some other single-strand binding protein, RAD51 is very inefficient at carrying out strand exchange, a process necessary for both gene conversion and reciprocal recombination. These localization experiments raise the possibility that only after RPA binds to DNA at MNs, is RAD51 capable of participating in a homology search (Plug et al. 1997b, 1998). If this is the case, the prolonged leptotene-zygotene interval during which RAD51 is present on asynapsed axes may represent a "search inactive" state (Plug et al. 1996).

Polβ. Polymerase β is a DNA polymerase. Although it is not required for DNA replication, it has been shown to be involved in base excision repair (Clairmont and Sweasy 1996; Sobol et al. 1996) and it can complete lagging strand replication in *Xenopus* oocytes and nuclear extracts (Jenkins et al. 1992). It shows a high level of expression in mammalian testes (Novak et al. 1990; Alcivar et al. 1992), and in meiotic prophase spermatocytes antibodies to Polβ localize to sites along newly formed SCs, and later in pachynema to the ends of the bivalents (Plug et al. 1997a). In addition, Polβ antibodies localize to nucleoli and the dense body (Plug et al. 1997a).

ATM. The *ATM* gene is mutated in the human autosomal recessive disease ataxia-telangiectasia (A-T). Human cells in culture normally halt cell cycle progression in response to DNA strand breaks until these breaks are repaired (Hartwell and Kastan 1994). Cells from A-T homozygous individuals do not arrest normally, but replicate with lesions (Kastan et al. 1992; Khanna and Lavin 1993; Xu and Baltimore 1996). The *ATM* defect appears to be involved in detection of DNA damage and the transmission of a damage signal to the cell cycle checkpoints that regulate the G1-S transition, S phase progression, and G2/M transition. In meiotic prophase spermatocytes, antibodies against ATM show a remarkable series of colocalization reactions with RPA. ATM not only colocalizes with RPA on postsynaptic MNs (Plug et al. 1997b), but the two proteins also show indistinguishable localization patterns at sites between homologous regions exhibiting delayed synapsis (Plug et al. 1998). The observation that cells from patients with A-T are unable to phosphorylate RPA (Liu and Bryant 1993) suggests the possibility of a direct interaction between the two proteins.

In mice homozygous for an *Atm* gene disrupted in the kinase domain, early meiotic prophase (leptonema and early zygonema) appears normal. However, the SCs of synapsed regions of bivalents soon begin to fragment (Xu et al. 1996) with fragmentation occurring preferentially at RPA sites (Plug et al. 1997b), i.e. within postsynaptic MNs (Plug et al. 1998). Since RPA is only known to directly associate with ssDNA, the correlation between fragment ends and RPA sites suggests that the observed breaks in the SCs also involve breaks in the DNA. The time of appearance of these breaks is intriguing. Fragments are only observed *after* homologues synapse, and even then, fragmentation is not immediate. ATM is reputed to detect, *not induce* breaks (see Shiloh 1995; Hoekstra 1997). While there may be a delay between the time breaks occur and the time they become cytologically detectable, the longest intact pieces in spermatocytes undergoing extensive fragmentation are always asynapsed axial elements with ATR foci (Plug et al. 1997b). This pattern suggests that breaks are confined to synapsed axes. These observations are consistent with the supposition that programmed breaks in DNA in meiotic prophase of mammals are delayed until after homology has been checked and homologous synapsis of chromosomes has already occurred (Plug et al. 1997b).

CHK1. CHK1 is a protein kinase involved in cell cycle control. In *S. pombe* it affects mitotic arrest following DNA damage, but is not required for normal S phase progression (Walworth et al. 1993; Al-Khodairy et al. 1994; Walworth and Bernards 1996). CHK1 plays a similar role in mammals (Sanchez et al. 1997). Antibody localization studies show that it begins to localize to discrete sites along synapsed autosomal axes in zygonema, as homologous chromosomes synapse, then accumulates on the synapsed axes during pachynema (Flaggs et al. 1997). During early pachynema it is not present on the asynapsed axis of the X chromosome. However, later it is seen first as foci, then as an accumulation on the X axis. The time of appearance of CHK1 foci on the X roughly coincides with the time of disappearance of RAD51 foci and appearance of RPA at sites along the X axis (Flaggs et al. 1997). In *Atm*$^{-/-}$ mice there is no CHK1 localization, an observation which suggests that CHK1 acts downstream of ATM (Flaggs et al. 1997).

MLH1. Mlh1 is a mismatch repair gene that is involved in the detection and correction of DNA mismatches during replication and DNA repair/recombination (for reviews see Modrich 1991; Kolodner 1996; Modrich and Lahue 1996). The mechanisms of activity and functions of the mismatch repair genes are discussed in more detail later in this chapter. Mlh1 has been shown to play a role in crossing over in *S. cerevisiae* (Hunter and Borts 1997). In mice, a monoclonal antibody to MLH1 localizes to sites along homologously synapsed bivalents in both spermatocytes and oocytes of mice (Baker et al. 1996) and the patterns of localization reflect what is known about recombination in mice (Polani and Jagiello 1976). For example, there is an "obligatory" crossover between the X and Y chromosomes in the pseudoautosomal region of shared homology (Rouyer et al. 1986; Petit et al. 1988; Rappold and Lehrach 1988) and an MLH1 focus is consistently observed at this location in early pachytene mouse spermatocytes (Baker et al. 1996). In males there is a high frequency of recombination near autosomal chromosome ends, despite an overall lower frequency of recombination in males compared with females (1.02–1.25 for males vs. 1.34–1.43 for females depending on the strain; Polani and Jagiello 1976). This difference is reflected in distribution of MLH1 sites (1.2 per autosomal bivalent in spermatocytes vs. 1.5 in oocytes; Baker et al. 1996). The close correlation between the MLH1 sites and chiasmata suggests that MLH1 is a component of RNs (Baker et al. 1996). In spermatocytes of mice in which the gene has been disrupted, homologous chromosomes synapse completely normally, but the bivalents fall apart as the homologues begin to desynapse and repel one another in diplonema of meiotic prophase. In spermatocytes this failure to form chiasmata between homologous chromosomes results in a meiotic arrest in metaphase I and subsequent male sterility (Baker et al. 1996). Edelmann et al. (1996) have suggested that the male sterility of *Mlh1*$^{-/-}$ mice occurred during pachynema. While the meiotic error obviously occurs during this stage, the presence of univalents at metaphase I attests to the fact that actual arrest is delayed.

2 Sterility from a "Process-Oriented" Perspective

At multiple steps in the meiotic process, errors can occur which result in meiotic arrest and apoptosis of the spermatocytes (or oocytes). If enough spermatocytes are involved, this abrupt halt in gamete production results in sterility. Consequently there is a large descriptive literature on cytologically detectable errors that result in meiotic arrest (for review see de Boer and de Jong 1989; Speed 1989). These observable errors can be subdivided into (1) errors in synapsis, (2) errors in recombination, or (3) errors in migration or orientation of the chromosomes on the metaphase I or II plates that can lead to defects during either the first or second meiotic division.

2.1 Errors in Synapsis

2.1.1 *The Asynaptic Phenotype and Sterility*

It is likely that asynapsis (synaptic failure) can be linked to more cases of male sterility than any other aberrant meiotic defect (Chandley 1979; Tiepolo et al. 1981; Chandley 1984a). In turn, asynapsis arises under a variety of situations: chromosome rearrangements such as translocations, aneuploidy (lack of a homologous pairing partner), and genetic mutations. In addition, asynapsis is the primary cause of sterility in F1 hybrids in mammals (Benirschke 1967; King 1993). Although asynapsis also affects oogenesis, the consequence is generally less severe and usually results in a shorter reproductive life span, rather than immediate sterility (Burgoyne et al. 1985; Mahadevaiah and Mittwoch 1986; see also de Boer and de Jong 1989; Speed 1989).

Varying degrees of asynapsis occur in association with a wide range of chromosome aberrations (for reviews on mouse and human respectively see de Boer and de Jong 1989; Speed 1989). For example, in meiotic prophase some degree of asynapsis is usually evident around the converging arms of the translocation ring quadrivalent of autosome-autosome translocation heterozygotes (see de Boer and de Jong 1989; Gabriel-Robez and Rumpler 1996) or at the base of the inversion loop of inversion heterozygotes (Poorman et al 1981; Gabriel-Robez and Rumpler 1994). This small degree of asynapsis seldom results in death of spermatocytes. However, in some autosome-autosome translocations an entire translocated chromosomal segment may fail to synapse giving rise to a chain quadrivalent, as opposed to the ring quadrivalent configuration discussed above. In the laboratory mouse, a species with all acrocentric chromosomes, chain quadrivalents (one "arm" asynapsed) occur most frequently when the breakpoints of the translocation are very proximal on one chromosome and very distal on the other (Lyon and Meredith 1966). The result is an increase in the amount of asynapsis, and generally, as severity of asynapsis increases, so too does the degree of spermatogenetic impairment. These asynapsed axial elements often appear thickened and more electron dense in electron micrographs (for example, see de Boer and de Jong 1989). As we have already seen above, this thickening is likely to he the result of the accumulation of proteins (ATR and BRCA1, for example) on the asynapsed axes.

The correlation between degree of asynapsis and sterility is particularly evident in wild mice carrying Robertsonian translocations. Mice heterozygous for a single Robertsonian translocation form an incompletely synapsed trivalent during early meiotic prophase, but remain fertile despite the fact that they may have reduced testis weights, lower sperm counts and an increased frequency of misshapen spermatozoa (Cattanach and Moseley 1973; de Boer and de Jong 1989). During zygonema and early pachynema the Robertsonian arms begin synapsing with their acrocentric counterparts from each of the distal ends, but stop short of the centromeric regions. Later in pachynema the centromeric regions of the Robertsonian trivalent synapse, by synaptic adjustment, a process that is generally assumed to involve synapsis of nonhomologous sequences (see de Boer and de Jong 1989).

However, of more interest to the current discussion is the link between reduced male fertility and increased asynapsis in meiotic prophase of mice heterozygous for more than one Robertsonian translocation (Redi and Capanna 1988; Johannisson and Winking 1994; Everett et al. 1996). This effect is most obvious in mice heterozygous for different Robertsonian translocations with one progenitor acrocentric chromosome in common, a situation referred to as monobrachial homology. Monobrachial homology for a single arm produces a chain configuration which may be associated with mild meiotic disturbances. But as the number of participating Robertsonian translocation chromosomes with monobrachial homology increases, so too does the amount of asynapsis and the severity of the meiotic disruption. Mice carrying five participating chromosomes (three Robertsonian metacentrics and two acrocentrics) are always sterile (for discussion see de Boer 1986; de Boer and de Jong 1989).

Against this background, there is one major exception to the correlation between asynapsis and spermatogenetic impairment. The X chromosome in all male mammals has no homologue and, at best, only partially synapses with the Y during meiotic prophase. This perpetual asynapsis in males, as the heterogametic sex in mammals, has obviously not adversely affected male fertility nor mammalian evolution. Yet, as pachynema progresses, the axial elements of the X and Y exhibit the same type of thickenings as those seen in some of the chromosome aberrations mentioned above. These thickenings become more electron dense and often undergo various "differentiations" that are characteristic for the species (see Solari 1974). The X axis accumulates ATR and BRCA1, just like asynapsed autosomal axes (Keegan et al. 1996; Scully et al. 1997). However, there are a few mammalian species, such as the sand rat,

Psammomys obesus (Solari and Ashley 1977), and the common European field vole, *Microtus agrexis* (Ashley et al. 1989) in which the X and Y are totally asynaptic. Interestingly, the axial elements of the sex chromosomes in these species do not undergo the characteristic thickenings seen in spermatocytes of most other mammals.

While partial, or total asynapsis of the normal X (and Y) chromosome is tolerated in spermatocytes, reciprocal translocations involving the sex chromosomes seem to be particularly vulnerable to spermatogenic disruptions. All reciprocal translocations involving either the X or Y that have been described in humans (Speed 1989) or mouse (Russell 1983; de Boer and de Jong 1989; Gabriel-Robez and Rumpler 1996) are male sterile and meiotic arrest in most of these translocation heterozygotes occurs early in pachynema of meiotic prophase. As discussed below (Sect. 2.1.2), the extreme sensitivity of X-autosome translocations to spermatocyte death leads to the formulation of one theory of meiotic sterility.

Asynaptic mutants have been identified in a number of plant species (Golubovskaya 1979), but have not been reported in animals (see Handel 1988). As we have seen above, most of the proteins that have been identified as components of presynaptic MNs are embryonic lethals, as are mutations in the RPA complex. This may explain the dearth of asynaptic genes in mammals. However, the identification of meiotic proteins has just begun and few mammalian meiotic-specific proteins have yet been identified, so this conclusion may be premature.

Before ending this description of asynapsis and sterility it is important to note that asynapsis constitutes a major barrier to speciation, assuming the hybrid is viable. F1 sterility is usually detectable as global synaptic problems (Benirschke 1967; King 1993). In general, the greater the karyotypic differences (more chromosome rearrangements) between two species, the more severe the meiotic asynapsis and the greater the reduction in fertility (see King 1993). However, there are exceptions to this general rule. For example, *Bos taurus* and *B. javanicus* are chromosomally indistinguishable, but spermatocytes degenerate by pachynema (Pathak and Kieffer 1979). Considerable asynapsis in these F1 hybrids has been noted with asynapsis not necessarily confined to the chromosomes involved in any particular aberration (for discussion, see King 1993). These findings are similar to observations in some of the chromosome aberration heterozygous carriers discussed above which seem to have an increased frequency of asynapsis of chromosomes not involved in the rearrangement (A. W. Plug and T. Ashley, unpubl. obs). Although sequence divergence is often offered as an explanation for this link between asynapsis and the barrier to hybrid sterility, the slow rate of divergence at any specific locus seems insufficient to account for the globally observed asynapsis. We will return to a discussion of this issue in the context of mismatch repair genes.

2.1.2
Theoretical Links Between Asynapsis and Sterility

Two theories have been advanced to explain the correlation between asynapsis and sterility in higher eukaryotes in general, but male mammals in particular. For simplicity they will be referred to as the "unsaturated pairing sites" model proposed by Miklos (Miklos 1974) and the "interference with the inactivity of the sex chromosomes" model proposed to account for the sterility of X-autosome translocations by Lifschytz and Lindsley (Lifschytz and Lindsley 1972). The latter theory has been modified to account for autosomal chromosome rearrangements by Forejt (1984). A third theory here termed "mismatch repair and homology checking" will also be discussed below.

The "unsaturated pairing sites" model was developed by George Miklos to explain the difference in transmission of duplication/deficiency chromosomes in *Drosophila.* Miklos noted that the gamete carrying the chromosome with more extensive synapsis was more likely to survive than the one with less extensive synapsis. In flies heterozygous for a chromosome deficiency, a gamete containing a chromosome with a cytologically detectable deficiency was recovered more frequently than the normal homologue. The opposite was true in flies heterozygous for a duplication: the gamete with a duplication was recovered less frequently than the normal (unduplicated chromosome). It was therefore postulated that each meiotic chromosome had pairing sites along its length at which an unspecified event had to occur in order for the pairing site to become "saturated". Failure to achieve saturation of a site decreased the probability of survival of the gamete carrying such a site. The more sites that remained "unsaturated", the higher the risk of gamete loss. Although the pairing error in *Drosophilia* is incurred in meiotic prophase, the consequence is delayed until after the second meiotic division, a delay that allows recovery of one of the meiotic products.

In mammals, pairing defects have a more immediate consequence than in male flies – loss of the spermatocyte occurs during the first meiotic division, most often during meiotic prophase. Since death occurs before division, there is no differential survival of duplication vs. normal or deficient vs. normal chromosomes. However, Miklos (1974) suggested that the loss of spermatocytes can be traced to the same origin: an inability to saturate pairing sites. By extension, female mammals with chromosome aberrations experience similar synaptic problems. Although the consequences are often less severe, asynapsis in oocytes frequently results in reduced fertility or even sterility, leading Burgoyne and Baker (1984) to extend the hypothesis to females. In evaluating these differences in male and female sterility there are two things to keep in mind; (1) many oocytes in normal females fail to achieve full synapsis and this failure is most likely linked to the massive atresia that normally occurs during early prophase in females (Speed 1988), therefore, death of additional oocytes is merely the extreme of the normal process, and

(2) theoretically, fertility in females requires only a single functional egg in contrast to fertility in males which requires a minimum sperm count in the millions!

The unsaturated pairing site theory draws an interesting correlation between asynapsis and sterility, but offers no molecular explanation for this apparent relationship. However, the recent antibody studies on components of MNs may begin to provide a definition of pairing sites and a foundation for interpreting this phenomenon. It seems likely that a check for homology takes place at every MN (Albini and Jones 1987; Anderson and Stack 1988; Plug et al. 1996). Thus it is likely that an MN can be equated with a pairing site. RAD51 is a component of presynaptic MNs and is already present as axial elements begin to form in leptonema. Although it is not clear exactly what RAD51 is doing at this time, its preferred affinity for ssDNA makes it likely that RAD51 is already associated with ssDNA. As homologous autosomes synapse, RPA appears at these postsynaptic MNs, and RAD51 disappears from these sites shortly thereafter. This suggests that RAD51-associated DNA sequences are being "processed" in some manner during zygonema and pachynema, either by completion of zygDNA replication, or repair synthesis associated with gene conversion. Under the Miklos model "saturated pairing site," can be equated to "processing" of these sequences.

In this context, it is interesting to examine RAD51 and RPA antibody localization on the asynapsed X chromosome in spermatocytes. Under the gene conversion model of synaptic initiation discussed above, the eventual disappearance of the RAD51 foci from the axial element of the X chromosome can be explained by assuming that the RAD51-associated sequences on one sister chromatid invade the duplex DNA on the sister and repair the double-strand break via gene conversion. However, if this model is correct, multiple sites along the X contain unrepaired double-strand breaks for another 3 days after the autosomes synapse. (Recall the data from *Atm*$^{-/-}$ mice suggest that asynapsed axes do not contain double-strand breaks.) In contrast, the preselection model suggests that zygDNA at sites along the X axis must complete their replication (i.e. processing). The time of disappearance of RAD51 from the asynapsed X axis during mid pachynema roughly coincides with the time of appearance of RPA at these sites, consistent with replication of the RAD51-associated sequences at this time (Plug et al. 1998). Since the X chromosome in mammalian spermatocytes is late replicating (Kofman-Alfaro and Chandley 1970), a further de]ay in replication of X-associated zygDNA sequences relative to the autosomal sequences in spermatocytes would not be surprising. However, collaborative evidence for this prediction is needed.

There have been several suggestions that nonhomologous synapsis may be able to rescue a spermatocyte carrying a chromosome aberration from "death by asynapsis" (see de Boer and de Jong 1989). Two general types of nonhomologous synapsis have been described: heterologous synapsis that occurs during zygonema at the same time as homologous synapsis (Ashley and Russell 1986;

Ashley 1987), and heterologous synapsis after homologous synapsis has been completed, a process termed synaptic adjustment (Moses 1977). During meiotic prophase mice heterozygous for the T(1;13)70H and T(1;13)1Wa double translocation can form two heteromorphic bivalents, each with an unsynapsed axial loop (de Boer et al. 1986; Peters et al. 1997) and the loop regions can undergo both types of nonhomologous synapsis. When the heteromorphic bivalents initially synapse homologously, RAD51 foci remain on the asynapsed portions of the loops. However, as the loops shorten and nonhomologously synapse by synaptic adjustment, RAD51 foci disappear and RPA foci appear (Plug et al. 1998). These observations are consistent with the proposition that RAD51-associated sequences must undergo some type of processing, and provide an explanation, albeit incomplete, for spermatocyte arrest in the absence of this processing.

The second model suggests that sterility is linked to "interference with the inactivity of the sex chromosomes" (Lifschytz and Lindsley 1972). Lifschytz and Lindsley pointed out that the single X chromosome is inactive in the primary spermatocytes of all species in which the male is the heterogametic sex. They suggested that there is coordinated regulation of X chromosome transcription and that an X-autosome translocation disrupts this coordinated control. In normal mammalian spermatocytes, tritiated uridine studies show that the autosomes are transcriptionally active during pachynema, but that there is virtually no transcription in the X and Y chromatin within the sex body (Monesi 1965; Monesi et al. 1978). Lifschytz and Lindsley (1972) suggested that the autosomal material brought into the sex body as a result of X-autosome translocations leads to a reactivation of the normally transcriptionally silent X chromatin. This theory gained popular support since it suggests a meiotic modification of the well-described phenomenon of position-effect variegation of the X chromosomes in mammals (Lyon 1968; Russell 1983). However, the proposed mode of operation is the converse of the classic position effects. These are usually seen as the phenotypic results of transcriptional silencing (not activation) of normally active genes brought about by a chromosome rearrangement that repositions these genes near transcriptionally inert heterochromatin. Moreover, in the most direct test of this hypothesis so far, Jaafar et al. (1989) found little evidence of transcriptional reactivation of the X in a tritiated uridine study of T(X;16)16H translocation heterozygotes. (Tritiated uridine incorporation over the sex body was not significantly above background levels.)

In male mammals, the ends of asynapsed autosomes, or autosomal segments, tend to associate with the asynapsed ends of the X chromosome within the sex body (see reviews by Chandley 1984b; de Boer and de Jong 1989; Speed 1989). Such associations have been observed in autosome-autosome translocations, trisomy, partial trisomy and complex Robertsonian chains (see de Boer and de Jong 1989). These associations may be a manifestation of a more general phenomenon – a tendency of ends of partially or totally asynapsed chromosomes to associate with the ends of other asynapsed chromosomes, perhaps in search of a pairing partner (Ashley et al. 1981; Davission et al. 1981; de Boer and de Jong 1989). Since the X and Y chromosomes in male mammals have no homologues, they may be prime candidates for end-association partnerships. The tendency for unsynapsed autosomal axes to associate with the ends of the sex chromosomes led Forejt (Forejt et al. 1981; Forejt 1984) to extend the Lifschytz and Lindsley hypothesis to chromosome rearrangements involving the autosomes. However, this hypothesis is not easy to evaluate, since the autosomal association with the sex body is dependent on asynapsis. For example, in a small pericentric inversion in the sand rat, *Psammomys obesus*, the inverted region never forms an inversion loop (Ashley et al. 1981). As long as the heteromorphic bivalent remains asynaptic, it lies in close contact with the sex body, but as soon as it synapses nonhomologously via synaptic adjustment, the bivalent moves to the other side of the nucleus (Ashley et al. 1981). Moreover, Speed (Speed 1986) detected only a slight increase in transcription, evidenced by minimum tritiated uridine incorporation over the sex body in tertiary trisomic mice in which the extra chromosome was always associated with the sex body. Although this study and the Jaafar study mentioned above do not rule out the Lifschytz and Lindsley model, the support is tenuous at best. Any transcriptional activation would be expected to spread from the adjacent autosomal chromatin. Thus the model implies that transcription of any sequence on the X is enough to trigger cell death. As discussed in more detail below in section 3, there is a more straightforward explanation for the relationship between asynapsis and apoptosis.

2.1.2.3
Mismatch Repair and Mammalian Meiotic Sterility

Meiotic sterility, especially in the context of F1 interspecific hybrids might also be a consequence of sequence divergence. As species diverge, the accumulation of point mutations or small duplications/deletions leads to increasing differences in DNA sequences. Largely identical sequences between species are considered *homologous,* while more divergent sequences are referred to as *homeologous.* In general, the more divergent the sequences the greater the predicted impedance to gene flow between species. If homology must be verified during synapsis, lack of homology might be expected to result in synaptic arrest. Consistent with this assumption, a major barrier to F1 interspecific hybrid fertility is asynapsis. Since the responsibility for detecting sequence differences during both replication and repair/recombination falls on the mismatch repair (MMR) genes, it seems reasonable to presume that MMR genes might be a component of early meiotic nodules and act during meiotic synapsis to detect sequence divergence and halt meiotic progression. Indeed, MMR genes appear to play this role in bacteria and unicellular eukaryotes, such as yeast (for review, see Kolodner 1996). To evaluate their possible role in mammals, we must examine the individual MMR genes and their prokaryotic and yeast homologues.

The mismatch repair (MMR) system was first described in *E. coli,* where three genes *mutS, mutL,* and *mutH* play defining roles. The MutS protein detects mismatches (either mismatched bases, or small deletions and duplications) and binds to the site of the mismatch, while MutH, an endonuclease helps direct repair to the newly synthesized strand. The role of MutL appears to be the coupling of mismatch recognition by MutS to MutH activation. In addition to their role in maintaining replication fidelity, the mismatch repair proteins in *E. coli* play an important role in restricting recombination between homeologous sequences. Mutations in *mutS* lead to an increased rate of homeologous recombination between *E. coli* and *Salmonella typhymurium* (Rayssiquier et al. 1989). In vitro experiments suggest that the mode of action of MMR in repair/recombination is the blockage of RecA-catalyzed strand exchange between homeologous DNA by preventing branch migration of imperfect heteroduplexes by *MutS* (Worth et al. 1994). This activity, termed anti-recombination, appears to be the key to the impediment of gene flow between closely related species of bacteria. Eukaryotes have evolved multiple homologues of MutS and MutL. However, deciphering the role of these genes in meiotic prophase is proving to be difficult, especially since some of these genes appear to have evolved a meiotic role unrelated to mismatch repair. The following discussion will focus on this apparent separation of function.

MSH2 appears to play a key role in detection of mismatches (for review, see Kolodner 1996). In *S. cerevisiae,* Msh2 binds in vitro to heteroduplex DNA at mismatched base pairs in duplex DNA (Prolla et al. 1994b) and mutation in *Msh2* results in a dramatic increase in post meiotic segregation of linked genes. Alani et al. (1994) have presented evidence that one function of *Msh2 is* to prevent branch migration through homeologous sequences during meiotic recombination. These results are indicative of unrepaired heteroduplexes (Reenan and Kolodner 1992) and suggest a role in both gene conversion and reciprocal recombination. In mammals, MSH2 also binds to mismatches in mitotic cells where *MSH2* mutations are responsible for a majority of cases of hereditary non-polyposis colon cancers (Fishel et al. 1993; Leach et al. 1993). However, disruption of the *Msh2* gene in mice does not lead to meiotic sterility (de Wind et al. 1995; Reitmair et al. 1995) although, since the *Msh2*$^{-/-}$ mice were from an inbred background, lack of a mismatch repair phenotype may not be a conclusive test.

Two MutS homologues in *S. cerevisiae, MSH4* and *MSH5* are only expressed during meiosis (Ross-Macdonald and Roeder 1994; Hollingsworth et al. 1995). Although mutations in either *MSH4* (Ross-Macdonald and Roeder 1994) or *MSH5* (Hollingsworth et al. 1995) lead to a decrease in reciprocal recombination and increased levels of meiosis I chromosome nondisjunction, mutations in either gene appear to have no effect on gene conversion or meiotic synapsis, nor do they appear to be involved in mismatch detection and repair (Ross-Macdonald and Roeder 1994; Hollingsworth et al. 1995). As yet there are no reports on the effect of the targeted disruption of either of these genes in mammals. It has been suggested that, rather than binding to DNA mismatches, MSH4 and MSH5 may bind to a different DNA substrate, such as a recombination intermediate (Ross-Macdonald and Roeder 1994; Hollingsworth et al. 1995).

In *E. coli*, the MutS and MutL proteins are close partners in MMR. In *S. cerevisiae* two *mutL* homologues: *Pms1* and *Mlh1* (Williamson et al. 1985; Bishop et al. 1989; Kramer et al. 1989; Prolla et al. 1994a) that form a heterodimer before binding to the MutS complex at the site of the mismatch. Mutations in yeast *Pms1* result in a dramatic increase in post meiotic segregation of linked genes, indicative of unrepaired mismatches and suggestive of a role for *Pms1* in gene conversion, but not in reciprocal recombination (Williamson et al. 1985). The mammalian homologue of *Pms1 is* called *Pms2* (Nicolaides et al. 1994) and disruption of the mouse *Pms2* gene is associated with sterility. Spermatocytes of *Pms2*$^{-/-}$ mice exhibit a variety of pairing problems (Baker et al. 1995). However, interpreting the significance of this disruption is difficult since the defects were observed in an inbred strain of mice, where mismatches between homologous chromosomes should be minimum. Therefore, there is no clear link between disruption of *Pms2* and detection of mismatches.

Mlh1 is also a *mutL* homologue and as discussed above in the section on meiotic proteins, it participates in reciprocal recombination. However, both the antibody localization studies in spermatocytes, and the meiotic phenotype of the *Mlh1*$^{-/-}$ mice suggest a role for MLH1 during pachynema, not zygonema, the predicted time for detection of mismatches that would disrupt synapsis. Like *MSH4* and *MSH5* in *S. cerevisiae,* the role of *Mlh1* in meiosis seems to be restricted to reciprocal recombination (Baker et al. 1996).

In yeast, the MMR genes appear to play a role in meiotic sterility of interspecific hybrids. *Saccharomyces paradoxus* is closely related to *S. cerevisiae.* Although the karyotype ot the two species is similar (Naumov et al. 1992; Hunter et al. 1996) sequence divergence has been estimated at between 8 and 20% (Herbert et al. 1988; Adjiri et al. 1994). Mating between the two species is successful, but most meiotic spores are inviable (Hawthorne and Phillippsen 1994; Hunter et al. 1996). Recombination in the few (~1%) viable spores is low and the frequency of aneuploidy is high (Hunter et al. 1996). Mutations in *pms1* and *msh2* in these hybrids leads to increased recombination, decreased aneuploidy and improved spore viability (Chambers et al. 1996; Hunter et al. 1996). The increased recombination is attributed to the mutations relaxing the requirement for strict homology between sequences and allowing more homeologous recombination. Therefore, it appears that MMR genes do prevent gene flow between species in yeast, but the barrier seems to be inviable spores due to aneuploidy, not to meiotic arrest.

In mammals, a potential link between MMR genes and a general meiotic arrest associated with sequence divergence and asynapsis is even more tenuous. Although it is possible that asynapsis in mammalian F1 interspecific hybrids might be the result of an inability to complete a gene conversion event associated with synaptic initiation, this appears unlikely for several reasons: (1) to date no meiotic phenotype of a mutated mismatch repair gene has clearly linked mismatch detection to sterility, (2) the asynaptic phenotype of F1 interspecific hybrids appears to be similar to that observed in mice heterozygous for any number of other types of chromosome aberrations that are usually sterile on inbred backgrounds where mismatches are unlikely to be a contributing factor, and (3) as discussed above, the gene conversion model of synaptic initiation may not hold for multicellular eukaryotes. For example, a recent study of two mutants in *Drosophila* clearly shows that homologous synapsis in flies occurs in the absence of either gene conversion or reciprocal recombination (McKim et al. 1998).

2.2
Errors in the Meiotic Divisions (Metaphase I Through Anaphase II)

As meiotic prophase draws to a close and the SC disassembles, homologous chromosomes are held together only at chiasma sites where crossovers have occurred. In most organisms, at least one crossover per autosomal bivalent is essential to assure disjunction of homologues to opposite poles at anaphase I. Errors in the recombinational process itself might be expected to fall into two categories: (1) those that leave breaks in the DNA, and (2) those in which crossover between homologues never occurs or is not maintained. If there are breaks in the DNA, it is likely that the damage has already been detected and the path to apoptosis already triggered in prophase (see discussion on cell cycle control below). If a sufficient number of spermatocytes are involved, the result will be sterility. However, if individual chromosomes are intact, but homologues are not held together by chiasmata, as the spindle begins to form another type of error becomes apparent. Normally the bivalents move as units onto the metaphase I plate with the kinetochores of each homologue oriented toward opposite poles. Failure of homologues to recombine, or to maintain at least one chiasmata per bivalent will result in individual chromosomes moving onto the metaphase I plate as univalents. The presence of even one univalent (or pair of univalents) per spermatocyte is sufficient to trigger a cell cycle checkpoint and lead to arrest of the spermatocyte at metaphase I. As is the case with breaks in the DNA, errors in a sufficient number of spermatocytes will result in sterility.

This later type of meiotic sterility has been observed in both a homozygous knock-out mouse (*Mlh1*$^{-/-}$) and several types of chromosomal aberrations. Homologous chromosomes in *Mlh1*$^{-/-}$ mice synapse normally, but homologues either fail to negotiate reciprocal recombination or do not maintain chiasmata (Baker et al. 1996). As the bivalents begin to separate, they fall apart, although the chromosomes themselves appear to have remained intact. Chromosome condensation continues, but univalents, not bivalents accumulate on the metaphase I plate and all spermatocytes arrest at this point (Baker et al. 1996). The presence of a single univalent appears to be sufficient to cause arrest. For example, XO males have an X chromosome carrying a translocated piece of the Y chromosome that contains the male sex determining region (Sxr). In XSxr^{a}O males the single X chromosome often exhibits "fold-back" pairing (Chandley and Fletcher 1980; Mahadevaiah et al. 1988). This nonhomologous synapsis apparently prevents "death by asynapsis" and many spermatocytes with a univalent X reach metaphase I, where they arrest and lead to sterility.

There is no intervening interphase, or even a well defined prophase between the first and second divisions in mammalian spermatocytes. Aneuploid individuals have been identified that are the result of meiosis II nondisjunction errors (Lamb et al. 1996), suggesting that some cells in which sister chromatids separate prematurely can escape the metaphase II/anaphase II checkpoint, if one exists. However, no gene mutations have been identified that block the metaphase II to anaphase II transition.

3 Sterility from the Perspective of Cell Cycle Checkpoint Control

Cell cycle checkpoints are regulated gateways in the cell cycle. Coordination of cell cycle progression is controlled by a series of changes (phase transitions) in the cyclin-dependent kinases (CDKs). The active forms of the CDKs are a complex of at least two proteins, a protein kinase and a cyclin, but usually include other proteins whose function is generally less well understood (Hartwell and Weinert 1989). The dependence of cell proliferation on completion of some event (DNA synthesis or spindle assembly, for instance) is regulated by these checkpoint proteins. One function of cell cycle checkpoint proteins is to monitor genomic integrity by detecting breaks in the DNA and preventing cell-cycle progression until damage is repaired. Through phosphorylation/dephosphorylation these proteins form a signal transduction cascade that transmit the notification of damage to the CDKs, thereby coupling damage detection to cell cycle progression. Further regulation of CDKs and cyclins is achieved through ubiquitin targeted proteolysis. Detection of a lesion, or other type error can result in either cell cycle arrest or apoptosis. However, incapacitation of one of these genes by mutation generally relieves the nucleus of dependence on perfection and allows the cell to continue through the cycle despite errors. For more details on mitotic cell cycle checkpoints, the reader is referred to several recent reviews (Hartley et al. 1995; Paulovich and Hartwell 1995; Elledge 1996; Lydall et al. 1996; Weinert 1997).

Comprehensive reviews of cell cycle checkpoint genes and proteins that have been found in meiotic cells has been provided in two recent reviews (Hoekstra et al. 1991; Handel and Eppig 1998). Rather than recapitulate this information, this section is an extension of the above discussion of meiotic events reinterpreted in the context of cell cycle control, and will continue to focus on early meiotic prophase. We will examine how errors in synapsis and recombination may be detected and transmitted along specially evolved meiotic signal transduction cascades. Reviews relating mammalian sterility to asynapsis as recent as 1993 (Burgoyne and Mahadevaiah 1993) have not raised the issue of cell cycle control in mammalian meiosis. In retrospect this seems strange, since sterility in the situations we have been examining is the result of cell autonomous errors that result in apoptosis. However, there are probably several reasons for this delayed realization: (1) strictly speaking, meiosis does not involve a "cell cycle" since spermatocytes are terminally differentiated, (2) some aspects of meiosis, such as synapsis, have no counterpart in mitosis, therefore it is not intuitively obvious that these processes may be under meiotic cell cycle controls, (3) although meiosis may have borrowed cell cycle proteins from mitotic division, the unique aspects of meiotic division may have led to modifications of the more well-studied mitotic roles and additional controls may have evolved. Consequently, a one to one correspondence between mitotic and meiotic proteins and their respective roles may not hold, (4) until recently, mammalian meiotic studies have been primarily descriptive in nature and have not been designed to identify molecular components.

One of the events that has served to focus attention on cell cycle controls in meiosis has been the growing realization that mutations in many mitotic cell cycle genes in yeast also produce severe meiotic phenotypes (Hoekstra et al. 1991; Lydall et al. 1996). Detection of a meiotic phenotype requires survival of the organism until there *are* meiotic cells. In yeast this is easily accomplished experimentally through the use of temperature sensitive mutants; in mammals these are not available. Therefore, the immunolocalization of mitotic cell cycle proteins, such as ATM (Keegan et al. 1996) on mammalian meiotic prophase chromatin has helped lead to this important paradigm shift. Once the possibility has been raised, this perspective offers new insights into meiotic errors and their consequences.

Before beginning a reexamination of the meiotic process viewed from the perspective of cell cycle checkpoints, we should briefly review the defined mitotic checkpoints. Mitotic checkpoints control the G1/S transition, S phase progression, and the G2/M transition. The G1/S checkpoint prohibits entry into S phase of cells with unrepaired breaks in their DNA, since replication with either gaps or breaks would result in loss of genome integrity. A major regulator of G1/S is the tumor suppressor protein p53. However, p53 appears to play little part in meiosis (A.W. Plug and T. Ashley, unpubl. obs.). The S phase checkpoint monitors progression through S phase and, in the event of incurred lesions, halts progression of replication until the damage is repaired.

Proteins such as BRCA1, PCNA, and RPA appear to be important S phase effectors. The G2/M checkpoint plays a dual role. It monitors the status of the DNA and prevents cell division until replication is completed and DNA damage is repaired and also monitors the assembly of the spindle, and orientation and attachment of chromosomes to assure that every chromosome segregates sister chromatids to opposite poles at anaphase.

Proteins that are involved in mitotic cell cycle control are sometimes divided into two categories: those that monitor normal progression of cell cycle events (completion of S phase, for example) vs. those that detect damage and halt progression of the cell cycle until repair can be accomplished (Paulovich and Hartwell 1995). Both types funnel information into the cyclin/CDK checkpoint gateways. Several replication proteins, such as some of the polymerases (Navas et al. 1995), as well as several DNA repair/recombination proteins, such as Rad53 (Sanchez et al. 1996; Sun et al. 1996) have been implicated as early detectors in this signal transduction cascade. In meiosis, the distinction between "normal progression" and "damage detection" checkpoints may be much less distinct. For example, breakage in mitotic cells may be due to extrinsic damage or to repair errors, while in meiosis, double-strand breaks, at least in yeast, are a programmed step in meiotic recombination.

The use of the term "G2/M checkpoint", especially in reference to meiosis may be misleading, since this transition encompasses all of prophase and therefore bears more resemblance to a corridor than a gateway. The primary mitotic prophase "event" is chromatid condensation, while meiotic prophase encompasses the search and check for homology, synapsis, and recombination, as well as chromatin condensation. As we have seen above, these events may require completion of DNA replication and certainly involve extensive DNA repair associated with reciprocal recombination. Mitotic prophase is completed in less than an hour; meiotic prophase lasts mearly two weeks.

With these points in mind, it may be instructive to revisit the various meiotic processes and proteins we have discussed above and reexamine them in the context of meiotic cell cyclc controls. If we return to the synaptic process, it seems likely that asynapsis triggers one or more meiotic cell cycle checkpoints. But which one or ones'? First it may be helpful to compare meiotic prophase events in *S. cerevisiae* vs. mammals. Mutations such as *red1, hop1,* and *zip1* block or interfere with synapsis in *S. cerevisiae.* However when *red1, hop1,* or *zip1* mutants are combined with a *spo13* mutant, the defective *spo13* mutants permit a single round of chromosome segregation in which the chromosomes can divide either reductionally or equationally and nonrecombinant chromosomes are recovered (Klapholz and Esposito 1980; Hugerat and Simchen 1993). These observations suggest that asynapsis per se does not trigger a meiotic checkpoint in yeast. Although there are, as yet, no reports on the effect of the targeted disruption of COR1 (SCP3) in mammals, the cytogenetic evidence from chromosome aberrations suggests that autosomal asynapsis triggers a checkpoint that leads to apoptosis of spermatocytes during meiotic prophase. Interestingly, SCP3 is a phosphoprotein whose extent of phosphorylation changes during meiotic prophase (Lammers et al. 1995). As we have seen above, CHK1, a checkpoint protein whose activity involves phosphorylation, localizes to the SC. Whether SCP3, or some other, as yet unidentified protein is the target of this CHK1 phosphorylation remains to be determined.

Ubiquitination is also important in cell cycle control, especially in the mitotic G2/M checkpoint where it plays a role in sister chromatid separation during the anaphase transition (for a discussion, see Elledge 1998). As mentioned above, UBC9, a ubiquitin conjugating enzyme, also localizes to sites along the SC and shows many of the same localization patterns as COR1 /SCP3 (Kovalenko et al. 1996 and A.W. Plug and T. Ashley, unpubl. obs.). As a component of the axial element, COR1/SCP3 is located between sister chromatids in meiotic prophase chromosomes. The discovery that UBC9 interacts with both COR1/SCP3 and SYN1/SCP1 in a two hybrid screen (Tarsounas et al. 1997) raises the intriguing possibility that UBC9 may regulate various aspects of sister chromatid separation in meiosis.

In *S. cerevisiae, dmc1,* and *rad51* mutants can still complete low levels of recombination and sporulate (Klapholz and Esposito 1980; Bishop et al. 1992; Shinohara et al. 1992; Hugerat and Simchen 1993; Rockmill and Roeder 1994), suggesting that the recombinational defects resulting from the mutations do not trigger a yeast meiotic checkpoint arrest. Although the mammalian *Rad51* gene is embryonically lethal, preempting an examination of the meiotic effect, the continued presence of RAD51 foci at sites along asynapsed axes suggests an incomplete step. Whether the RAD51 foci observed on asynapsed axes are assisting with repair synthesis associated with DSBs, or are involved in some aspect of DNA replication, they are most likely bound to ssDNA. If so, they signal the existence of gaps or breaks in the DNA, the continued presence of which can be expected to trigger a checkpoint. However, RAD51 foci do not remain on the asynapsed axis of the X chromosome, but disappear at about the same time as RPA foci appear (Plug et al. 1998). This suggests that the RAD51-associated ssDNA on the X axis is replicated or repaired, releasing these sites from the asynaptic checkpoint and allowing spermatogenesis to proceed.

In the T70H/T1Wa translocation double heterozygote, the large heteromorphic bivalent almost always undergoes complete synaptic adjustment, at which time the RAD51 foci disappear from the nonhomologously synapsed axes (Plug et al. 1998). This observation supports the suggestion that nonhomologous synapsis can "rescue" a spermatocyte from "death by asynapsis" (de Boer and de Jong 1989). Sometimes the small heteromorphic bivalent also "adjusts" completely with the longer axis shortening to the length of the shorter one. In these nuclei both lateral elements of the small 1^{13} bivalent appear to be of equal thickness and a normal distance apart. However, in other nuclei, the smaller 1^{13} bivalent never completely synapses and the longer axis appears thicker than the shorter axis, creating a bent or slightly "horseshoe-shaped" 1^{13} bivalent (Peters et al. 1997). In an electron microscopic study, the "horseshoe" type 1^{13} bivalent (but not the completely synaptically adjusted 1^{13} bivalent) was found to associate with the sex body and the frequency of this association correlated with a corresponding drop in sperm count (Peters et al. 1997). RAD51 foci remain associated with the axes of these incompletely synapsed horseshoe-shaped 1^{13} bivalents (Plug et al. 1998). The preferred association of RAD51 with ssDNA suggests that these sites still contain ssDNA, whose continued presence constitutes gaps in the genome. If so, it is likely that it is these lesions, not the association of the bivalent with the sex body, that have triggered a cell cycle checkpoint leading to apoptosis of the affected spermatocytes.

While these observations provide a rational link between asynapsis, cell cycle control and sterility, they do not identify the checkpoint proteins involved in the signal transduction cascade, nor do they differentiate between an "S phase progression pathway" and a "damage control pathway". However, ATR, ATM and CHK1 are all cell cycle proteins. In *S. pombe rad3* (the ATR homologue), *chk1,* and *cdc2* form a damage detection cascade (see Carr 1997). Recent studies have provided evidence of a similar signal transduction cascade in mammals (Flaggs et al. 1997; Furnari et al. 1997; Sanchez et al. 1997). *rad3,* the *S. pombe* homologue of ATR (Bentley et al. 1996), is required for both the damage checkpoint and the S-M checkpoint pathways and Mec1 plays a similar dual role in *S. cerevisiae.* However, the downstream targets of *rad3*/Mec1 appear to differ in the damage control and the S phase progression pathways (see Carr 1997). In this regard it is interesting that rad3/Mec1 have two mammalian homologues: ATR and ATM. Perhaps ATR functions in the replication/normal progression pathway, while ATM acts in the damage detection pathway.

The first of the identified mammalian checkpoint proteins to make a meiotic appearance is ATR. Despite the fact that ATR is first evident as it colocalizes with RAD51 to sites along asynapsed axes in zygonema, its localization differs in a subtle way from that of RAD51. Unlike RAD51, ATR is not visible in leptonema as the axial elements are forming (Keegan et al. 1996); also A. W. Plug and T. Ashley (unpubl. obs.). In fact, it does not become evident on asynapsed axes until late zygonema, at which time it is generally seen only at sites along axes that are the last in the nucleus to synapse. If there is prolonged delay of synapsis, as is the case along the X chromosome axis or the asynapsed axes of translocation heterozygotes, ATR accumulates along the axes (Keegan et al. 1996). Might ATR accumulation be signaling asynapsis – perhaps in response to a delay in replication of RAD51-associated sequences? If the autosomal axes eventually synapse, whether or not that synapsis is homologous, ATR disappears from the axis (Plug et al. 1997b, 1998), supporting a role for ATR in signaling asynapsis. However, in the case of the X which never synapses, ATR remains associated with the axis (Plug et al. 1997b, 1998) even after the RAD51-associated sequences are "processed" and disappear. Since processing of these sequences seems to relieve the "asynaptic" checkpoint by fulfilling an event necessary for meiotic progression, the continued presence of ATR on the

X axis might seem to suggest that ATR is not involved in this checkpoint. However, recent experiments have shown that ATR localization occurs independently of the state of protein kinase activity (K. S. Keegan and M. F. Hoekstra, pers. comm.). Since phosphorylation/dephosphorylation is the major mechanism of signal transduction and since the current antibody localization studies do not distinguish between these forms, further experiments are necessary before a conclusion can be drawn regarding the role of ATR in signaling asynapsis.

As discussed above, another PIK kinase, ATM, colocalizes with RAD51 (and RPA) on synapsed homologous chromosomes (Plug et al. 1998). It is at these sites that the bivalents fragment in mice homozygous for the targeted disruption of the *Atm* gene (Plug et al. 1997b). A-T patients are hypersensitive to radiation and chemicals that induce breaks in the DNA (Hari et al. 1995; Morrow et al. 1995; Bentley et al. 1996; Hoekstra 1997; Sanchez et al. 1997). Since the *Atm* gene and its homologues in other organisms have been identified as checkpoint proteins, not repair proteins (Sedgwick and Boder 1991), the chromosome fragmentation associated with disruption of the *Atm* gene suggests that the defect is an inability to signal the presence of breaks. Yet in $Atm^{-/-}$ mice the fragmentation of chromosomes occurs, not at asynaptic RAD51/ATR sites, but at synapsed RPA/ATM sites (Plug et al. 1997b). Even in nuclei with 60 or more fragments, the longest intact axes are axial elements with ATR foci (Plug et al. 1997b). These observations suggest that the fatal lesions that involve both DNA and SCs in $Atm^{-/-}$ mice are incurred only after homologues synapse (Plug et al. 1997b).

The observations on chromosome aberrations invite an interesting comparison to those on the $Atm^{-/-}$ mice. In mice heterozygous for numerous aberrations, RAD51 foci remain associated with the asynapsed axes until the cells begin to undergo apoptosis several days into pachynema. However, neither these axial elements nor those present in other chromosome aberrations that result in univalency, or partial asynapsis, show any indication of fragmentation of the axes. This suggests a difference in the nature of the presumed interruptions (perhaps unreplicated DNA gaps?) associated with RAD51 foci at presynaptic MN vs. the interruptions (perhaps double-strand breaks?) at RPA foci at postsynaptic MN sites in $Atm^{-/-}$ mice. The detection of fragments in $Atm^{-/-}$ mice *after* synapsis may be telling us that breaks occur in mammalian meiotic prophase only after homology has been checked and chromosomes have synapsed. These differences seem to separate the apoptosis associated with asynapsis from the apoptosis associated with $Atm^{-/-}$ fragmentation. The association of ATR with asynapsed axes vs. the association of ATM with synapsed axes again suggests they may be functioning in different pathways, or different branches of the same pathway.

The molecular structure of ATM does not contain an obvious DNA binding domain, suggesting that damage detection may be indirect and require mediation by an additional protein. An example of this type of interaction is DNA-

PK, another PIK-kinase, which binds to the ku autoantigens that, in turn, bind directly to broken double-strand ends (Lees-Miller et al. 1990). Although the "DNA intermediary" for ATM has not been identified, RPA is an interesting candidate. Cultured cells from A-T patients are defective in DNA damage detection and in phosphorylation of RPA (Liu and Weaver 1993) and, not only do the meiotic localization patterns of RPA and ATM mimic one another (Plug et al. 1997b, 1998), but fragmentation of the meiotic bivalents in *Atm*$^{-/-}$ mice occurs at RPA sites (Plug et al. 1997b).

Atm$^{-/-}$ mice raise another interesting point. Many checkpoint proteins have been identified by the fact that mutations in the gene allow the cell to progress to the next step of the cell cycle without completing the previous step (progression checkpoints), or with unrepaired damage (damage control checkpoints). This has led to the assumption that mutations in, or elimination of, the checkpoint protein will always result in continued progression of the cell through the cell cycle. However, there is a fallacy in this assumption, especially in relation to meiosis. It fails to take into account that damage which escapes detection by a checkpoint and remains unrepaired will eventually be lethal, and that progression through mammalian meiotic prophase requires days, rather than hours. Consequently, lethal effects that might be masked for one or more cycles in mitotically dividing cells, can be expected to become evident during meiotic prophase during that period.

As discussed above, DNA damage is transmitted to the CDK/cyclin complex through a signal transduction cascade. In *S. pombe, chk1* acts downstream of *rad3* (the ATR homologue) in this cascade, but only transmits signals in the damage control pathway, not the S-M progression pathway (Al-Khodairy et al. 1994). Human CHK1, like its *S. pombe* homologue, is modified in response to DNA damage and in turn phosphorylates human CDC25C (Sanchez et al. 1997). It has been proposed that this phosphorylation inhibits the ability of CDC25C to dephosphorylate and activate the CDC2-cyclin B complex, thereby blocking entry into mitotic cell division, until repair is accomplished (Sanchez et al. 1997). CHK1 localizes to synapsed SCs (Flaggs et al. 1997), but unlike RPA and ATM, its localization does not remain focal in nature. It begins to coat the SCs, although not as extensively as does ATR on the asynapsed axes. As might be predicted for a protein that only associates with synapsed autosomes, CHK1 is not found on the asynapsed X chromosomal axis in early pachytene nuclei. However, at about the time the RPA and ATM foci appear, then disappear, CHK1 can be seen first as foci, then as a partial coating of the X axis. If CHK1 accumulation signals completion of processing of RAD51-associated sequences, its presence on the X axis may be signaling an "OK-go ahead" message.

The relationship between normal synapsis and meiotic progression is further strengthened by the observation that CHK1 localization is dependent on the presence of a functionally active *Atm* gene (Flaggs et al. 1997). If ATM is signaling the presence of DNA damage, failure to transmit that signal interrupts the signal transduction cascade and blocks even the localization of CHK1. In the mitotic cycle CHK1 phosphorylates CDC25C in response to damage (Furnari et al. 1997; Sanchez et al. 1997). A model has been proposed (Sanchez et al. 1997) suggesting that CDC25C phosphorylation prevents CDC25C from activating the CDC2-cyclin B complex and mitotic entry. Although CDC25C has not yet been localized to mammalian meiotic chromosomes, the *Cdc25* gene is expressed in pachytene spermatocytes (Wu and Wolgemuth 1995).

The heat shock protein, HSP70-2, mentioned above, also appears to be involved in the meiotic signal transduction cascade in spermatocytes. HSP70-2 localizes to SCs in pachytene spermatocytes, but not oocytes (Allen et al. 1996). When the *Hsp70-2* gene is disrupted, SC formation appears normal, but fragmentation of the axes occurs by mid-pachynema and is followed by apoptosis (Dix et al. 1996b). The cytological description of this disruption sounds similar to that observed in the *Atm*$^{-/-}$ mice. Zhu et al. (1997) have shown that HSP70-2 interacts with CDC2 in mouse testis and appears to be required for the correct assembly of the CDC2/cyclin B complex in mouse spermatocytes. Destruction of the CDC2/cyclin B complex triggers the anaphase I transition, making it likely that HSP70-2 acts downstream of ATM and CHK1. HSP70-2 localizes to the SC and may be involved in late prophase meiotic events. Since the CDC2/cyclin B complex is believed to be the cell division "gate keeper", it appears that we are making progress toward identifying the protein components of a meiotic signal transduction cascade that links completion of synapsis and reciprocal recombination to the metaphase I/anaphase I transition.

Hawn et al. (1995) have proposed a link between the mismatch repair pathway and the G2 cell cycle checkpoint in mitotic cells. A similar link seems to exist in meiotic cell division. It can be predicted that detection of reciprocal recombination and chiasma formation is also under checkpoint control. Failure of recombination could have either of two outcomes. It could leave an unrepaired break, or result in two intact univalents. The targeted disruption of the *Mlh1* gene illustrates the separation of these two events. As discussed above, meiosis in *Mlh1*$^{-/-}$ mice appears to proceed normally until homologous chromosomes begin to separate at diplonema. At that time, the previously synapsed bivalents fall apart as an obvious consequence of a failure to form chiasmata. Nonetheless, intact univalents congregate on the metaphase I plate and the cells arrest then eventually die (Baker et al. 1996). This meiotic checkpoint may detect tension on the spindle, since attachment of homologous kinetochores to opposite spindle poles and a pulling force from both poles appear to provide the trigger necessary for the onset of anaphase (Nicklas and Kock 1969; Nicklas and Kubai 1985). More recent studies with a phosphoepitope-specific antibody, 3F3/2, have shown that this tension results in phosphorylation of an unidentified kinetochore protein (Nicklas et al. 1995). The protein that phosphorylates this kinetochore protein also remains to be identified.

The function of the meiotic spindle checkpoint is to inhibit the metaphase I to anaphase I transition if there are univalents present. Recent evidence suggests that this checkpoint is more stringent in spermatocytes than oocytes (see Hunt and LeMaire-Adkins 1998). If this sex-specific effect is confirmed, it may help to explain the prevalence of maternally vs. paternally derived aneuploidy (see Hassold 1998).

4 Summary and Conclusions

The many events of meiotic prophase can now be viewed as a series of specialized incidents that are monitored by meiotic checkpoints, some of which are similar to their mitotic counterparts, and some of which are probably unique to meiosis. This shift in perspective means that meiotic sterility in mammals must be reexamined and viewed as the result of errors subject to meiotic checkpoint controls. Like their mitotic counterparts, the meiotic checkpoints detect defects and halt normal progression until these mistakes can be repaired. Some of these checkpoints utilize mitotic checkpoint proteins, others may involve meiotic-specific proteins, or splice forms. If repair is impossible, the checkpoints then either trigger immediate apoptosis or cause an arrest of meiotic progression followed by eventual cell death. If a sufficient number of spermatocytes are involved, either alternative results in sterility. Identification of these meiotic checkpoints and delineation of the signal transduction cascades involved has only just begun. While yeast, or other model organisms, may provide clues to some of these pathways, others appear to have arisen during vertebrate evolution. The study of mammalian meiosis has entered a new era and the foundations are being laid for a growing understanding of the many problems that may contribute to sterility.

Acknowledgments. I would like to thank Drs. Lorinda Anderson, Peter de Boer, Merl Hoekstra, and Annemieke Plug for their comments and suggestions and Adelle Hack for her meticulous editing. The work discussed above from the author's laboratory was supported by NIH grants GM49779 and GM55300.

Note added in proof: Since this chapter was written, the targeted disruption of the mammalian meiotic mismatch repair gene *Msh5*, has been shown to prevent synapsis between homologous chromosomes (de Vries et al. 1999; Edelmann et al. 1999). The phenotype appears to be similar to that of *Dmc1*$^{-/-}$ mice (Pittman et al. 1998; Yoshida et al. 1998), also described since the completion of this article. In both cases, meiosis is arrested at a stage equivalent to around mid-pachynema. De Vreies et al. (1999) have suggested that this arrest defines the time of a meiotic checkpoint.

de Vries SS, Baart EB, Dekker M, Siezen A, de Rooij DG, de Boer P, te Riele, H (1999) Mouse MutS-like protein MSH5 is required for proper chromosome synapsis in male and female meiosis. Genes Dev 13:523–531

Edelmann W, Cohen PE, Kneitz B, Winand N, Lia M, Heyer J, Kolodner R, Pollard JW, Kucherlapati R (1999) Mammalian MutS homologue 5 is required for chromosome pairing in meiosis. Nat Genet 21:123–127

Pittman DL, Cobb J, Schimenti KJ, Wilson LA, Cooper DM, Brignull E, Handel MA, Schimenti JC (1998) Meiotic prophase arrest with failure of chromosome synapsis in mice deficient for *Dmc1,* a germline-specific RecA homolog. Molec Cell 1:697–705

Yoshida K, Kondoh G, Matsuda Y, Habu T, Nishimune Y, Marita T (1998) The mouse RecA-like gene DMC1 is required for homologous chromosome synapsis during meiosis. Molec Cell 1:707–718

References

Adjiri A, Chanet R, Mezard C, Fabre F (1994) Sequence comparison of the *ARG4* chromosomal regions from the two related yeast, *Saccharomyces cerevisiae* and *Saccharomyces douglasii.* Yeast 10:309–317

Alani E, Reenan RAG, Kolodner R (1994) Interactions between mismatch repair and genetic recombination in *Saccharomyces cerevisiae.* Genetics 137:19–39

Albini SM, Jones GH (1987) Synaptonemal complex spreading in *Allium cepa* and *A. fistulosum* I. The initiation and sequence of pairing. Chromosoma 95:324–338

Alcivar AA, Hake LE, Hecht NB (1992) DNA polymerase β and poly(ADP)ribose polymerase mRNAs are differentially expressed during the development of male germinal cells. Biol Reprod 46:201–207

Al-Khodairy F, Fotou E, Sheldrick KS, Griffiths DJF, Lehman AR, Carr AM (1994) Identification and characterization of new elements involved in checkpoints and feedback controls in fission yeast. Mol Biol Cell 5:147–160

Al-Khodairy F, Enoch T, Hagan IM, Carr AM (1995) The *Schizosaccharomyces pombe hus5* gene encodes a ubiquitin conjugating enzyme required for normal mitosis. J Cell Sci 108: 475–486

Allen JW, Dix DJ, Collins BW, Merrick BA, He C, Selkirk JK, Poorman-Allen P, Dresser ME, Eddy EM (1996) HSP70-2 is part of the synaptonemal complex in mouse and hamster spermatocytes. Chromosoma 104:414–421

Anderson LK, Stack SM (1988) Nodules associated with axial cores and synaptonemal complexes during zygotene in *Psilotum nudum.* Chromosoma 97:96–100

Ashley T (1987) Nonhomologous synapsis of the XY during early pachynema in male mice. Genetica 72:81–84

Ashley T, Russell LB (1986) A new type of nonhomologous synapsis in T(X;4)1R1 translocation mice. Cytogen Cell Genet 43:194–200

Ashley T, Moses MJ, Solari AJ (1981) Fine structure and behavior of a pericentric inversion in the sand rat, *Psammomys obesus.* J Cell Sci 50:105–119

Ashley T, Jaarola M, Fredga K (1989) The behavior during pachynema of a normal and inverted Y chromosome in *Microtus agrestis.* Hereditas 111:281–294

Ashley T, Plug AW, Xu J, Solari AJ, Reddy G, Golub EI, Ward DC (1995) Dynamic changes in Rad51 distribution on chromatin during meiosis in male and female vertebrates. Chromosoma 104:19–28

Baker SM, Bronner CE, Zhang L, Plug AW, Robatzek M, Warren G, Elliott EA, Yu J, Ashley T, Arnheim N, Flavell RA, Liskay RM (1995) Male mice defective in the DNA mismatch repair gene PMS2 exhibit abnormal chromosome synapsis in meiosis. Cell 82:309–319

Baker SM, Plug AW, Prolla TA, Bronner CE, Harris AC, Yao X, Christie D-M, Monell C, Arnheim N, Bradley A, Ashley T, Liskay RM (1996) Involvement of mouse *Mlh1* in DNA mismatch repair and meiotic crossing over. Nat Genet 13:336–342

Benirschke K (1967) Sterility and fertility of interspecific mammalian hybrids. In: Benirschke K (ed) Comparative Aspects of Reproductive Failure. Springer, Berlin Heidelberg New York

Bentley NJ, Holtzman D, Flaggs G, Keegan KS, DeMaggio AJ, Ford JC, Hoekstra MF, Carr AM (1996) The *Schizosaccharomyces pombe* rad3 checkpoint gene. EMBO J 15:6641–6651

Bishop DK (1994) RecA homologues Dmc1 and Rad51 interact to form multiple nuclear complexes prior to meiotic chromosome synapsis. Cell 79:1081–1092

Bishop DK, Andersen J, Kolodner RD (1989) Specificity of mismatch repair following transformation of *Saccharomyces cerevisiae* with heteroduplex plasmid DNA. Proc Natl Acad Sci USA 86:3713–3717

Bishop DK, Park D, Xu L, Kleckner N (1992) DMC1: a meiotic specific yeast homolog of *E. coli* recA required for recombination, synaptonemal complex formation and cell cycle progression. Cell 69:439–456

Burgoyne PS, Baker TG (1984) Meiotic pairing and gametogenic failure. In: Evans CW, Dickenson HG (eds) Controlling events in meiosis. Company of Biologists, Ltd, Cambridge, pp 349–362

Burgoyne PS, Mahadevaiah SK (1993) Unpaired sex chromosomes and gametogenic failure. In: Sumner AT, Chandley AC (eds) Chromosomes Today, vol 11. Chapman and Hall, London, pp 243–263

Burgoyne PS, Mahadevaiah S, Mittwoch U (1985) A reciprocal autosomal translocation which causes male sterility in the mouse also impairs oogenesis. J Reprod Fertil 75:647–652

Carpenter ATC (1984) Meiotic roles of crossing-over and of gene conversion. Cold Spring Harbor Symp Quant Biol 49:23–29

Carpenter ATC (1987) Gene conversion, recombination nodules, and the initiation of meiotic synapsis. Bioessays 6:232–236

Carr AM (1996) Checkpoints take the next step. Science 271:314–315

Carr AM (1997) Control of cell cycle arrest by the Mec1sc/Rad3sp DNA structure checkpoint pathway. Curr Opin Genet Dev 7:93–98

Cattanach BM, Moseley H (1973) Non-disjunction and reduced fertility caused by tobacco mouse metacentric chromosomes. Cytogenet Cell Genet 12:264–287

Chambers SR, Hunter N, Louis EJ, Borts RH (1996) The mismatch repair system reduces meiotic homeologous recombination and stimulates recombination-dependent chromosome loss. Mol Cell Biol 16:6110–6120

Chandley AC (1979) The chromosomal basis of human infertility. Br Med Bull 35:181–186

Chandley AC (1984a) Infertility and chromosome abnormalities. In: Clarke JR (ed) Oxford Reviews of Reproductive Biology, vol 6. Oxford Univ. Press, Oxford, pp 1–46

Chandley AC (1984b) On the nature and extent of XY pairing at meiotic prophase in man. Cytogenet Cell Genet 38:241–247

Chandley AC, Fletcher JM (1980) Meiosis in Sxr male mice. I. Does a Y-autosome rearrangement exist in sex-reversed (Sxr) mice? Chromosoma 81:9–17

Chandley AC, McBeath S (1987) DNase I hypersensitive sites along the XY bivalent at meiosis in man include the XpYp pairing region. Cytogenet Cell Genet 44:22–31

Cimprich K, Shin TB, Keith CT, Schreiber SL (1996) cDNA cloning and gene mapping of a candidate human cell cycle checkpoint protein. Proc Natl Acad Sci USA 93:2850–2855

Clairmont CL, Sweasy JB (1996) Dominant negative rat DNA Polymerase β mutants interfere with base excision repair in *Saccharomyces cerevisiae.* J Bacteriol 178:656–661

Coverley D, Kenny MK, Munn M, Rupp WD, Lane DP, Wood RD (1991) Requirement for the replication protein SSB in human DNA excision repair. Nature 349:538–541

Coverley D, Kenney MK, Lane DP, Wood RD (1992) A role for the human single-stranded DNA binding protein HSSB/RPA in an early stage of nucleotide excision repair. Nucleic Acids Res 20:3873–3880

Davisson MT, Poorman PA, Roderick TH, Moses MJ (1981) A pericentric inversion in the mouse. Cytogenet Cell Genet 30:70–76

de Boer P (1986) Chromosomal causes for fertility reduction in mammals. In: de Serres FJ (ed) Chemical Mutagens, vol 10. Plenum, New York, pp 427–467

de Boer P, de Jong JH (1989) Chromosome pairing and fertility in mice. In: Gillies CB (ed) Fertility and chromosome pairing: recent studies in plants and animals. CRC Press, Boca Raton, Florida, pp 37–76

de Boer P, Searl AG, van der Hoeven FA, de Rooij DG, Beechey CV (1986) Male pachytene pairing in single and double translocation heterozygotes and spermatogenic impairment in the mouse. Chromosoma 93:326–336

de Wind N, Dekker M, Berns A, Radman M, te Riele H (1995) Inactivation of the mouse Msh2 gene results in mismatch repair deficiency, methylation tolerance, hyperrecombination, and predisposition to cancer. Cell 82:321–330

Dix DJ, Allen JW, Collins BW, Mori C, Nakamura N, Poorman-Allen P, Goulding EH, Eddy EM (1996a) Targeted gene disruption of *Hsp70-2* results in failed meiosis, germ cell apoptosis, and male infertility. Proc Natl Acad Sci USA 93:3264–3268

Dix DJ, Rosario-Herrle M, Gotoh H, Mori C, Goulding EH, Barrett CV, Eddy EM (1996b) Developmentally regulated expression of *Hsp70-2* and a *Hsp70-2/lacZ* transgene during spermatogenesis. Dev Biol 174:310–321

Dobson MJ, Pearlman RE, Karaiskakis A, Spyropoulos B, Moens PB (1994) Synaptonemal complex proteins, epitope mapping and chromosome disjunction. J Cell Sci 107:2749–2760

Dresser ME, Moses MJ (1980) Synaptonemal complex karyotyping in spermatocytes of the Chinese hamster *(Cricetulus griseus)*. IV. Light and electron microscopy of synapsis and nucleolar development by silver staining. Chromosoma 76:1–22

Dresser ME, Ewing DJ, Conrad MN, Dominguez AM, Barstead R, Jiang H, Kodadek T (1997) *Dmc1* functions in a *Saccharomyces cerevisiae* meiotic pathway that is largely independent of the *RAD51* pathway. Genetics 147:533–544

Edelmann W, Cohen PE, Kane M, Lau K, Morrow B, Bennett S, Umar A, Kunkel T, Cattoretti G, Chaganti R, Pollard JW, Kolodner RD, Kucherlapati R (1996) Meiotic pachytene arrest in *Mlh1*-deficient mice. Cell 85:1125–1134

Elgin SCR (1988) The formation and function of DNAase I hypersensitive sites in the process of gene activation. J Biol Chem 263:19259–19262

Elledge SJ (1996) Cell cycle checkpoints: preventing an identity crisis. Science 274:1664–1672

Elledge SJ (1998) Mitotic arrest: Mad2 prevents Sleepy from waking up the APC. Science 279:999–1000

Everett CA, Searle JB, Wallace BMN (1996) A study of meiotic pairing, nondisjunction and germ cell death in laboratory mice carrying Robertsonian translocations. Genet Res 67:239–247

Fan QQ, Petes TD (1996) Relationship between nuclease-hypersensitive sites and meiotic recombination hotspot activity at the HIS4 locus of *Saccharomyces cerevisiae*. Mol Cell Biol 16:2037–2043

Fawcett DW (1956) The fine structure of chromosomes in the meiotic prophase of vertebrate spermatocytes. J Biophys Biochem Cytol 2:403–406

Finley D, Chau V (1991) Ubiquitination. Annu Rev Cell Biol 7:25–69

Fishel R, Lescoe MK, Rao MRS, Copeland NG, Jenkins NA, Garber J, Kane M, Kolodner R (1993) The human mutator gene homolog MSH2 and its association with hereditary nonpolyposis colon cancer. Cell 75:1027–1038

Flaggs G, Plug AW, Dunks KM, Mundt KE, Ford JC, Quiggle MRE, Taylor EM, Westphal CH, Ashley T, Hoekstra MF, Carr AM (1997) ATM-dependent interactions of a mammalian CHK1 homolog with meiotic chromosomes. Curr Biol 7:977–986

Forejt J (1984) X-inactivation and its role in male sterility. In: Bennett MD, Gropp A, Wolf U (eds) Chromosomes Today, vol 8. George Allen and Unwin, London, pp 117–127

Forejt JS, Gregorova S, Goetz P (1981) XY pair associates with the synaptonemal complex of autosomal male-sterile translocations in pachytene spermatocytes of the mouse *(Mus musculus).* Chromosoma 82:41–53

Furnari B, Rhind N, Russell P (1997) Cdc25 mitotic inducer targeted by Chkl DNA damage checkpoint kinase. Science 277:1495–1497

Gabriel-Robez O, Rumpler Y (1994) The meiotic pairing behavior in human spermatocyte carriers of chromosome anomalies and their repercussions on reproductive fitness. I: Inversions and insertions. A European Study. Ann Genet 37:3–10

Gabriel-Robez O, Rumpler Y (1996) The meiotic pairing behaviour in human spermatocyte carriers of chromosome anomalies and their repercussion on reproductive fitness. II. Robertsonian and reciprocal translocations. A European collaborative study. Ann Genet 39:17–25

Golubovskaya IN (1979) Genetical control of meiosis. Int Rev Cytol 52:247–290

Gowen LC, Johnson BL, Latour AM, Sulik KK, Koller BH (1996) *Brca1* deficiency results in early embryonic lethality characterized by neuroepithelial abnormalities. Nat Genet 12:191–194

Greenwell PW, Kronmal SL, Porter SE, Gassenhuber J, Obermaier B, Petes TD (1995) Tel1, a gene involved in controlling telomere length in S. *cerevisiae,* is homologous to the human ataxia telangiectasia gene. Cell 82:823–829

Haaf T, Golub EI, Reddy G, Radding CM, Ward DC (1995) Nuclear foci of mammalian Rad51 recombination protein in somatic cells after DNA damage and its localization in synaptonemal complexes. Proc Natl Acad Sci USA 92:2298–2302

Hakem R, de la Pomba JL, Surard C, Mo R, Woo M, Hakem A, Wakeman A, Potter J, Reitmair A, Billia F, Firpo E, Hui C, Roberts J, Rossant J, Mak TW (1996) The tumor suppressor gene *Brca1 is* required for embryonic cellular proliferation in the mouse. Cell 85:1009–1023

Handel MA (1988) Genetic control of spermatogenesis in mice. In: Hennig W (ed) Results and problems in cell differentiation, spermatogenesis: genetic aspects., vol 15. Springer, Berlin Heidelberg New York, pp 1–62

Handel MA, Eppig JJ (1998) Sexual dimorphism in the regulation of mammalian meiosis. In: Handel MA (ed) Meiosis and gametogenesis. Academic Press, San Diego, pp 333–358

Hari KL, Santerre A, Sekelsky JJ, McKim KS, Boyd JB, Hawley RS (1995) The mei-41 gene of *D. melanogaster* is a structural and functional homologue of the human ataxia telangiectasia gene. Cell 82:815–821

Hartley KO, Gell D, Smith GCM, Zhang H, Divecha N, Connelly MA, Admon A, Lees-Miller SP, Anderson CW, Jackson SP (1995) DNA-dependent protein kinase catalytic subunit: a relative of phosphatidylinositol 3-kinase and the ataxia telangiectasia gene product. Cell 82:849–856

Hartwell LH, Kastan MB (1994) Cell cycle control and cancer. Science 266:1821–1828

Hartwell LH, Weinert TA (1989) Checkpoints: controls that ensure the order of cell cycle events. Science 249:629–634

Hassold TJ (1998) Nondisjunction in the human male. In: Handel MA (ed) Meiosis and gametogenesis. Academic Press, San Diego, pp 383–406

Hawn MT, Umar A, Carethers JM, Marra G, Kunkel TA, Boland CR, Koi M (1995) Evidence for a connection between the mismatch repair system and the G2 cell cycle checkpoint. Cancer Res 55:3721–3725

Hawthorne D, Phillippsen P (1994) Genetic and molecular analysis of hybrids in the genus *Saccharomyces* involving *S. cerevisiae, S. uvarum* and a new species, *S. douglasii.* Yeast 10:1885–1896

He Z, Henrickson LA, Wold MS, Ingles CJ (1995) RPA involvement in the damage-recognition and incision steps of nucleotide excision repair. Nature 374:566–569

Herbert CJ, Dujardin D, Labouesse M, Slonimski PP (1988) Divergence of the mitochondrial leucyl tRNA synthetase gene in two closely related yeasts: *Saccharomyces cerevisiae* and *S. douglasii:* a paradigm of incipient evolution. Mol Gen Genet 213:297–309

Heyer W-D, Rao MRS, Erdile LF, Kelley TJ, Kolodner RD (1990) An essential *Saccharomyces cerevisiae* single-stranded DNA binding protein is homologous to the large subunit of human RP-A. EMBO J 9:2321–2329

Hoekstra MF (1997) Responses to DNA damage and regulation of cell cycle checkpoints by the ATM protein kinase family. Curr Opin Gen Dev 7:170–175

Hoekstra MF, Demaggio AJ, Dhillon N (1991) Genetically identified protein kinases in yeast. II: DNA metabolism and meiosis. TIG 7:293–298

Hollingsworth NM, Byers B (1989) *Hop1:* a yeast meiotic pairing gene. Genetics 121:445–462

Hollingsworth NM, Goetsch L, Byers B (1990) The *Hop1* gene encodes a meiosis-specific component of yeast chromosomes. Cell 61:73–84

Hollingsworth NM, Ponte L, Halsey C (1995) MSH5, a novel MutS homolog, facilitates meiotic reciprocal recombination between homologs in *Saccharomyces cerevisiae* but not mismatch repair. Genes Dev 9:1728–1739

Hotta Y, Stern H (1971) Analysis of DNA synthesis during meiotic prophase in *Lilium.* J Mol Biol 55:337–355

Hotta Y, Ito M, Stern H (1966) Synthesis of DNA during meiosis. Proc Natl Acad Sci USA 56: 1184–1191

Hotta Y, Tabata S, Stubbs L, Stern H (1985) Meiotic-specific transcripts of a DNA component replicated during chromosome pairing: homology across the phylogenetic spectrum. Cell 40:785–793

Hugerat Y, Simchen G (1993) Mixed segregation and recombination of chromosomes and YACs during single-division meiosis in *spo13* mutant strains of *Saccharomyces cerevisiae.* Genetics 135:297–308

Hunt PA, LeMaire-Adkins R (1998) Genetic control of mammalian female meiosis. In: Handel MA (ed) Meiosis and gametogenesis. Academic Press, San Diego, pp 359–381

Hunter N, Borts RH (1997) Mlh1 is unique among mismatch repair proteins in its ability to promote crossing-over during meiosis. Genes Dev 11:1573–1582

Hunter N, Chambers SR, Louis EJ, Borts RH (1996) The mismatch repair system contributes to meiotic sterility in an interspecific yeast hybrid. EMBO J 15:1726–1733

Ito M, Hotta Y, Stern H (1967) Studies of meiosis in vitro. II. Effect of inhibiting DNA synthesis during meiotic prophase on chromosome structure and behavior. Dev Biol 16:54–77

Jaafar H, Gabriel-Robez O, Rumpler Y (1989) Pattern of ribonucleic acid synthesis in vitro in primary spermatocytes from mouse testis carrying an X-autosome translocation. Chromosoma 98:330–334

Jenkins TM, Saxena JK, Kumar A, Wilson SH, Ackerman EJ (1992) DNA polymerase β and DNA synthesis in *Xenopus* oocytes and in a nuclear extract. Science 258:475–478

Jentsch S (1992) The ubiquitin-conjugation system. Annu Rev Genet 26:179–207

Johannisson R, Winking H (1994) Synaptonemal complexes of chains and rings in mice heterozygous for multiple Robertsonian translocations. Chromosome Res 2:137–145

Kastan MB, Zhan Q, El-Deiry WS, Carrier F, Jacks T, Walsh WV, Plunkett BS, Vogelstein B, Fornace JA Jr (1992) A mammalian cell cycle checkpoint pathway utilizing p53 and GADD45 is defective in ataxia-telangiectasia. Cell 71:587–597

Keegan KS, Holtzman DA, Plug AW, Christenson ER, Brainerd EE, Flaggs G, Bentley NJ, Taylor EM, Meyn MS, Moss SB, Carr AM, Ashley T, Hoekstra MF (1996) The ATR and ATM protein kinases associate with different sites along meiotically pairing chromosomes. Genes Dev 10: 2423–2437

Khanna KK, Lavin ML (1993) Ionizing radiation and UV induction of p53 protein by different pathways in ataxia telangietasia cells. Oncogene 8:3307–3312

King M (1993) Species evolution: the role of chromosome change. Cambridge University Press, Cambridge

Klapholz S, Esposito RE (1980) Recombination and chromosome segregation during the single division meiosis in *spo12-1* and *spo13-1* diploids. Genetics 96:589–611

Kleckner N (1996) Meiosis: how could it work? Proc Natl Acad Sci USA 93:8167–8174

Kleckner N, Weiner BM (1993) Potential advantages of unstable interactions for pairing of chromosomes in mitotic, somatic and premeiotic cells. Cold Spring Harbor Symp Quant Biol 58:553–565

Kleckner N, Padmore R, Bishop DK (1991) Meiotic chromosome metabolism: one view. Cold Spring Harbor Symp Quant 56:729–743

Kofman-Alfaro S, Chandley AC (1970) Meiosis in the male mouse. An autoradiographic investigation. Chromosoma 31:404–420

Kolodner R (1996) Biochemistry and genetics of eukaryotic mismatch repair. Genes Dev 10: 1433–1442

Kornberg RD, Lorch Y (1995) Interplay between chromatin structure and transcription. Curr Opin Cell Biol 7:371–375

Kovalenko OV, Plug AW, Haaf T, Gonda DK, Ashley T, Ward DC, Radding CM, Golub EI (1996) Mammalian ubiquitin-conjugating enzyme, UBC9, interacts with RAD51 recombination protein and localizes in synaptonemal complexes. Proc Natl Acad Sci USA 93:2958–2963

Kowalczykowski SC, Dixon DA, Eggleston AK, Lauder SD, Rehrauer WM (1994) Biochemistry of homologous recombination in *Escherichia coli.* Microbiol Rev 58:401–465

Kramer B, Kramer W, Williamson MS, Fogel S (1989) Heteroduplex DNA correction in *Saccharomyces cerevisiae* is mismatch specific and requires functional *PMS* genes. Mol Cell Biol 9: 4432–4440

Krude T, Elgin SCR (1996) Chromatin: pushing nucleosomes around. Curr Biol 6:511–515

Lamb N, Freeman S, Savage-Austin A, Pettay D, Taft D, Hersey J, Gu Y, Shen J, Saker D, May K, Avamopoulos D, Petersen M, Hallberg A, Mikkelsen M, Hassold T, Sherman S (1996) Susceptible chiasmata configurations of chromosome 21 predispose to non-disjunction in both maternal meiosis I and meiosis II. Nat Genet 14:400–405

Lammers JHM, Offenberg HH, van Aalderen M, Vink ACG, Dietrich AJJ, Heyting C (1994) The gene encoding a major component of the lateral element of the synaptonemal complex of the rat is related to X-linked lymphocyte-regulated genes. Mol Cell Biol 14:1137–1146

Lammers JHM, van Aalderen M, Peters AHFM, van Pelt AAM, Gaemmers IC, de Rooij DG, de Boer P, Offenberg HH, Dietrich AJJ, Heyting C (1995) A change in the phosphorylation pattern of the 30 000–33 000 Mr synaptonemal complex proteins of the rat between early and mid-pachytene. Chromosoma 104:154–163

Leach FS, Nicolaides NC, Papadopoulos N, Liu B, Jen J, Parsons R, Peltomaki P, Sistonen P, Aaltonen LA, Nystrom-Lahti M, Guan X-Y, Zhang J, Meltzer PS, Yu J-W, Kao F-T, Chen DJ, Cerosaletti KM, Fournier REK, Todd S, Lewis T, Leach RJ, Naylor SL, Weissenbach J, Mecklin J-P, Jarvinen H, Petersen GM, Hamilton SR, Green J, Jass J, Watson P, Lynch HT, Trent JM, de la Chapelle A, Kinzler KW, Vogelstein B (1993) Mutations of a mutS homolog in hereditary nonpolyposis colorectal cancer. Cell 75:1215–1225

Lees-Miller SP, Chen Y-R, Anderson CW (1990) Human cells contain a DNA-activated protein kinase that phosphorylates simian virus 40 T antigen, mouse p53, and the human Ku autoantigen. Mol Cell Biol 10:6472–6481

Lehmann AR, Walicka M, Griffiths JF, Murray JM, Watt FZ, Mc Cready S, Carr AM (1995) The *rad18* gene of *Schizosaccharomyces pombe* defines a new subgroup of the SMC superfamily involved in DNA repair. Mol Cell Biol 15:7067–7080

Lifschytz E, Lindsley DL (1972) The role of X-chromosome activation during spermatogenesis. Proc Natl Acad Sci USA 69:182–186

Lim D-S, Hasty P (1996) A mutation in mouse rad51 results in an early embryonic lethal that is suppressed by a p53 mutation. Mol Cell Biol 16:7133–7143

Liu C-Y, Flesken-Nikitin A, Li S, Zeng Y, Lee W-H (1996a) Inactivation of the mouse *Brca1* gene leads to failure in the morphogenesis of the egg cylinder in early postimplantation development. Genes Dev 10:1835–1843

Liu J-G, Yuan L, Brundell E, Bjorkroth B, Daneholt B, Hoog C (1996b) Localization of the N-terminus of SCP1 to the central element of the synaptonemal complex and evidence for direct interactions between the N-termini of SCP1 molecules organized head-to-head. Exp Cell Res 226:11–19

Liu N, Bryant PE (1993) Response of ataxia-telangiectasia cells to restriction endonuclease induced DNA double-strand breaks: I. Cytogenetic characterization. Mutagenesis 8:503–510

Liu VF, Weaver DT (1993) The ionizing radiation-induced replication protein A phosphorylation response differs between ataxia telangiectasia and normal human cells. Mol Cell Biol 13:7222–7231

Lydall D, Nikolsky Y, Bishop DK, Weinert T (1996) A meiotic recombination checkpoint controlled by mitotic checkpoint genes. Nature 383:840–843

Lyon MF (1968) Chromosomal and subchromosomal inactivation. Annu Rev Genet 2:31–52

Lyon MF, Meredith R (1966) Autosomal translocations causing male sterility and viable aneuploidy in the mouse. Cytogenetics 5:335–354

Maguire MP (1978) A possible role for the synaptonemal complex in chiasma maintenance. Exp Cell Res 112:297–308

Maguire MP (1993) Sister chromatid association in meiosis. Madica 38:93–106

Mahadevaiah S, Mittwoch U (1986) Synaptonemal complex analysis in spermatocytes and oocytes of tertiary trisomic Ts(5–12)31H mice with male sterility. Cytogenet Cell Genet 41: 169–176

Mahadevaiah SK, Setterfield LA, Mittwoch U (1988) Univalent sex chromosomes in spermatocytes of Sxr-carrying mice. Chromosoma 97:145–153

Masai H, Arai K (1997) Frpo: a novel single-strand DNA promoter for transcription and for primer RNA synthesis of DNA replication. Cell 89:897–907

McGhee JD, Felsenfeld G (1980) Nucleosome structure. Annu Rev Biochem 49:1115–1156

McKim KS, Green-Marroquin BL, Sekelsky JJ, Chin G, Steinberg C, Khodosh R, Hawley RS (1998) Meiotic synapsis in the absence of recombination. Science 279:876–878

Meuwissen RLJ, Offenberg HH, Dietrich AJJ, Riesewijk A, van Iersel M, Heyting C (1992) A coiled-coil related protein specific for synapsed regions of meiotic prophase chromosomes. EMBO J 11:5091–5100

Meuwissen RLJ, Meerts I, Hoovers JMN, Leschot NJ, Heyting C (1997) Human synaptonemal complex protein 1 (SCP1): isolation and characterization of the cDNA and chromosomal localization of the gene. Genomics 39:377–384

Miklos GLG (1974) Sex-chromosome pairing and male fertility. Cytogenet Cell Genet 13: 558–577

Modrich P (1991) Mechanisms and biological effects of mismatch repair. Annu Rev Genet 25:229–253

Modrich P, Lahue R (1996) Mismatch repair in replication, fidelity, genetic recombination, and cancer biology. Annu Rev Biochem 65:101–133

Moens PB, Chen DJ, Shen Z, Kolas N, Tarsounas M, Heng HHQ, Spyropoulos B (1997) RAD51 immunocytology in rat and mouse spermatocytes and oocytes. Chromosoma 106:207–215

Monesi V (1965) Differential rate of ribonucleic acid synthesis in the autosomes and sex chromosomes during male meiosis in the mouse. Chromosoma 17:11–21

Monesi V, Geremia R, D'Agostino A, Boitani C (1978) Biochemistry of male germ cell differentiation in mammals: RNA synthesis in meiotic and post meiotic cells. Curr Top Dev. Biol 12: pp 11–36

Moore DP, Orr-Weaver TC (1998) Chromosome segregation during meiosis: building an unambivalent bivalent. In: Handel MA (ed) Meiosis and gametogenesis. Academic Press, San Diego, pp 264–299

Morrow DM, Tagle DA, Shiloh Y, Collins FS, Hieter P (1995) TEL1, an *S. cerevisiae* homolog of the human gene mutated in ataxia telangiectasia, is functionally related to the yeast checkpoint gene MEC1. Cell 82:831–840

Moses MJ (1956) Chromosomal structures in crayfish spermatocytes. J Biophys Biochem Cytol 2:215–218

Moses MJ (1977) Microspreading and the synaptonemal complex in cytogenetic studies. In: de la Chapelle A, Sorsa M (eds) Chromosomes today, vol 6. Elsevier/North Holland, Amsterdam, pp 72–82

Moses MJ (1980) New cytogenetic studies on mammalian meiosis. In: Serio M, Martini L (eds) Animal models in human reproduction. Raven Press, New York, pp 169–190

Murray AW, Hunt T (1993) The cell cycle: an introduction. Oxford University Press, New York
Naumov GI, Naumova ES, Lantto RA, Louis EJ, Korhola M (1992) Genetic homology between *Saccharomyces cerevisiae* with its sibling species *S. paradoxus* and *S. bayanus:* electrophoretic karyotypes. Yeast 8:599–612
Navas TA, Zhou Z, Elledge SJ (1995) DNA polymerase ε links the DNA replication machinery to the S phase checkpoint. Cell 80:29–39
Nicklas RB, Kock CA (1969) Chromosome micromanipulation III spindle fiber tension and the reorientation of mal-orientated chromosomes. J Cell Biol 43:40–50
Nicklas RB, Kubai DF (1985) Microtubules, chromosome movement, and reorientation after chromosomes are detached from the spindle by micromanipulation. Chromosoma 92: 313–324
Nicklas RB, Ward SC, Gorbsky GJ (1995) Kinetochore chemistry is sensitive to tension and may link mitotic forces to a cell cycle checkpoint. J Cell Biol 130:929–939
Nicolaides NC, Papadopoulos N, Liu B, Wei Y, Carter KC, Ruben SM, Rosen CA, Haseltine WA, Fleischmann RD, Frasser CM, Adams MD, Venter JC, Dunlop MG, Hamilton SR, Petersen GM, de la Chapelle A, Vogelstein B, Kinzler K (1994) Mutations of two PMS homologues in hereditary nonpolyposis colon cancer. Nature 371:75–80
Novak R, Woszczynski M, Siedecki JA (1990) Changes in the DNA polymerase β gene expression during development of lung, brain and testis suggest an involvement of the enzyme in DNA recombination. Exp Cell Res 191:51–56
Ohta K, Shibata T, Nicolas N (1994) Changes in chromatin structure at recombination initiation sites during yeast meiosis. EMBO J 13:5754–5763
Pathak S, Kieffer N (1979) Sterility in hybrid cattle. I Distribution of constitutive heterochromatin and nucleolus organizer regions in somatic and meiotic chromosomes. Cytogenet Cell Genet 24:42–52
Paulovich AG, Hartwell L (1995) A checkpoint regulates the rate of progression through S phase in *S. cerevisiae* in response to DNA damage. Cell 82:841–847
Peters AHFM, Plug AW, de Boer P (1997) Meiosis in carriers of heteromorphic bivalents: sex differences and implications for male fertility. Chromosome Res 5:313–324
Petit C, Levillers J, Weissenbach J (1988) Physical mapping of the pseudoautosomal region: comparison with genetic linkage map. EMBO J 7:2369–2379
Plug AW, Xu J, Reedy G, Golub EI, Ashley T (1996) Presynaptic association of RAD51 protein with selected sites in meiotic chromatin. Proc Natl Acad Sci USA 93:5920–5924
Plug AW, Clairmont CA, Sapi E, Ashley T, Sweasy JB (1997a) Evidence for a role for DNA polymerase β in mammalian meiosis. Proc Natl Acad Sci USA 94:1327–1331
Plug AW, Peters AHFM, Xu Y, Keegan KS, Hoekstra MF, Baltimore D, deBoer P, Ashley T (1997b) ATM and RPA in meiotic chromosome synapsis and recombination. Nat Genet 17: 457–461
Plug AW, Peters AHFM, van Breuklen B, Keegan KS, Hoekstra M, de Boer P, Ashley T (1998) Changes in protein composition of meiotic nodules during mammalian meiosis. J Cell Sci 111:413–423
Polani PE, Jagiello GM (1976) Chiasmata, meiotic univalents, and age in relation to aneuploid imbalance in mice. Cytogenet Cell Genet 16:505–529
Poorman PA, Moses MJ, Davisson MT, Roderick TH (1981) Synaptonemal complex analysis of mouse chromosome rearrangements. III. Cytogenetic observations on two paracentric inversions. Chromosoma 83:419–429
Prolla TA, Christie D-M, Liskay RM (1994a) Dual requirement in yeast DNA mismatch repair for MLH1 and PMS1, two homologs of the bacterial mutL gene. Mol Cell Biol 14: 407–415
Prolla TA, Pang Q, Alani E, Kolodner RD, Liskay RM (1994b) MLH1, PMS1, and MSH2 interactions during the initiation of DNA mismatch repair in yeast. Science 265:1091–1093
Rappold GA, Lehrach H (1988) A long-range restriction map of the pseudoautosomal region by partial digest PFGE analysis from the telomere. Nucleic Acids Res 16:5361–5377

Rayssiquier C, Thaler DS, Radman M (1989) The barrier to recombination between *Escherichia coli* and *Salmonella typhimurium* is disrupted in mismatch repair mutants. Nature 342: 396–401

Redi CA, Capanna E (1988) Robertsonian heterozygotes in the house mouse and the fate of their germ cell. In: Daniel A (ed) The cytogenetics of mammalian autosomal rearrangements. Alan R Liss, New York, pp 315–359

Reenan RA, Kolodner RD (1992) Isolation and characterization of two *Saccharomyces cerevisiae* genes encoding homologs of the bacterial HexX and MutS mismatch repair proteins. Genetics 132:963–973

Reitmair AH, Schmits R, Ewel A, Bapat B, Redston M, Mitri A, Waterhouse P, Mittrucker H-W, Wakeham A, Liu B, Thomason A, Griesser H, Gallinger S, Ballhausen WG, Fishel R, Mak TW (1995) *MSH2* deficient mice are viable and susceptible to lymphoid tumours. Nat Genet 11: 64–70

Rockmill B, Roeder GS (1990) Meiosis in asynaptic yeast. Genetics 126:563–574

Rockmill B, Roeder GS (1994) The yeast med1 mutant undergoes both meiotic homolog nondisjunction and precocious separation of sister chromatids. Genetics 136:65–74

Rockmill B, Sym M, Scherthan H, Roeder GS (1995) Roles for two RecA homologs in promoting meiotic chromosome synapsis. Genes Dev 9:2684–2695

Roeder GS (1997) Meiotic chromosomes: it takes two to tango. Genes Dev 11:2600–2621

Ross-Macdonald P, Roeder GS (1994) Mutation of meiosis-specific MutS homolog decreases crossing over but not mismatch correction. Cell 79:1069–1080

Roth TF, Ito M (1967) DNA-dependent formation of the synaptonemal complex at meiotic prophase. J Cell Biol 35:247–255

Rouyer F, Simmler M-C, Johnson C, Vergnaud G, Cooke HJ, Weissenbach J (1986) A gradient of sex linkage in the pseudoautosomal region of the human sex chromosomes. Nature 319: 291–295

Russell LB (1983) X-chromosome translocations in the mouse: their characterization and use as tools to investigate gene inactivation and gene action. In: Sandberg AA (ed) Cytogenetics of the mammalian X chromosome. Part A: basic mechanisms of X chromosome behavior. Alan R Liss, New York, pp 205–250

Sanchez Y, Desany BA, Jones WJ, Liu Q, Wang B, Elledge SJ (1996) Regulation of *RAD53* by the *ATM*-like kinases *Mec1* and *TEL1* in yeast cell cycle checkpoint pathways. Science 271: 357–360

Sanchez Y, Wong C, Thoma RS, Richman R, Wu Z, Piwnica-Worms H, Elledge SJ (1997) Conservation of the Chk1 checkpoint pathway in mammals: linkage of DNA damage to Cdk regulation through Cdc25. Science 277:1497–1501

Savitsky K, Bar-Shira A, Gilad S, Rotman G, Ziv Y, Vanagaite L, Tagle DA, Smith S, Uziel T, Sfez S, Ashkenazi M, Pecker I, Frydman M, Harnik R, Patanjali SR, Simmons A, Clines GA, Sartiel A, Gatti RA, Chessa L, Sanal O, Lavin MF, Jaspers NGJ, Taylor AMR, Arlett CF, Miki T, Weissman SM, Lovett M, Collins FS, Shiloh Y (1995) A single ataxia telangiectasia gene with a product similar to PI-3 kinase. Science 268:1749–1753

Scherthan H, Weich S, Schweger H, Heyting C, Harle M, Cremer T (1996) Centromere and telomere movements during early meiotic prophase of mouse and man are associated with the onset of chromosome pairing. J Cell Biol 134:1109–1125

Scully R, Chen J, Plug A, Xiao Y, Weaver D, Feunteun J, Ashley T, Livingston DM (1997) Association of BRCA1 with RAD51 in mitotic and meiotic cells. Cell 88:265–275

Seaton BL, Yucel J, Sunnerhagen P, Subramani S (1992) Isolation and characterization of *Schizosaccharomyces pombe* rad3 gene, involved in the DNA damage and DNA synthesis checkpoints. Gene 119:83–89

Sedgwick PP, Boder E (1991) Ataxia-telangiectasia. In: deJong JMBV (ed) Hereditary neuropathies and spinocerebellar atrophies. Elsevier, New York, pp 347–423

Seufert W, Futcher B, Jentsch S (1995) Role of a ubiquitin-conjugating enzyme in degradation of S- and M-phase cyclins. Nature 373:71–81

Shiloh Y (1995) Ataxia-telangiectasia: closer to unraveling the mystery. Eur J Hum Genet 3: 116–138
Shinohara A, Ogawa H, Ogawa T (1992) Rad51 protein involved in repair and recombination in *S. cerevisiae* is a RecA-like protein. Cell 69:457–470
Smith AV, Roeder GS (1997) The yeast Red1 protein localizes to the cores of meiotic chromosomes. J Cell Biol 136:957–967
Sobol RW, Horton JK, Kuhn R, Gu H, Singhal RK, Prasad R, Rajewsky K, Wilson SH (1996) Requirement of mammalian DNA polymerase β in base excision repair. Nature 379:183–186
Solari AJ (1974) The behavior of the XY pair in mammals. Int Rev Cytol 38:273–317
Solari AJ, Ashley T (1977) Ultrastructure and behavior of the X and Y chromosomes of the achiasmatic, telosynaptic XY pair of the sand rat *(Psammomys obesus)*. Chromosoma 69:319–336
Speed RM (1986) Abnormal RNA synthesis in sex vesicles of tertiary trisomic male mice. Chromosoma 93:267–270
Speed RM (1988) The possible role of meiotic pairing anomalies in the atresia of human fetal oocytes. Hum Genet 78:260–266
Speed RM (1989) Heterologous pairing and fertility in humans. In: Gillies CB (ed) Fertility and chromosome pairing: recent studies in plants and animals. CRC Press, Boca Raton, Florida, pp 1–36
Stack S, Sherman J, Anderson L, Herickhoff L (1993) Meiotic nodules in vascular plants. In: Summer A, Chandley A (eds) Chromosomes today, vol 11. Chapman & Hall, London, pp 301–311
Stern H, Hotta Y (1985) Molecular biology of meiosis: synapsis-associated phenomena. In: Dellarco V, Voytek PE, Hollaender A (eds) Aneuploidy: etiology and mechanisms. Plenum, New York, pp 305–316
Sun Z, Fay F, Marini F, Foiani M, Stern DF (1996) Spk1/Rad53 is regulated by Mec1-dependent protein phosphorylation in DNA replication and damage checkpoint pathways. Genes Dev 10:395–406
Svejstrup JQ, Wang Z, Feaver WJ, Wu X, Bushnell DA, Donahue TF, Friedberg EC, Kornberg RD (1995) Different forms of TFIIH for transcription and DNA repair: holo-TFIIH and a nucleotide excision repairosome. Cell 80:21–28
Sym M, Engebrecht J, Roeder GS (1993) Zip1 is a synaptonemal complex protein required for meiotic chromosome synapsis. Cell 72:365–378
Sym M, Roeder GS (1994) Crossover interference is abolished in the absence of a synaptonemal complex protein. Cell 79:283–292
Tarsounas M, Pearlman RE, Gasser PJ, Park MS, Moens PB (1997) Protein-protein interactions in the synaptonemal complex. Mol Biol Cell 8:1405–1414
Teruhisa T, Fujii Y, Sakumi K, Tominaga Y, Nakao K, Sekiguchi M, Matsushiro A, Yoshimura Y, Morita T (1996) Targeted disruption of the Rad51 gene leads to lethality in embryonic mice. Proc Natl Acad Sci USA 93:6236–6240
Tiepolo L, Zuffardi O, Fraccaro M, Giarola A (1981) Chromosome abnormalities and male infertility. In: Frajase G, Hafez ESE, Conti C, Fabbrini A (eds) Oligozoospermia: recent progress in andrology. Raven Press, New York, pp 233–246
Walworth N, Bernards R (1996) rad-dependent response of the chk1-encoded protein kinase at the DNA damage checkpoint. Science 271:353–356
Walworth N, Davey S, Beach D (1993) Fission yeast chk1 protein kinase links the rad checkpoint pathway to cdc2. Nature 363:368–371
Weinert T (1997) A DNA damage checkpoint meets the cell cycle engine. Science 277:1450–1451
Weith A, Traut W (1980) Synaptonemal complexes with associated chromatin in a moth, *Ephestia kuehniella Z.* Chromosoma 78:275–291
West SC (1992) Enzymes and molecular mechanisms of genetic recombination. Annu Rev Biochem 61:603–640
White MA, Dominska M, Petes TD (1993) Transcription factors are required for the meiotic recombination hotspot at the HIS4 locus in *Saccharomyces cerevisiae.* Proc Natl Acad Sci USA 90:6621–6625

Williamson M, Game J, Fogel S (1985) Meiotic gene conversion mutants in *Saccharomyces cerevisiae.* I. Isolation and characterization of *pms1-1 and pms1-2.* Genetics 110:609–646

Wold MS, Kelly T (1988) Purification and characterization of replication protein A, a cellular protein required for in vitro replication of simian virus 40 DNA. Proc Natl Acad Sci USA 85:2523–2527

Worth LJ, Clark S, Radman M, Modrich P (1994) Mismatch repair proteins MutS and MutL inhibit RecA-catalyzed strand transfer between diverged DNAs. Proc Natl Acad Sci USA 91:3238–3241

Wu S, Wolgemuth DJ (1995) The distinct and developmentally regulated patterns of expression of members of the mouse *Cdc25* gene family suggest differential functions during spermatogenesis. Dev Biol 179:195–206

Wu T-C, Lichten M (1994) Meiosis-induced double-strand break sites determined by yeast chromatin structure. Science 263:515–518

Wu T-C, Lichten M (1995) Factors that affect the location and frequency of meiosis-induced double-strand breaks in *Saccharomyces cerevisiae.* Genetics 140:55–66

Xu Y, Baltimore D (1996) Dual roles of ATM in the cellular response to radiation and in cell growth control. Genes Dev 10:2401–2410

Xu Y, Ashley T, Brainerd EE, Bronson RT, Meyn MS, Baltimore D (1996) Targeted disruption of ATM leads to growth retardation, chromosomal fragmentation during meiosis, immune defects, and thymic lymphoma. Genes Dev 10:2411–2422

Zakeri ZF, Wolgemuth DJ, Hunt CR (1988) Identification and sequence analysis of a new member of the mouse HSP70 gene family and characterization of its unique cellular and developmental pattern of expression in the male germ line. Mol Cell Biol 8:2925–2932

Zhu D, Dix DJ, Eddy EM (1997) HSP70-2 is required for CDC2 kinase activity in meiosis I of mouse spermatocytes. Development 124:3007–3014

Mutations of the Cystic Fibrosis Gene and Congenital Absence of the Vas Deferens

Pasquale Patrizio[1] and Debra G. B. Leonard[2]

1 Introduction

It is estimated that about 30 to 40% of couples seeking fertility treatments are diagnosed with male factor infertility. These males have a range of gonadal dysfunctions which include azoospermia (i.e., no sperm in the ejaculate); oligozoospermia (i.e. sperm count less than 20 million/ml), asthenozoospermia (i.e.sperm motility less than 50%) and teratozoospermia (i.e. sperm with normal morphology less than 30%). The group of patients with azoospermia represent about 25% of the total and, of these, about 30% have an obstructive process (obstructive azoospermia) while the remaining have a primary testicular failure (non-obstructive azoospermia). In the obstructive azoospermia group, 25% of males have congenital bilateral absence of the vas deferens (CBAVD), while the incidence among all infertile males is about 2%. In the USA, it is estimated that approximately 16000 males are affected by CBAVD. Anatomically, CBAVD is a disorder characterized by regression bilaterally of variable portions of the epididymis, vas deferens, and, in about 80% of cases, absence of the seminal vesicles. In about 10 to 20% of the patients, a renal anomaly is also present. These anatomical hallmarks are so strikingly similar to those observed in men with cystic fibrosis (CF) that, as early as 1971 (Holsclaw et al. 1971), these two apparently unrelated disorders were hypothesized to have the same genetic origin. The hypothesis was proven when mutations in the Cystic Fibrosis Transmembrane conductance Regulator (CFTR) gene (Kerem et al. 1989; Riordan et al. 1989;Rommens et al. 1989) were found in patients with CF as well patients with isolated CBAVD (Dumur et al. 1990; Anguiano et al. 1992; Patrizio et al. 1993).

The clinical link between CF and CBAVD has also been strengthened by documenting the presence of subclinical CF symptoms in the majority of men

[1] Director, Male Infertility Program, University of Pennsylvania Health System, Department of Obstetrics & Gynecology, Division of Human Reproduction, Philadelphia, PA 19104-4283, USA

[2] Director, Molecular Pathology Laboratory, University of Pennsylvania Health System, Department of Pathology and Laboratory Medicine, Philadelphia, PA 19104-4283, USA

Results and Problems in Cell Differentiation, Vol. 28
McElreavey (Ed.): The Genetic Basis of Male Infertility

with CBAVD, including mild elevations of sweat chloride concentrations, nasal polyps, and chronic sinusitis. In a recent study, more than 70% of CBAVD patients manifested at least one mild CF symptom (Durieu et al. 1995), giving further credit to the concept that CBAVD is indeed a mild or incomplete form of CF.

The diagnosis of CBAVD is relatively simple to make. It relies on a semen analysis showing a volume of <1.5 ml, acidic pH, absence of fructose, and azoospermia. On physical examination, the palpation of the spermatic cord reveals absence of the vas deferens. Since patients with CBAVD have usually normal or only slightly reduced spermatogenesis (Silber et al. 1990), these patients can benefit from the use of assisted reproductive techniques, including epididymal sperm aspiration, in vitro fertilization, and, recently, intracytoplasmic sperm injection. Indeed, many pregnancies with live births have been reported (Patrizio et al. 1988; Tournaye et al. 1994; Silber et al. 1995).

This chapter focuses on the evidence that isolated CBAVD is a mild form of CF, with the presence of mutations causing a milder phenotype than the mutations seen in CF. The most common CBAVD genotype consists of a severe CFTR mutation on one allele and a mild mutation on the second CFTR allele. Others have either a mild or severe CFTR mutation in combination with a 5T-tract splice variant of intron 8 on the other CFTR allele.

2 CF and CBAVD: A Common Genetic Background

Cystic Fibrosis is a fatal, autosomal recessive disorder with an incidence ranging from 1 in 2000 to 1 in 2500 live births. Classical symptoms of CF are obstructive pulmonary disease, pancreatic exocrine insufficiency, gastrointestinal obstruction, failure to thrive, elevated concentrations of electrolytes in sweat and, in males, infertility due to absence of the vas deferens. However, the spectrum of the presenting symptoms is very wide and the severity of the disease cannot be predicted by knowing the genotype (Estivill, 1996). The gene for CF was identified in 1989 (Rommens et al. 1989) and mapped to the long arm of chromosome 7 (7q31.2) (Riordan et al. 1989; Kerem et al. 1989); it spans approximately 230 kilobases and consists of 27 exons. It is called Cystic Fibrosis Transmembrane conductance Regulator gene (CFTR) because the encoded protein of 1480 amino acids is a multidomain glycoprotein functioning as a low-voltage, cyclic-AMP-regulated chloride channel. This protein is present at high levels in the apical membranes of various epithelial cells within the organs commonly affected by CF. The presence of two transmembrane domains, each containing six hydrophobic segments and two sets of nucleotide binding folds, led to the classification of CFTR as a member of the ATP-binding cassette (ABC) family of transporter proteins. Because CF is a recessive disease, both copies of the CFTR gene must be mutated in order to express the CF phenotype.

Currently, there are over 700 published mutations in the CFTR gene associated with disease. In addition, there are 68 coding region and 55 non-coding region polymorphisms, not associated with disease. (Source: CF Genetic Analysis Consortium; http://www.genet.sickkids.on.ca/).

Following cloning of the CF gene, it was finally possible to test the hypothesis that CF and CBAVD shared a common genetic cause. Dumur et al. (1990), reported 17 patients with CBAVD of which 7 were found positive to ΔF508; six patients, including 2 with no detectable mutations were also found to have elevated sweat chloride concentrations. These preliminary results were soon confirmed by other investigators or studies (Anguiano et al. 1992; Patrizio et al. 1993).

The frequency of mutated CFTR genes in men with infertility due to CBAVD is 20 times higher than the carrier frequency detected in the general population. In general, according to the molecular mechanisms, the mutations identified in the CFTR gene are divided into four classes:

Class I mutations cause defective production of the CFTR protein by introducing translation termination signals in the CFTR mRNA. Examples are insertions or deletions, missense mutations or splice site variants that shift the reading frame and create a stop codon or other truncations of the protein. In these cases very little or no CFTR protein is produced.

Class II mutations cause defective processing of the CFTR protein. The major example is the ΔF508 mutation. Incompletely processed CFTR protein becomes trapped in the endoplasmic reticulum and never reaches the cell surface.

Class III mutations alter the regulation of the CFTR protein function resulting in decreased chloride conductance.

Class IV mutations cause a defect in the chloride ion conductance function of the CFTR protein.

The most severe CF phenotypes are associated with mutations which result in an absence or severe decrease in the amount of CFTR protein expressed on the cell surface. Mutations with a milder phenotype alter the amount or the function of CFTR less severely. CBAVD patients usually have a severe and a mild mutation or two mild mutations. The mutations more frequently associated with CBAVD belong to the class II, defective processing of the CFTR protein, which can cause a CF phenotype, such as pancreatic insufficiency, when in combination with another severe mutation. The most common of this group is the ΔF508, which accounts for about 70% of the total CF mutations in patients. Other mutations frequently found in CBAVD are G542X, R553X, W1282X (all nonsense mutations where X symbolizes a stop codon), N1303K (missense), 1717-1G→A (splicing error), 2184delA (frameshift) and R1162X (nonsense). Other CFTR mutations associated with a mild phenotype in CF patients tend to occur more frequently in CBAVD than in CF. One example is the missense mutation R117H located in exon 4. This mutation is peculiar because when found in patients with CF, it is invariably associated with the

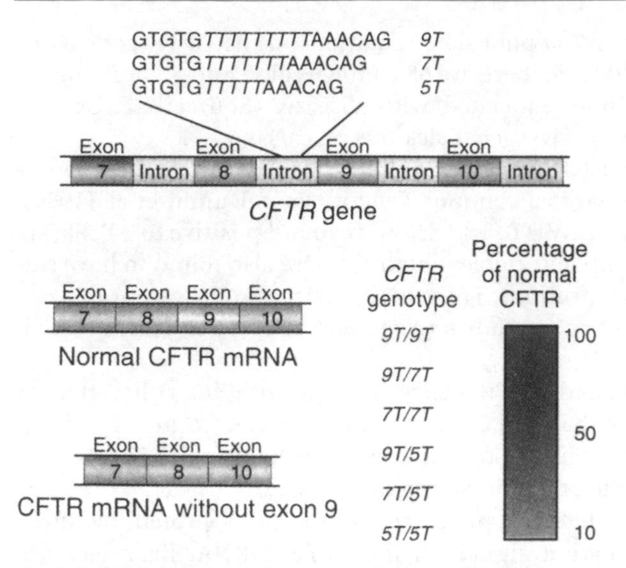

Fig. 1. Splicing variants of exon 9 in the cystic fibrosis transmembrane conductance regulator gene (CFTR). 9T, 7T or 5T refers to the number of thymydine bases. The 5T-tract variant is associated with lower expression of normal, full-length CFTR mRNA than the 7T and 9T-tract variants. The homozygote 5T/5T produces the lowest amount of normal CFTR

pancreatic-sufficient (mild) form of CF, even when there are severe mutations of the second allele. In men with isolated CBAVD, R117H has a variable intragenic background, associated with either of two intragenic variants differing in the splicing efficiency of exon 9. The site of variation is a polypyrimidine tract within the acceptor splice site of intron 8 (IVS8) of the CFTR gene (see Fig. 1). There are three reported variants: IVS8-5T, IVS8-7T and IVS8-9T, consisting of 5, 7 and 9 thymidines, respectively. The length of the T-tract affects the splicing efficiency of exon 9 and thus the percent of normal CFTR mRNA. The 9 T-tract is the most efficient while the 5 T-tract is the least efficient, allowing about 8-10% of CFTR mRNA to be completed. Since CFTR transcribed from mRNA that lacks exon 9 is non-functional as a cyclic-AMP-activated Cl^- channel, the low splicing efficiency of the 5T variant influences the phenotypic effect of the R117H mutation on the same allele.

The intriguing question of why so many men with CBAVD have only one detectable mutation was answered when the polythymidine tract of IVS8 was included in the genetic testing. Although a second uncommon mutation was suspected in patients with CBAVD, only recently the screening for variant 5T,7T and 9T in the T-tract of IVS8 revealed the five- to six-fold increase in the frequency of the 5T variant among CBAVD chromosomes. The presence of 5T variant significantly reduces the levels of functional CFTR. The total amount of available normal CFTR protein may cause the difference in phenotype between development of CF or CBAVD (see Table 1). In particular,the male

Table 1. Quantification of functional CFTR in relation to the severity of the phenotype

Amount of Functional CFTR	Organs Affected	Phenotype
0–10	Lungs, sweat glands, pancreas, vas deferens	CF
10–50	Vas deferens	CBAVD
50–100	None	Normal

genital tract seems to be more sensitive to CFTR deficiency than other organs involved in CF. It is also important to mention that the frequency of a particular mutation is also influenced by the ethnic composition of the population analyzed. In Ashkenazi Jewish patients, for example, the mutation W1282X is particularly frequent, while in German ethnic groups R117H is more frequent.

Other mutations, found exclusively in men with CBAVD or extremely rarely in CF, result in conservative amino acid changes and/or occur in segments of the CFTR less critical for its function as a chloride channel. To this end, a number of rare substitutions (G576A, R668C, R75Q), previously classified as benign CF variations, resurfaced in CBAVD patients (Patrizio and Zielenski, 1996).

2.1 Mutation Analysis Results and the 5T-Tract Variant

The inclusion of the T-tract variants as CBAVD mutations has changed the proportion of CBAVD patients with two identifiable mutations. The 5T-tract variant is associated with many milder CF phenotypes, and is now considered a mild mutation rather than a polymorphism. The 5T-tract variant has also been associated with atypical CF (Kerem et al. 1997), atypical sinopulmonary disease (Friedman et al. 1997), and with hypertrypsinemia and abnormal sweat tests in newborns (Castellani et al. 1997).

Table 2 combines data from mutation analyses of the CFTR gene with and without the 5T-tract variant (modified from Lissens et al. 1996). The majority (60%) of patients with CBAVD showed two mutations when the 5T-tract was included as a mutation, while the remaining had either one mutation (23%) or no mutations (17%). It is not clear that all the patients in the first five studies listed in Table 2 were without evidence of other congenital malformations (renal), nor whether the entire CFTR gene was analyzed for mutations.

Two of the most recent studies (the last two of Table 2), evaluated men with CBAVD using a conformational analysis of the exons, flanking intron sequences and promoter regions of the CFTR gene followed by sequence analysis of abnormal regions (Dork et al. 1997; Bienvenu et al. 1997). Neither study included those CBAVD patients with other congenital abnormalities such as renal agenesis which is not associated with CFTR mutations. In the German study by Dork et al. (1997), only 6 of the 92 males (7%) had no identified mutation

Table 2. Survey of CFTR mutations in CBAVD

No. of patients	5T Variant	CFTR genotype (% of patients studied) 2 Mutations	1 Mutation	No mutation	Reference
102	–	19	53	28	Chillon et al. (1995)
	+	53	25	22	
70	–	13	57	30	Zielenski et al. (1995)
	+	59	20	21	
45	–	33	56	11	Costes et al. (1995)
	+	80	9	5	
38	–	16	39	45	Dumur et al. (1996)
	+	40	26	34	
25	–	16	20	64	Jarvi et al. (1995)
	+	24	56	20	
92	–	58	32	10	Dork et al. (1997)
	+	82	11	7	
64	–	24	59	17	Bienvenu et al. (1997)
	+	47	36	17	
Total	–	28	47	25	
436	+	60	23	17	

and another 10 (11%) had only one mutation in the CFTR gene. This study resulted in the identification of 15 new CFTR gene mutations associated with CBAVD. In the French study by Bienvenu et al. (1997), 11 of 64 men (17%) had no identifiable mutation, 23 (36%) had one CFTR gene mutation, and 30(47%) had two CFTR mutations. These last two studies used very thorough methods to identify mutations in the CFTR gene in CBAVD patients, and found more than 80% to have one or two identifiable CFTR gene mutations, when including the 5T variant.

These data clearly support the concept that the 5T-tract variant should be considered a mild CFTR mutation rather than a polymorphism and convincingly demonstrate that CFTR mutations account for the majority of CBAVD.

3 Pathogenesis of CBAVD

The pathogenesis of the male genital abnormalities characteristic for CBAVD remains unclear. The components of the male genital tract (testes, epididymis and vas deferens) are well developed by 16-18 weeks of gestation (Harris and Coleman 1989). It is known that the CF gene is already expressed in some tissues of the mid trimester fetus, specifically, pancreas, lungs, and male genital tract. Indeed, CFTR mRNA detected by in situ hybridization is present in the genital tract at 18 weeks of gestation and is maintained throughout postnatal life. One report has described a fetus with CF that showed extensive fibrosis and regression of the epididymis (Harris and Coleman 1989). In a study (Patrizio and Salameh 1998), epididymal biopsies obtained from men with failed vasectomy reversal operations or with CBAVD undergoing epididymal sperm aspiration procedures, FISH analysis demonstrated the presence of CFTR mRNA in the columnar epithelium of adult human epididymis. The most intense CFTR signal was in the most proximal portion of the epididymis (caput). This study, however, could not conclusively demonstrate that CFTR mRNA expression in epididymis of men with CBAVD was reduced when compared to men with vasectomy because in both groups of patients, the obstructive process had created sloughing of epididymal epithelial cells, creating a non-homogeneous background signal. The presence of CFTR mRNA in the columnar epithelium of the epididymis has been confirmed by other authors (Tizzano et al. 1994) who also demonstrated high levels of CFTR expression in the head of the epididymis, low in the body and tail of the epididymis and none in the seminal vesicles. Taken together these data favor an early obstructive process in the pathogenesis of CBAVD. The most plausible hypothesis to explain the absence of the vas deferens and of variable portions of the epididymis supposes a progressive obstruction of the deferent ductal system by excessive and thick mucus secretions in the epididymal lumen. This excessive accumulation of abnormal secretions may lead to the damage and subsequent atrophy and regression of these structures. Sodium and water reabsorption from the epididymal lumen far outweighs the limited chloride secretion, and thus, the epididymal fluid become viscous and thickened, thus impairing the proper development of the most distal portions of the epididymis and of the remaining derivatives of the wollffian ducts, vas deferens, and seminal vesicles. A similar mechanism is also responsible for the blockage of exocrine outflow from the pancreas and for the accumulation of thick, dehydrated, mucous in the pulmonary airways.

4
Spermatogenesis and Epididymal Length

Men with CBAVD have either normal or slightly reduced spermatogenesis as demonstrated by quantitative analysis of testicular biopsies (Silber et al. 1990). In contrast to spermatogenesis, the development of the epididymis and of the vas deferens is severely abnormal in CBAVD. The epididymis is of variable length in CBAVD patients. The most proximal segment of the epididymis, the caput, is always present with the same length as seen in normal men of about 0.5 cm. The majority (65%) of patients with CBAVD have an epididymis represented by the caput and the proximal portion of the corpus with a total length between 2 and 4 cm. Very few (15%) have the epididymis displaying the entire caput, corpus and tail (Patrizio et al. 1994). In an attempt to explain why there should be such variability in length, a correlation with the presence of specific CFTR mutation was attempted, but no association with any particular genotype was found (Patrizio et al. 1994).

Embryologically, the proximal portion of the caput, also known as globus major, comprises the vasa efferentia and along with the testes, originates from the genital ridge (Charny and Gillenwater, 1965). Further, in this area the epithelial principal cells lining the lumen are abundant in CFTR mRNA.

The morphological defects seen in the male genital tract might also be sustained by the lack of the paracrine effects of testosterone. The Wolffian ducts must receive testosterone and dihydrotestosterone directly from the nearby Leydig cells to sustain normal development. In the absence of this paracrine effect the secretions of the epididymis are too thick to allow normal intraluminal flow rate, and the development of the Wolffian ducts cannot be completed. In these instances, only the most proximal portion of the caput, of different embryological origin, rich in CFTR mRNA and close to the testicular outflow, can sustain normal development.

In about 10–15% of patients with CBAVD there is also an associated renal anomaly (agenesis).

However, these cases are probably not related to CFTR gene mutations since extensive screening of the gene sequence has always failed to detect mutations. In these instances, a different mechanism is probably responsible for both kidney and vas deferens maldevelopment.

5 Remaining Questions

Mutations in the CFTR gene cause CBAVD. However, the inability to find CF mutations in some patients with CBAVD (about 15%) may indicate a subpopulation of men with other factors involved in the reproductive anomaly or a further genetic heterogeneity of the CFTR gene. Some questions that need unequivocal answers are the following:

1. What are the molecular mechanisms behind the observed tissue- and organ-specific differences in response to the same defect in the CFTR gene? Are extragenic factors responsible for the variable expressivity of CBAVD as illustrated by the clinical discordance in siblings with the same CFTR genotype?

The variable phenotype for the same genotype in the spectrum of CF from CBAVD to complete clinical CF with pancreatic insufficiency is theorized to be caused by variability at loci other than the CFTR gene. Even though no human data supports this hypothesis, it has been tested in a mouse model of CF (Rozmahel et al. 1996; reviewed by Estivill, 1996). Three different phenotypes are observed in mice homozygous for CFTR gene disruption. One phenotype died by 5 weeks of age due to intestinal obstruction and rupture. The second phenotype died at weaning, and the third phenotype survived for more than 6 weeks. Investigation of inheritance patterns at other loci throughout the mouse genome identified an allele type at a locus on mouse chromosome 7 which was linked to the milder phenotype. The milder phenotype was associated with the presence of a calcium-activated Cl- conductance which was not present in the mice with the severe phenotype.

The mouse chromosome 7 region is a conserved synteny with human chromosome 19q13, which is the location of several genes with the potential to modulate the CF disease phenotype. These genes include the γ subunit of protein kinase C (PKC), the α3 subunit of the type 1 Na+/K+ exchanging ATPase (Atp1a3), and the sodium channel, type 1, β subunit (Scn1b). Further work in the mouse model will be required to identify a CF phenotype modulator gene, which can then be tested in humans. Estivill (1996) points out that even monogenic diseases like CF will become more complex as we understand more about phenotypic differences which cannot be explained by the CFTR genotype alone.

2. Is it possible that the CBAVD patients without CFTR gene mutations represent a separate disease process and isolated congenital malformation of the vas deferens has a cause similar to CBAVD patients with renal agenesis?

Either there are as yet unidentified CFTR mutations which are not identified by the methods currently used, or additional genetic loci other than the CFTR gene can result in CBAVD.

3. Are women with genotypes characteristic of men with CBAVD asymptomatic?

A recent study demonstrated that CFTR gene expression occurs in endocervical cells throughout the menstrual cycle (Hayslip et al. 1997). Therefore, the absence of CFTR gene expression due to CFTR mutations has the potential to alter female reproduction. A case report of a woman with two brothers diagnosed with CBAVD has been described (Gervais et al. 1996). All three were compound heterozygotes for CFTR mutations (Delta F508/R1070W). The woman had bronchiectasis of the lingula, an intermediate sweat chloride level, and tenacious endocervical mucus which did not vary according to the menstrual cycle and contained no motile sperm on three separate assessments. She had been unable to conceive after 5 years of unprotected sexual intercourse. Four pregnancies were subsequently obtained by intrauterine insemination after ovulation induction. Hypofertility has been observed in women with CF, often related to delayed menarche or secondary amenorrhea due to the patient's clinical condition. However, thickened cervical mucus has also been observed. A thorough investigation of the CFTR gene in hypofertile or infertile women will be required to determine the significance of CF mutations in female reproduction.

6 Conclusions

In conclusion, advances in molecular biology and molecular genetics are continously improving our understanding of many forms of male infertility previously classified as idiopathic. These discoveries are important today more than ever because with the use of assisted reproductive technologies, it is possible to offer reproductive hopes to men once considered irreversibly sterile. If mendelian genetic anomalies are the cause of infertility, then there is an increased risk of transmitting a genetic defect to offspring when assisted reproductive technologies are used.

A typical example is infertility due to CBAVD where the majority (83%) of these men have a defect in both copies (60%) or in one copy (23%) of the CFTR gene and they represent, beyond any doubt, an incomplete phenotypic form of cystic fibrosis. As a consequence, CF testing has been implemented as a screening test for each couple where CBAVD is the reason for their infertility prior to epididymal sperm aspiration procedures. The high frequency of 5T-variants found in CBAVD needs to be taken into account when performing mutation analysis. If CF mutations are detected in the couple, then the risk for the offspring to have CF or CBAVD will be increased. The wide range of mutations and their different degree of penetrance makes genetic counselling and the risk estimates quite difficult. The advice of genetic counselors with expertise in CF, CBAVD and other atypical CF expressions should be made available to every couple with infertility due to CBAVD before progressing to assisted reproductive techniques.

References

Anguiano A, Oates RD, Amos JA et al (1992) Congenital bilateral absence of the vas deferens. A primarily genital form of cystic fibrosis. J Am Med Assoc 267:1794–1797

Bienvenu T, Adjiman M, Thiounn N et al (1995) Molecular diagnosis of congenital bilateral absence of vas deferens: analyses of the CFTR gene in 64 French patients. Ann Genet 40:5–9

Castellani C, Bonizzato A, Mastella G (1997) CFTR mutations and IVS8-5T variant in newborns with hypertrypsinaemia and normal sweat test. J Med Genet 34(4):297–301

Charny CW and Gillenwater JY (1965) Congenital absence of the vas deferens. J Urol 93:399–402

Chillon M, Casals T, Mercier B et al (1995) Mutations in the cystic fibrosis gene in patients with congenital absence of the vas deferens. N Engl J Med 332:1475–1480

Costes B, Girodon E, Ghanem N et al (1995) Frequent occurrence of the CFTR intron 8 (TG) 5T allele in men with congenital bilateral absence of the vas deferens. Eur J Hum Genet 3: 285–293

Dork T, Dworniczak B, Aulehla-Scholz C et al (1997) Distinct spectrum of CFTR gene mutations in congenital absence of vas deferens. Hum Genet 100:365–377

Durieu I, Bey-Omar F, Rollet J et al (1995) Diagnostic criteria for cystic fibrosis in men with congenital absence of the vas deferens. Medicine 74:42–47

Dumur V, Gervais R, Rigot JM et al (1990) Abnormal distribution of CF ΔF508 allele in azoospermic men with congenital aplasia of the epididymis and vas deferens. Lancet 336:512 (letter)

Dumur V, Gervais R, Rigot JM et al (1996) Congenital bilateral absence of the vas deferens (CBAVD) and cystic fibrosis transmembrane regulator (CFTR): correlation between genotype and phenotype. Hum Genet 97:7–10

Estivill X (1996) Complexity in a monogenic disease. Nature Genet 12:348–350

Friedman KJ, Heim RA, Knowles MR, Silverman LM (1997) Rapid characterization of the variable length polythymidine tract in the cystic fibrosis (CFTR) gene: association of the 5T allele with selected CFTR mutations and its incidence in atypical sinopulmonary disease. Hum Mutat 10(2):108–115

Gervais R, Dumur V, Letombe B et al (1996) Hypofertility with thick cervical mucus: another mild form of cystic fibrosis? J Am Med Assoc 276:1638 (letter)

Harris A and Coleman L (1989) Ductal epithelial cells cultured from human foetal epididymis and vas deferens: relevance to sterility in cystic fibrosis. J Cell Sci 92:687–690

Hayslip CC, Hao E, Usala SJ (1997) The cystic fibrosis transmembrane regulator gene is expressed in the human endocervix throughout the menstrual cycle. Fertil Steril 67(4): 636–640

Holsclaw DS, Lobel B, Jockin H, Schwachman H (1971) Genital abnormalities in male patients with cystic fibrosis. J Urol 106:568–574

Jarvi K, Zieleinski J, Wilschanski M et al (1995) Cystic fibrosis transmembrane conductance regulator and obstructive azoospermia. Lancet 345:1578 (letter)

Kerem B, Rommens JM, Buchanan JA et al (1989)Identification of the cystic fibrosis gene: genetic analysis. Science 245:1073–1080

Kerem E, Rave-Harel N, Augarten A et al (1997) A cystic fibrosis transmembrane regulator splice variant with partial penetrance associated with variable cystic fibrosis presentations. Am J Resp & Crit Care Med 155(6):1914–1920

Lissens W, Mercier B, Tournaye H et al (1996) Cystic fibrosis and infertility caused by congenital absence of the vas deferens and related clinical entities. Hum Reprod 11 (suppl 4):55–78

Patrizio P, Silber SJ, Ord T et al (1988) Two births after microsurgical epididymal sperm aspiration in congenital absence of the vas deferens. Lancet 2:1364 (letter)

Patrizio P, Asch RH, Handelin B,Silber SJ (1993) Aetiology of congenital absence of the vas deferens: genetic study of three generations. Hum Reprod 8:215–220

Patrizio P, Ord T, Silber SJ, Asch RH (1994) Correlation between epididymal length and fertilization rate in men with congenital absence of the vas deferens. Fertil Steril 61 (2):265–268

Patrizio P and Zielenski J (1996) Congenital absence of the vas deferens: A mild form of cystic fibrosis. Mol Med Today 2(1):24–31

Patrizio P and Salameh WA (1998) Expression of cystic fibrosis transmembrane conductance regulator (CFTR) mRNA in natural and pathological adult human epididymis. J Reprod Fertil (53):261–270

Rozmahel R, Wilschanski M, Matin A et al (1996) Modulation of disease severity in cystic fibrosis transmembrane conductance regulator deficient mice by a secondary genetic factor. Nature Genet 12:280–287

Riordan JR, Rommens JM, Kerem BS et al (1989) Identification of the cystic fibrosis gene: cloning and characterization of complementary DNA. Science 245:1066–1073

Rommens JM, Iannuzzi, MC, Kerem BS et al (1989) Identification of the cystic fibrosis gene: chromosome walking and jumping. Science 245:1059–1065

Silber SJ,Patrizio P, Asch RH (1990) Quantitative evaluation of spermatogenesis by testicular histology in men with congenital absence of the vas deferens. Hum Reprod 5:89–93

Silber SJ, Nagy Z, Liu J et al (1995) The use of epididymal and testicular spermatozoa for intracytoplasmic sperm injection: the genetic implications for male infertility. Hum Reprod 10: 2031–2043

Tizzano EF, Silver MM, Chitayat D et al (1994) Differential cellular expression of cystic fibrosis transmembrane regulator in human reproductive tissues. Clues for the infertility in patients with cystic fibrosis. Am J Pathol 144:906–914

Tournaye H, Devroey P, Liu J et al (1994) Microsurgical epididymal sperm aspiration and intracytoplasmic sperm injection: a new effective approach to infertility as a result of congenital bilateral absence of the vas deferens. Fertil Steril 61:1045–1051

Zielenski J, Patrizio P, Corey M et al (1995) CFTR gene variant for patients with congenital absence of the vas deferens. Am J Hum Genet 57:958–960

Mitochondrial Function and Male Infertility

Thomas Bourgeron

1 Introduction

During the recent years, alteration of mitochondrial function has been associated with an increasing numbers of phenotypes (Wallace 1992; Munnich et al. 1996). Although alteration of mtDNA is only rarely identified in patients, mtDNA has been claimed to be at the origin of aging (Wallace et al. 1995), several degenerative disorders (Wallace et al. 1995), and more recently male infertility (Cummins et al. 1994; St. Johns et al. 1997). Thus, some authors have suggested that some forms of infertility may be explained as premature aging of the testis (Cummins et al. 1994). Frank and Hurst (1996) have also hypothesized that a germ-line mutation of the mtDNA with severe effects on males but only mild effects on females might increase to a relatively high frequency because natural selection of mitochondria occurs only in females. Thus, the strictly maternal inheritance of mtDNA could create an important male-female asymmetry in the expected severity of mitochondrial disease. However, although in plants there is a direct role of mtDNA mutations in the phenotype of cytoplasmic male sterility (Hanson 1991), in mammals the evidence is not so convincing.

In the mammalian testis, mitochondria change markedly during spermatogenesis, passing through distinct metabolic states. First, germ cells require energy mainly for biosynthetic processes but not for motility. Then, in mature spermatozoa, the ATP-demand shifts from biosynthesis to motility (Kamp et al. 1996). Mutations of mitochondrial or nuclear genes encoding proteins involved in the function of the respiratory chain (RC) should decrease or suppress the oxidative phosphorylation (OXPHOS) with dramatic consequences on sperm cell metabolism. Taking into account the importance of mitochondria for the spermatozoa, testing mitochondrial functions could be one component of a male fertility diagnostic, possibly enhancing the routine semen analysis index (Bartoov et al. 1994).

Laboratoire d'Immunogénétique Humaine, INSERM U276, Institut Pasteur,
25, rue du docteur Roux; 75724 Paris Cedex 15, France

Results and Problems in Cell Differentiation, Vol. 28
McElreavey (Ed.): The Genetic Basis of Male Infertility

Since spermatozoa are (1) clearly dependent on their mitochondria, (2) specially sensitive to lipid peroxydation (Aitken and Fisher 1994) and ATP depletion (de Lamirande and Gagnon 1992b; Vigue et al. 1992) and that (3) mitochondrial mutations have occasionally been found in the germ line of males (Kao et al. 1995; Reynier et al. 1997), we should consider the possibility that RC deficiencies could affect the testis and the spermatozoa and have an influence on the male fertility (Cummins et al. 1994). However, there has been no clear indication for a common role of the mitochondria in male infertility up to now.

Finally, the possibility that spermatozoa could harbor mutant mtDNA molecules and the availability of new techniques of medically assisted procreation such as intra cytoplasmic sperm injection (ICSI; Palermo et al. 1992) have raised the important question of mtDNA inheritance in humans (Ankel-Simons and Cummins 1996). If spermatozoa accumulate mtDNA mutations and there is a leak of paternal mtDNA in the zygote, this could have a dramatic effect on the embryo bearing mutated mtDNA at the first stage of its development.

2 Mitochondrial Diseases

Mitochondrial diseases include an increasing number of different disorders with different phenotypes, genetic mutations and heredity (Tyler 1992; Wallace 1992). Particularly in childhood, the clinical presentation of mitochondrial disorders is quite varied including: neuromuscular, liver, cardiac,

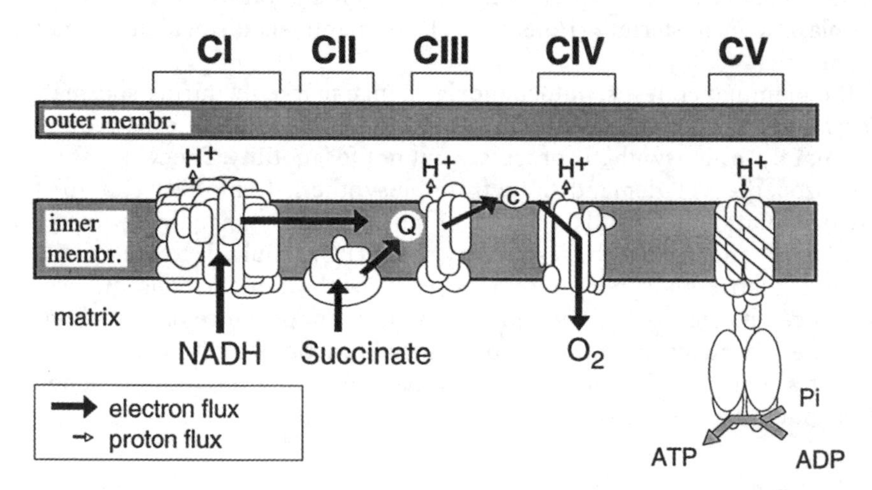

Fig. 1. The respiratory chain. *CI, CII, CIII, CIV, CV* various complexes of the respiratory chain, *Q* quinones, *c* cytochrome *c*

renal, gastrointestinal, endocrine, hematological, facial dimorphism and dermatological symptoms (Munnich et al. 1996). The molecular bases of the mitochondrial disorders also differ, involving nuclear genes (Bourgeron et al. 1995), the mitochondrial genome (Wallace 1992) or the combination of both genomes (Suomalainen et al. 1995). At least three metabolic pathways have been involved in mitochondrial disorders: fatty acid oxidation (Tyler 1992), the Krebs cycle (Rustin et al. 1997), and the respiratory chain (Fig. 1; Wallace 1992).

In humans, the mitochondrial genome is very compact (16569 bp) and carries 37 genes: 22 tRNA, 2 rRNA, and 13 genes encoding for the catalytic subunits of the respiratory chain (Fig. 2; Tyler 1992). Almost all of the regulatory sequences are located in the deplacement loop (D-loop), including the replica-

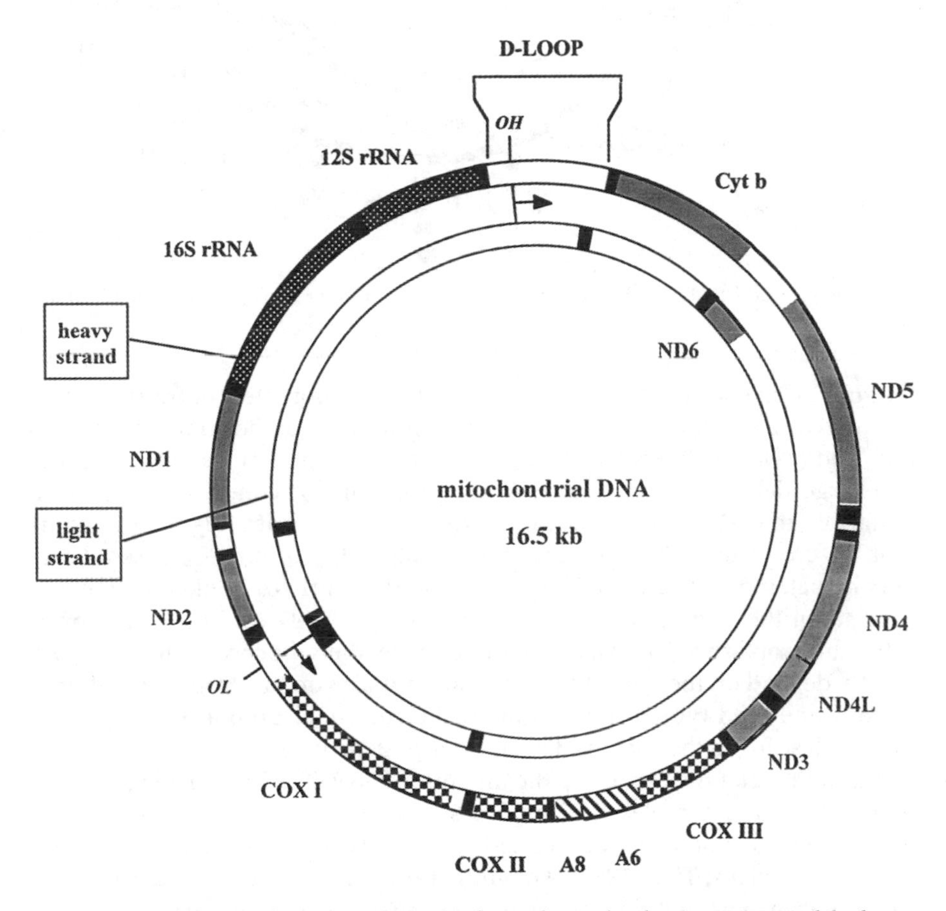

Fig. 2. The organization of human mtDNA. *Cyt b* cytochrome b subunit, *ND* NADH dehydrogenase subunit, *COX* cytochrome *c* oxidase subunit, *A* ATPase subunit, *D-LOOP* deplacement loop, *OH* and *OL* replication origins for the heavy and the light strands respectively

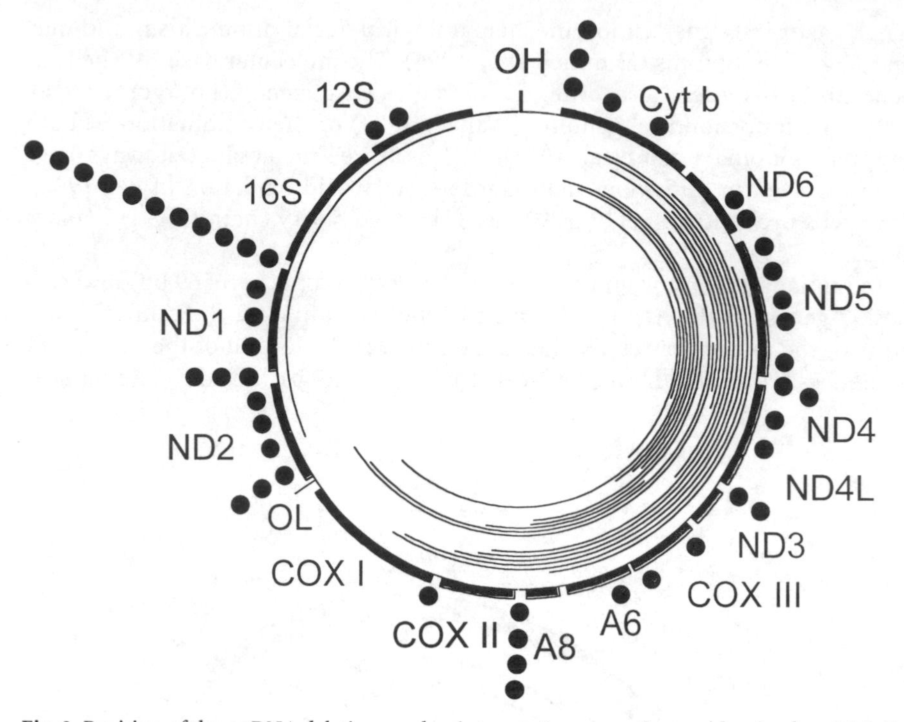

Fig. 3. Position of the mtDNA deletions and point mutations in patients with mitochondrial diseases

tion initiation site of the heavy strand and the two promoters of transcription. An increasing number of mtDNA point mutations or deletions have been reported (Tyler 1992; Wallace 1992). However, point mutations in tRNA genes and large deletions of the mtDNA are predominant compared with point mutations in the rRNAs and the RC enzyme encoding genes (Fig. 3). An important feature of the genotype/phenotype relationship in mtDNA based disorders is that, in most cases, only a part of the total mtDNA pool is mutant and coexists in the same cell or tissue with wild-type mtDNA. This condition is called heteroplasmy (Fig. 4). As a consequence, the mitochondrial defect will greatly depend on the type of the mutation but also on the proportion of both mutant and wild-type mtDNA. Finally, the threshold proportion of mutant mtDNA leading to the enzyme defect will also depend on the type of the mutation and the cell type harboring the mutant mtDNA. The RC is made up of five multi-enzymatic complexes (Complexes I–V) containing both nuclear and mtDNA encoded polypeptides (except complex II, which is exclusively encoded by the nucleus; Tyler 1992). The mitochondrial RC ensures the process of electron transport from reducing equivalents (e.g. NADH, succinate) to molecular oxygen with a very large loss of free energy, much of which is conserved by the phosphorylation of ADP to yield ATP, in the process of OXPHOS.

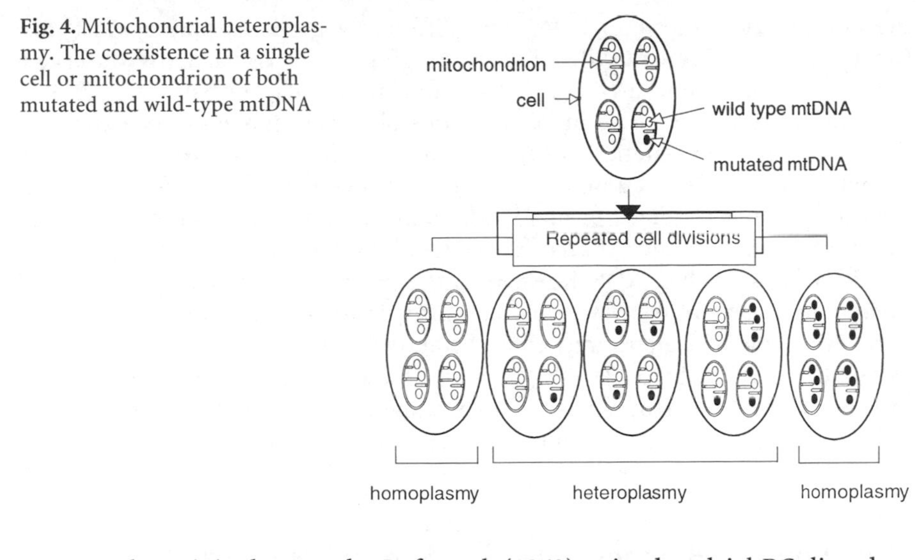

Fig. 4. Mitochondrial heteroplasmy. The coexistence in a single cell or mitochondrion of both mutated and wild-type mtDNA

Since the original report by Luft et al. (1962), mitochondrial RC disorders have been essentially regarded as neuromuscular diseases. However, the process of OXPHOS is not restricted to muscles and nerves. Consequently, defects of the mitochondrial RC might potentially affect any organ or tissue, including testes or spermatozoa. At least three hypotheses might explain the specific involvement of one given organ or tissue. First, it might result from the occurrence of mutations in nuclear genes coding for tissue-specific isoforms. These may be specific subunits of the RC complexes or proteins involved in specialized mitochondrial biogenesis. On the other hand, organ specific involvement might result from mtDNA heteroplasmy. Indeed, variable proportions of both mutated and normal mitochondria genomes, with tissue to tissue and cell to cell differences, are often observed in affected patients (Rötig et al. 1990). Finally, organ-specific involvement might originate from organ specific regulation of electron fluxes in the RC, or from organ-specific requirements of energy supply.

3 Mitochondrial Function and Aging

Apart from energy loss, another consequence of RC deficiency is the possible accumulation of reactive oxygen species (ROS) able to actively peroxidize lipids, proteins or nucleic acids (Aitken and Fisher 1994). This ability of the RC to produce ROS was used to support the mtDNA somatic mutation theory of aging (Wallace et al. 1995). According to this theory, mtDNA accumulates somatic mutations and when the proportion of mutations is above a certain threshold (specific to a given tissue), this leads to a decrease in OXPHOS causing degeneration and aging. However, different arguments play against this theory: (1) decrease of OXPHOS during aging is not consistently observed (Chretien et al. 1998); (2) using mitochondrial cybrids, Hayashi et al. (1994) have shown that a decrease in RC activities observed in aged fibroblasts was due to the nuclear genetic background instead of the mtDNA; (3) it is well known that mtDNA accumulates about ten times more mutations than the nuclear genome (Wallace et al. 1995). Such an elevated mutation rate was explained by the absence of an effective repair system and of the proximity of the RC able to produce ROS. However, if the pyrimidine dimer repair mechanism appears absent from mammalian mitochondria, recent work has shown that several mtDNA alterations could be actively repaired (Thyagarajan et al. 1996; Shadel and Clayton 1997); (4) finally, the proportion of mutant mtDNA in the somatic tissues of elderly people is always far below the known threshold of mutant mtDNA leading to a decrease of OXPHOS.

4 Mitochondrial Biogenesis During Spermatogenesis

In mammalian testis, mitochondria change markedly during spermatogenesis (De Martino et al. 1979; Hecht and Liem 1984). During the initial phase of spermatogonial renewal and proliferation (in prepuberal animals), the testis contains type A spermatogonia, intermediate spermatogonia, type B spermatogonia, and a group of supporting somatic cells. The mitochondria of the germ cells at this stage appear structurally similar to those found in somatic tissues. During meiosis, however, mitochondria with diffuse and vacuolated matrices appear. These mitochondria ultimately constitute 80% of the testicular mitochondrial pool in the sexually mature rat (De Martino et al. 1979).

As replication of nuclear DNA occurs in spermatogonia and preleptotene spermatocytes, labeling studies using ^{3}H thymidine suggest that testicular mtDNA is synthesized at specific intervals during spermatogenesis and some of the mitochondrial genomes must have half-lives of at least 2 weeks (Hecht and Liem 1984). The peak of mtDNA synthesis has been detected at the end of meiosis and into early spermiogenesis. Mitochondrial transcription factor A (mtTFA) is a key activator of transcription in mammals (Shadel and Clayton 1997) and also has a role in mtDNA replication, since transcription generates an RNA primer necessary for the initiation of mtDNA replication. In the mouse, testis-specific mtTFA transcripts encode a protein isoform that is imported to the nucleus rather than into the mitochondria of spermatocytes and elongating spermatids (Larson et al. 1996). Similarly to the mouse, in humans, Larsson et al. (1997) have identified abundant testis-specific transcript isoforms generated by alternate transcription initiation sites. However, none of the testis specific transcripts predicts a nuclear protein isoform, and western blot analysis identified only the mitochondrial form of mtTFA in human testis. mtTFA protein and mtDNA exhibit parallel gradients with high levels in undifferentiated male germ cells and low levels or an absence in differentiated male germ cells. Testis-specific mtTFA transcripts exhibit an opposite pattern, suggesting that in both humans and mice, the presence of testis-specific transcripts down-regulates mtTFA protein levels in mitochondria (Larson et al. 1997). The mechanism involved is still not known.

Comparison of total testicular RNA preparation from prepuberal and sexually mature mice revealed no major qualitative or quantitative differences in the levels of the mitochondrial transcripts (Alcivar et al. 1989). Similar results were observed from enriched preparations of type A and B spermatogonia and interstitial cells obtained from the testes of 8-day-old mice. However, transcripts for COX I, COXII, ATPase 6 or ND1 were reduced in amount in the enriched preparation of pachytene spermatocytes, round spermatids, and residual bodies when compared with the amount in total testis or liver RNA (Alcivar et al. 1989).

During spermiogenesis, impressive organelle rearrangements occur as spermatids develop into the highly specialized spermatozoa (De Martino et al. 1979). A distinct form of mitochondria with a condensed matrix and the absence of an expanded intracrystal matrix are found in spermatozoa. At this stage, the mammalian spermatozoon contains approximately 72 mitochondria in the mitochondrial sheath of the midpiece (Hecht and Liem 1984). Quantitation of the number of mitochondrial genomes between meiosis and the end of spermatogenesis in mice has revealed an eight-to-ten fold decrease per haploid genome with spermatozoa containing approximately one mitochondrial genome per mitochondrion (Hecht and Liem 1984). Finally, during sperm maturation in the epididymis, mitochondria undergo further modifications. One well-established change is that the outer mitochondrial membrane becomes extensively cross-linked by disulfide bonds which account for its exceptional resistance to solubilization by anionic detergents. Sonication or nitrogen cavitation do not disrupt the mitochondrial sheath and the outer and inner mitochondrial membranes as well as the mitochondrial matrix appear intact (Olson and Winfrey 1992).

5 Mitochondrial Organization in the Spermatozoon

Mitochondria of the mammalian spermatozoon are restricted to the midpiece of the flagellum. The elongate mitochondria wrap in a helical fashion around the outer dense fiber-axoneme complex to form the cylinder-shaped mitochondrial sheath (Olson and Winfrey 1992). Within the sheath, adjacent mitochondria associate both end to end and along their lateral surfaces. This positioning of a concentrated array of mitochondria adjacent to the flagellum is believed to represent an efficient mechanism for the provision of energy required for flagellar motility (Olson and Winfrey 1992).

The mechanisms that generate and maintain this arrangement of mitochondria around the midpiece are poorly understood. Mitochondria aggregate around the spermatid midpiece following the migration of the annulus from its original position at the neck to its final destination at the midpiece-principal piece junction (Phillips 1977). Two structural specializations have been identified in mature sperm which could maintain the arrangement of mitochondria within the sheath. First, bridge-like connections between adjacent mitochondria have been noted, and second, a midpiece-specific cytoskeletal complex associated with the adaxial surface of the mitochondrial sheath has been identified (Olson and Winfrey 1992). Thus, the mitochondrial surface is comprised of at least three structurally distinct domains. These three surface domains include first, the outer-facing surface of the mitochondria that is oriented toward the plasma membrane. The second surface domain consists of the inner-facing surface of the mitochondria that is oriented toward the central outer dense fiber-axoneme complex; this surface is further distinguished by its association with a midpiece-specific cytoskeletal network, the submitochondrial reticulum (Olson and Winfrey 1990). The third surface domain includes the lateral surfaces and ends of the mitochondria that are in closed apposition with the neighboring mitochondria; typically the outer membranes of adjacent mitochondria are separated by a space of about 5–6 nm (Olson and Winfrey 1992).

Pallini et al. (1979) have reported that the bull sperm mitochondria capsule consists essentially of three polypeptide chains having molecular weights of 31, 29, and 20 kDa. The 20-kDa protein was rich in cysteine (17.9%), proline (26.5%), and contains selenium (Pallini et al. 1979). The identification of the selenoprotein in mammalian sperm is important because spermatozoa contain the highest concentration of selenium of any tissue in the mammalian body; ^{75}Se is incorporated selectively into the midpiece and purified mitochondrial capsules of sperm. Selenium was shown to play a role in the organization of the mitochondria around the sperm tail dense fibers in dietary experiments. Rats on a selenium-deficient diet show reduced sperm motility and disorganization of the sperm mitochondria (Wu et al. 1979; Wallace et al. 1983). In humans, both high and low sperm selenium concentrations were reported to have a negative influence on the number of spermatozoa and their (Hansen and Deguchi 1996). It has been suggested that a seminal Se concentration of between 40 and 70 μg/l is optimal (Bleau et al. 1984).

Two proteins containing selenium have been identified in the rat testis: the mitochondrial capsule selenoprotein (MCS; Calvin et al. 1981) and the phospholipid hydroperoxide glutathione peroxidase (Roveri et al. 1992). However, the originally cloned MCS does not appear to be a selenoprotein (Cataldo et al. 1996): (1) MCS is not labeled with ^{75}Se-selenite in seminiferous tubule culture; (2) the translation start site in the mouse mRNA seems to be located downstream of the potential UGA selenocysteine codons and (3) finally, the reading frame encoding the cysteine-rich protein in rat lacks in-phase UGA selenocysteine codons (Adham et al. 1996).

In situ hybridization reveals that the MCS mRNA begins to be transcribed in step 3 spermatids and persists through step 16 spermatids in mice (Shih and Kleene 1992). Using light and electron microscopy immunocytochemistry, Cataldo et al. (1996) could detect this cysteine-rich protein first during step 11 of spermiogenesis in the mouse demonstrating that the cysteine-rich protein mRNA is under translational control. It has been shown that MCS mRNA is translationally repressed with long homogenous poly(A) tracts in round spermatids and is translationally active with shortened heterogenous poly(A) tracts in elongating spermatids (Kleene 1989).The protein is distributed in the cytoplasm in spermatids in steps 11 through early step 16 in the mouse, and is associated with the outer mitochondrial membranes of spermatids in late step 16 and epididymal spermatozoa (Cataldo et al. 1996). In step 16, the assembly of the mitochondrial sheath is complete producing a structure in which elongate, crescent-shaped mitochondria touch end-to-end every third gyre, creating five equidistant longitudinal lines within the sheath (Otani et al. 1988). The observation that the immunoreactive MCS is homogenously distributed in the spermatid cytoplasm during step 13 to early step 16, when the mitochondrial sheath is forming, implies that MCS does not participate directly in the formation of the mitochondrial sheath (Cataldo et al. 1996). MCS could function in one or more of three ways: by stabilizing the mitochondrial capsule, by cementing the mitochondria together in the sheath, and by attaching the mitochondria to the outer dense fibers (Cataldo et al. 1996).

When spermatogenesis is complete, the precise helicoidal structure of the spermatozoa mitochondria should allow the best environment for the production and release of energy from the RC to the dynein arm of the flagellon.

6 Regulation of Oxidative Phosphorylation in Mitochondria

Spermatozoa pass through distinct metabolic states. During spermiogenesis (cell proliferation and differentiation) they require energy mainly for biosynthetic processes but not for motility (Kamp et al. 1996). On the other hand, mature spermatozoa have lost their biosynthetic potential, the nucleus is condensed and the Golgi-apparatus has been extruded together with most of the endoplasmic reticulum. Hence, ATP-demand has shifted from biosynthesis to motility. These two periods may be separated by a period with a very low metabolic rate (Kamp et al. 1996).

The contribution of oxidative phosphorylation to the generation of ATP in human spermatozoa is unclear. These cells exhibit a high rate of aerobic glycolysis and substrate level phosphorylation can completely fulfil their energy needs when glucose or fructose is available (McLeod 1941; Peterson and Freund 1970; Ford and Harrison 1981). The principal mechanism by which oxidative phosphorylation in isolated sperm mitochondria is regulated via the extramitochondrial ATP/ADP ratio is the same as that demonstrated for other isolated mitochondria (Halangk et al. 1987). Respiration was found to depend on the extramitochondrial ATP/ADP ratio in the range of 1–100. The contribution of the adenine nucleotide translocator to this dependence was determined by titration with the irreversible inhibitor carboxyatractyloside in the presence of ADP. Using lactate plus malate as the substrate, the active state respiration was controlled to about 30% by the translocator, whereas 12 and 4% were determined in the presence of glycerol-3-phosphate and malate alone, respectively (Halangk et al. 1987). However, production of ATP by mitochondria in the midpiece and ATP consumption by dynein ATPase along the whole sperm tail might lead to higher ATP/ADP ratios surrounding the mitochondria than that indicated by the average value calculated for the cytosolic compartment (Halangk et al. 1987).

Cardullo and Baltz (1991) have shown that mitochondrial volume determines sperm length and flagellar beat frequency and that there is a morphometric relationship between the dimensions of the midpiece and the length of the flagellum. The length of the midpiece varies approximately as the 3/2 power of the flagellar length although the proportionality constant is different for eutherian and marsupial sperm. The flagellum, which propels the cell, is the main consumer of this energy. Thus, the source and the destination of the metabolic energy of sperm are spatially separated. Consequently, one important question concerns the transport of energy-rich phosphate from the mitochondria to the distal dynein-ATPases along the flagella. In sea urchin this transport is carried out by the phosphocreatine (PCr) and creatine kinase (CK, EC 2.7.3.2) phosphagen shuttle (Tombes and Shapiro 1985). According to this model, ATP is generated in the sperm mitochondria and the energy is delivered to the sperm tail via phosphorylated creatine. The CK isoenzyme system catalyses the reversible transfer of the N-phosphoryl group from phosphorylcreatine (PCr) to ADP in order to generate ATP (Wallimann and Hemmer 1994). Creatine phosphate serves as a donor for rephosphorylation of ADP to ATP, which supports the flagellar dynein-adenosine triphosphatase and sperm motility. The creatine diffuses back to the mitochondria, completing the shuttle.

Different isoenzymes of CK are present and in most tissues, cytosolic as well as mitochondrial CK isoenzymes are co-expressed. The mitochondrial CK (Mi-CK) isoenzymes are associated with the outer side of the inner mitochondrial membrane and form both dimeric and octameric molecules that are readily interconvertible (Jacobus and Lehninger 1973). Studies on isolated mitochondria have shown that mitochondrial oxidative phosphorylation and the Mi-CK reaction are functionally coupled, that is, Mi-CK preferentially utilizes the ATP synthesized through oxidative phosphorylation for PCr synthesis (Walliman and Hemmer 1994).

In contrast to sea urchin sperm which thrive exclusively on mitochondrial fatty acid oxidation (Tombes and Shapiro 1985), mammalian sperm must also remain motile under the low-oxygen conditions prevalent in the female tract, deriving chemical energy from glycolytic pathways or from external sources, in the form of glucose, fructose, mannose or even PCr, substances all shown to be present in seminal plasma (Hammerstedt and Lardy 1983). Therefore, Mi-CK may not be absolutely required for sperm motility and fertilization in mammals. For instance, a transgenic mouse mutant lacking Mi-CK, which is present as the main Mi-CK isoenzyme in sperm of normal mice, was shown to produce motile sperm and to be fertile (Steeghs et al. 1995). The importance of the CK system for proper sperm function and fertility may range from essential (sea urchin) to marginal (some mammals). In mammalian sperm, no significant phosphocreatine shuttle has been identified. Boar sperm contain several mitochondria in the middle piece but no PCr and no CK activity (Kamp et al. 1996). Hence, transport of energy-rich phosphate from mitochondria to the distal dynein-ATPases by the phosphagen system, as proposed to be important for spermatozoa of various species, is not necessary for boar sperm motility. Schoff et al. (1989) have identified adenylate kinase activity in bovine sperm flagella which may act to regulate ATP levels in motile sperm. If such a system is operating, adenylate kinase could use the energy of two high-energy phosphate bonds in each molecule of ATP. Finally, another alternative to a phosphogen shuttle derives from the observation that spermatozoa without or with only low CK activities have relatively high activities of glycolitic enzyme, such as the glyceraldehyde-3-phosphate dehydrogenase and lactate dehydrogenase, but low activities of enzymes representing aerobic pathways such as hydroxyacyl-CoA dehydrogenase and citrate synthase (Kamp et al. 1996). On the other hand, CK-rich spermatozoa show considerable capacities for oxidation (turkey, carp, and lugworm spermatozoa). Sea urchin spermatozoa appear to be completely dependent on mitochondrial ATP production from fat and show high CK activities (Mita and Yasumasu 1983). If ATP in the distal part of the flagellum was preferentially produced glycolytically, an energy shuttle from the mitochondria would not be necessary. This organization would, however, require the regeneration of glycolytic NADH in the distal part of the flagellum. In addition to the typical shuttle systems for the reoxidation of cytosolic NADH (glycerophosphate shuttle, malate/aspartate shuttle), mammalian spermatozoa seem to contain another system. In several mammalian spermatozoa, a mitochondrial-specific LDH-C has been found (Gallina et al. 1994). It appears that lactate produced in the distal part of the flagellum is transferred to the midpiece where pyruvate plus NADH are formed which can be oxidized in the mitochondria.

Another difference between sea urchin sperm and other sperms lies in the coupling between the motility and ATP production. In the sea urchin and tunicate sperm, O_2 consumption is markedly decreased upon decreasing the flagellar beat frequency with increasing solution viscosity (Brokaw 1967; Brokaw and Benedict 1968). However, sperm from other species including rat (Cardullo and Cone 1986) and *Xenopus laevi* (Bernadini et al. 1988) show no such coupling between motility and metabolism. In fact, these species continue to consume oxygen at the same high rate even though their flagella have been mechanically immobilized.

7 Abnormal Mitochondria and Infertility

Sperm motility is one of the major determinants of male fertility. The capability for active flagellar motion is essential for the functional competence of sperm and is crucial for their successful transport through the female reproductive tract to the oolemma. Thus, spermatozoa with poor motility cannot penetrate through mucus-filled cervices and arrive at the site of fertilization. Various factors may cause impairment of sperm motility leading to infertility: genital infections, maturational abnormalities in the epididymis, factor in the plasma, intrinsic defects in the flagellar axonemes or defects in metabolism (Olds-Clarke 1996). Absence of or abnormal mitochondria have been described in asthenozoospermic or akinetozoospermic men (Pedersen et al. 1971; Ross et al. 1973; Alexandre et al. 1978; McClure et al. 1983; Wilton et al. 1992; Gopalkrishnan et al. 1995; Mundy et al. 1995). The spectrum of midpiece abnormalities ranges from irregular and disorganized mitochondria to a decreased number or absence of mitochondria. For instance, spermatozoa with shorter midpieces were found in asthenozoospermic subjects, while midpieces width and tail length were comparable to controls (Mundy et al. 1995). At the ultrastructural level, the asthenozoospermic subjects demonstrated significantly fewer mitochondrial gyres than their fertile counterparts. Different hypotheses have been proposed to explain this anomaly. It may be that fewer mitochondria are initially formed or there is a faulty migration system where an excess of mitochondria are lost as the peripheral cytoplasm is shed. On the other hand, since sperm mitochondria are further modified during sperm maturation in the epididymis, the disordered arrangement of the midpiece mitochondria could represent a failure of this process. Finally, it is also possible that both a reduction in mitochondria and a failure of bond formation in the midpiece play a part in the poor quality of movement of spermatozoa in these subjects.

Interestingly, asthenozoospermia due to a complete absence of mitochondria has been reported (Ross et al. 1973; Alexandre et al. 1978; McClure et al. 1983). In testicular biopsies from one set of patients, spermatozoa had normal mitochondria and it was suggested that the mitochondria have degenerated during epididymal passage (Ross et al. 1973). The anomalies could also have a genetic origin since absence of sperm mitochondria has been described in two infertile brothers (Alexandre et al. 1978).

8
Mitochondrial Respiratory Chain, mtDNA and Infertility

Considering the importance of the mitochondrial functions for complete spermatogenesis, many groups have studied the properties of mitochondria from infertile or elderly men and hypothesize that some forms of infertility may be explained as premature aging of the testis (Cummins et al. 1994). It should be considered that, in the somatic mtDNA mutation theory, the best candidate cells in the testis are the Leydig and the Sertoli cells. For instance, in Leydig cells, the testicular testosterone biosynthesis consists of two major components: mitochondrial steroidogenesis, which produces pregnenolone from cholesterol, and microsomal steroidogenesis, which produces testosterone from pregnolone supplied by the mitochondria. Studying the decline in sexual function that accompanies the aging process in men, Takahashi et al. (1983) have found that in the testes of elderly men with prostatic cancer, there is a decreased supply of pregnolone from the mitochondria, resulting in a decline in testicular steroidogenesis. However, it is not clear if the origin of this defect is directly due to the mitochondrial pregnolone synthesis or if it is an indirect effect due to the alteration of another cellular pathway.

Apart from this, most of those studying mitochondria have searched for mtDNA mutations or measured the mitochondrial RC and OXPHOS activities. For instance, Hsu et al. (1986) have studied the RC activities in rats with varicocele. Varicocele is considered to be one of the causes of male infertility but the exact pathogenesis of the disease is still an enigma. Several reports have shown that blood flow and temperature are elevated in both testes following experimentally induced varicocele (Saypol et al. 1981). In addition the concentration and the motility of the spermatozoa in the caudal epididymis is decreased (Hurt et al. 1986). In the experimentally induced varicocele-bearing rats, Hsu et al. (1986) found a decrease in the RC enzyme activities of NADH cytochrome *c* reductase (complex I + III), succinate cytochrome *c* reductase (complex II + III), and cytochrome *c* oxidase (complex IV). In addition, the energy charge of the adenine nucleotides, as well as the RC enzyme activities, were significantly decreased after inducing a varicocele in the rat, and their level never recovered (Hsu et al. 1986).

On the other hand, studying patients with mitochondrial disorders, Folgero et al. (1993) have found one infertile patient with asthenozoospermia. Electron microscopy of spermatozoa from the patient revealed mitochondria with increased matrix, thickening of membranes, parallelization of cristae and lipid inclusions, which are characteristic findings in mitochondrial disorders. Abnormal mitochondria were also found in spermatids, suggesting that the ultrastructural changes of the mitochondria are primary rather than secondary to the degeneration of the spermatozoa. This patient was a member of a four-generation family with mitochondrial encephalomyopathy and abnormal mitochondria were also found in his muscle. Mutation screening of the mtDNA revealed a mutation in the tRNA Leu (UUR) in the family. More recently, Reynier et al. (1997) have found a significant proportion of multiple deletions of the mtDNA in the spermatozoa from an infertile patient with oligo-asthenozoospermia. These deletions were accompanied with an increase of the total mtDNA pool in the spermatozoa. In this case, the patient also presented ptosis, a clinical sign often associated with mitochondrial diseases. Histology of the deltoid muscle revealed the presence of ragged red fibers, cytochrome *c* oxidase negative fibers and multiple deletions were also present in the muscle. However, if some patients suffering from mitochondrial disorders could have decreased fertility, to our knowledge, no other decrease of male or female fertility was reported in families with MERRF (myoclonus epilepsy associated with ragged red fibers) or MELAS (mitochondrial myopathy, encephalopathy, lactic acidosis, and stroke-like episodes) harboring mtDNA mutations. Moreover, Huang et al. (1994) did not find altered spermatozoa motility in a healthy patient harboring the MELAS mutation of the tRNA Leu (UUR). These apparently contradictive results could be explained by a different proportion of mutant mtDNA present in the testis or in the spermatozoa of the patients.

Finally, mtDNA deletions were also found in the spermatozoa of infertile men with no apparent sign of mitochondrial disease. Kao et al. (1995) have examined the accumulation of the "common" 4977-bp mtDNA deletion in spermatozoa obtained from patients with infertility or subfertility. This deletion removes 4977 bp (seven genes and five tRNAs) from the wild-type mtDNA and is found in different mitochondrial disorders. The highest frequency of occurrence of the 4977-bp mtDNA deletion was found in sperm in the fraction with the lowest motility and a higher incidence of the deletion was found in patients with asthenozoospermia, oligozoospermia and primary infertility compared with normal individuals. However, the proportion of the 4977-bp mtDNA was less than 0.1% of the total mtDNA pool. It should be noted that this proportion of mutant mtDNA is certainly not sufficient to cause a general decrease of the OXPHOS in spermatozoa. Two arguments have been used to support the role of such small amounts of mtDNA mutations in the aging process or in the case of human infertility. First, only one type of deletion has been looked for and spermatozoa might harbor a huge number of other mutations. Thus, the proportion of the 4977 bp deletion could only be the "tip of the iceberg" but this remains to be shown. Here, we should recall that in muscle, the threshold of mutant mtDNA leading to the RC defect is over 80 to 90% of the total mtDNA pool, depending on the type of mutation (Shoubridge et al. 1990). B lymphoblastoid cell lines harboring 80% of deleted mtDNA (4977 bp deletion) have normal RC activities (Bourgeron et al. 1993). Secondly, only 10% of chloramphenicol resistant mtDNA are sufficient to protect a heteroplasmic cell line from the inhibition of mitochondrial translation by chloramphenicol (Wallace 1986). The second argument, as already mentioned, stipulates that the consequence of an isolated defective RC in a single mitochondrion could, by the defective RC production of ROS, be able to create damage first in the cell and then in the tissue. To our knowledge, in humans, the production of ROS by a defective RC has not yet been shown, and it is very important to clarify this point since ROS plays such an important role in many aspects of cell physiology.

9
Reactive Oxygen Species Generation and Human Spermatozoa

Human spermatozoa have been shown to have the capacity to generate ROS and the reactivity, short half-life and limited diffusion of these molecules is consistent with their possible significance as second messengers in such cells (Aitken and Fisher 1994). Free radicals have been defined as species containing one or more unpaired electrons in a given atomic or molecular orbital. The possession of unpaired electrons is the feature that bestows upon these molecules their immense reactivity, and thus, their biological significance. In the case of human spermatozoa, a physiological role for ROS is supported by the involvement of these molecules in the induction of key biological events such as the acrosome reaction and hyperactivated motility. The reactivity of a free radical is inversely related to its stability. Examples of free radicals are the superoxide anion, O_2^-, the protonated form of this molecule, the hydroperoxyl radical HO_2, and the hydroxy radical OH°. In biological terms, a number of non-radical derivatives of oxygen, including hydrogen peroxide (H_2O_2) or hypochlorous acid (HOCl), are also extremely important and are grouped under the collective term, reactive oxygen species (Aitken and Fisher 1994).

For spermatozoa, the risk of manufacturing reactive oxygen metabolites is considerable because these cells are particularly vulnerable to lipid peroxidation (Aitken and Fisher 1994). This susceptibility stems from the fact that during their differentiation, spermatozoa discard most of their cytoplasm during the final stages of spermiogenesis. As a consequence, these cells tend to lose some of their antioxidant cytoplasmic enzymes, such as glutathione peroxidase, superoxide dismutase or catalase that protect most cell types from peroxidative damage. Moreover, spermatozoa are especially susceptible to peroxydative damage because their plasma membranes are enriched with unsaturated fatty acids that are particularly vulnerable to free radical attack (Jones et al. 1979). To some extent the lack of cytoplasmic defensive enzymes is compensated for by the antioxidant properties of seminal plasma, which is well endowed with a variety of factors that are designed to protect the spermatozoa against oxidative stress (Jones et al. 1979). These factors include small molecular weight scavengers such as ascorbic and uric acids, chain-breaking antioxidants such as α-tocopherol, and antioxidant enzymes such as superoxide dismutase and catalase (Fraga et al. 1991; Zinni et al. 1993). Thus, as long as spermatozoa are suspended in seminal plasma they benefit from some protection against oxidative stress (Aitken and Fisher 1994).

Work on bovine spermatozoa has suggested that ROS could be generated via the oxidative deamination of aromatic amino acids (Tosic and Walton 1950). In contrast, the ROS generated by rabbit spermatozoa appear to derive from the leakage of electrons from the mitochondrial RC (Killian et al. 1985). Neither of these mechanisms appear to apply to the generation of ROS by human spermatozoa, since neither aromatic amino acids nor mitochondrial inhibitors influence the rate at which ROS are produced by these cells. In human spermatozoa, the mechanism could be similar to that present in leukocytes involving an NADPH-oxidase complex (Aitken and Fisher 1994).

ROS can affect sperm axonemes but the possibility of a direct action is unlikely, since neither hydrogen peroxide nor xanthine oxidase + xanthine had any significant effect on the motility parameters when added to control demembranated-reactivated spermatozoa (de Lamirande and Gragnon 1992a). The inhibitory action of ROS on sperm motility may not be direct, but could be mediated by other substances. Three lines of evidence suggested that ATP depletion induced by ROS treatment is responsible for the effect observed in spermatozoa (de Lamirande and Gragnon 1992b). First, the rapid decrease in intracellular ATP observed after ROS treatment is closely followed by a decrease in beat frequency, loss of intact sperm motility, and axonemal damage due to insufficient phosphorylation. Second, incubation of spermatozoa with the combination pyruvate-lactate allows maintenance of sperm ATP at a normal level and prevents the effects of ROS; furthermore, spermatozoa immobilized after ROS treatment, then supplemented with pyruvate-lactate, are able to reinitiate motility in parallel with an increase in their ATP level. Third, rotenone, an inhibitor of OXPHOS, caused ATP depletion, sperm immobilization, and axonemal alterations similar to those observed after ROS treatment.

The mechanism of ATP depletion in spermatozoa is not yet known. Since ROS can pass through cell membranes, ATP depletion could result from the action of ROS on glyceraldehyde 3-phosphate dehydrogenase, one of the key enzymes for the generation of ATP in the cell. Low concentrations of these substances (1 to 100 μmol/l) have been shown to inhibit a large number of cellular enzymes and functions including mitochondrial functions, anaerobic glycolysis, DNA, RNA and protein synthesis, plasma membrane adenylate cyclase (Comporti 1989) and sperm motility (Selley et al. 1991).

10 ATP Concentration, Creatine Kinase Activity and Infertility

The lack of reliable methods for assessing sperm fertilizing potential is a problem for infertile couples and for their physicians. Because of this, a lot of work has been done to identify cellular markers of sperm quality that, in addition to their diagnostic value, may also facilitate the identification of specific deficiencies in sperm function. For instance, mitochondrial functions could be one component of a male fertility diagnostic profile based on quantitative parameters to enhance the routine semen analysis index (Bartoov et al. 1994). Because one of the most appreciable properties of spermatozoa is ATP-fueled motility, the possible value of semen and sperm ATP concentrations for predicting male fertility has been studied extensively. The clinical value of sperm or semen ATP concentrations for the evaluation of male fertility has been promoted by some investigators, whereas others have questioned the utility of this approach. In support, Vigue et al. (1992) found that a single measurement of the ATP content per living spermatozon is not directly related to sperm motility. However, a significant decrease of ATP concentration in relation to the number of spermatozoa that have survived after a couple of hours indicates good sperm movement and may add information to the routinely used subjective motility assessment (Vigue et al. 1992).

In normozoospermic, oligozoospermic and asthenozoospermic samples the per sperm ATP concentrations, ATP/ADP ratios, or the total contents of ATP and ADP were found to be similar (Vigue et al. 1992). When measuring CK activity in the normozoospermic and oligozoospermic samples Vigue et al. (1992) have also shown that a fourfold difference in per sperm CK activity does not change the ATP concentrations or ATP/ADP ratios. This indicates that the increased capacity for ATP production did not actually cause higher sperm ATP concentration. Also, it raises the possibility that pockets of extra cytoplasm demonstrated by CK immunocytochemistry, which apparently account for the increased CK concentrations in immature sperm, may be segregated or metabolically inactive (Vigue et al. 1992). An inverse correlation between per sperm CK activity and sperm concentration has been measured in normozoospermic and oligozoospermic specimens and in fertile and infertile oligospermic men, as well as consistently lower per sperm CK activity in swim-up fractions (Huszar et al. 1990). However, these higher sperm CK concentrations in oligozoospermic patients is mostly related to an incomplete loss of cytoplasm before the release of sperm from the Sertoli cell during spermatogenesis; thus the CK activity differences are markers of incomplete sperm maturity rather than a cause of diminished fertilizing potential due to metabolic factors (Huszar et al. 1990). This index has been refined with the measurement of CK isoenzyme ratios rather than the per sperm CK activity (Huszar et al. 1992).

In conclusion, sperm ATP concentrations or ATP/ADP ratios do not seem to be predictive of male fertility, and higher sperm ATP concentrations do not reflect an increase in the fertilizing potential of men (Gottlieb et al. 1991; Vigue et al. 1992). However, samples with ATP concentrations of less than 40 pmol/10 million sperm had limited success in in vitro fertilization (Morshedi 1990). On bovine sperm, the apparent Km for force generation by dynein was 24 µM ATP (Halangk et al. 1985). Assuming that ATP is evenly distributed, the typical level of 100 pmol ATP/10 million sperm would yield an intra-sperm ATP concentration of 30 µM, which is well onto the physiological plateau.

11 Mitochondrial Inheritance

The possibility of mtDNA mutations in the spermatozoa and the availability of new techniques of medically assisted procreation such as ICSI have raised the important question of mtDNA inheritance in humans (Ankel-Simon and Cummins 1996). If spermatozoa accumulate mtDNA mutations and if there is a leak of paternal mtDNA in the zygote, this could have the dramatic effect of the embryo harboring mutated mtDNA at the first stage of its development. Ankel-Simons and Cummins (1996) have recently claimed that strictly maternal inheritance of mtDNA was a misconception and that paternal mtDNA inheritance could occur at a low level. This assumption was supported by recent findings of biparental transmission of mtDNA in different species (Kondo et al. 1990; Gyllensten et al.1991; Zouros et al. 1994). The new evidence suggested that biparental transmission does occur in species in which intact sperm enter the egg cytoplasm during fertilization (Ankel-Simons and Cummins 1996). In almost all mammal species the entire sperm, including the midpiece mitochondrial sheath, enters the egg at fertilization. Thus, tail and midpiece structures can be traced for several division cycles (Sutovsky et al. 1996). As already mentioned, the typical mammalian sperm midpiece contains approximately 72 mitochondria with one copy of mtDNA in each. In contrast, the mammalian oocyte contains from 10^5 to 10^8 mitochondria, and the human oocyte in particular is estimated to contain 10^5 copies of mtDNA (Tyler 1992). Thus the oocyte's mtDNA copy number exceeds that of the sperm by a factor of at least 1000. This dilution effect could account for the observation of strictly maternal inheritance of the mtDNA using low resolution techniques such as the Southern blot. For example, Gyllensten et al. (1991) have detected paternal mtDNA by PCR in interspecific mitochondrial congenic mice derived from backcrosses between *Mus musculus* and *Mus spretus*. Because the paternal contribution was only 0.01–0.1%, these authors suggested that earlier failures to detect paternal mtDNA were due to the low sensitivity of the applied assay.

However, there is also strong evidence for the presence of an active process to eliminate paternal mitochondria and mtDNA during the first stage of embryogenesis (Kaneda et al. 1995). To examine whether mtDNA is uni- or bi-parentally transmitted in mice, Kaneda et al. (1995) developed an assay that can detect sperm mtDNA in a single mouse embryo. In intraspecific hybrids of *Mus musculus*, paternal mtDNA was detected only through the early pronucleus stage, and its disappearance coincided with loss of membrane potential in sperm-derived mitochondria. By contrast, in interspecific hybrids between *Mus musculus* and *Mus spretus*, paternal mtDNA was detected throughout development from pronucleus stage to neonates. The authors propose that oocyte cytoplasm has a species-specific mechanism that recognizes and eliminates sperm mitochondria and mtDNA. This mechanism should recognize nuclearly encoded proteins in the sperm midpiece, and not the mtDNA or the protein it encodes, because sperm mitochondria from the congenic strain B6.mtspr, which carries *Mus spretus* mtDNA on a background of *Mus musculus* (B6) nuclear genes, were eliminated early by B6 oocytes as in intraspecific crosses (Kaneda et al. 1995). Interestingly, in lower plants with isogametes, such as *Chlamidomonas*, a gene involved in the active degradation of the chloroplast and the mitochondrial genomes from one parent has recently been cloned (Armbrust et al. 1993).

Using immature spermatozoa during fertilization, the risk of a leakage of paternal mtDNA in the zygote increases and several groups have studied the mitochondrial inheritance in children born after ICSI. However, the paternal mtDNA was never detected in the blood of the children (Houshmand et al. 1997). All these results shown that mtDNA is transmitted maternally in intraspecific crosses in mammals and that a leakage of parental mtDNA is limited to interspecific crosses. Thus, strictly maternal transmission could help to prevent the spread of deleterious mitochondrial genomes from the spermatozoa.

Allen (1996) has proposed the theory that separation of sexes could be linked to the separation of mitochondrial function and replication between male and female germ lines. In males, the gametes maximize energy production for motility by sacrificing mtDNA to electron transfer and its mutagenic byproducts, while in females, the gametes, which are non-motile, repress mitochondrial OXPHOS, protecting mtDNA for faithful transmission between generations. Thus, male gametes make no contribution to the mitochondrial genome of the zygote and mitochondria are maternally inherited. If Allen's hypothesis is correct, the primary difference between sexes may be defined as follows: "male" is that sex in which germ-line mitochondria perform OXPHOS; "female" is that sex in which germ-line mitochondria do not perform OXPHOS (Allen 1996).

12 Conclusion

Germ cell differentiation is an extremely complex process involving a huge number of genes and specific regulations. Spermatogenesis is also very susceptible to the environment. Accordingly, male infertility is a complex and very heterogenous trait. Hitherto, in humans, there has been no experimental evidence to accuse mitochondria and their mtDNA of being a major cause of male infertility. However, the presence of mutant mtDNA in spermatozoa should force us to be vigilant when using new techniques for medically assisted procreation.

Acknowledgments. I would like to thank Dr. Pierre Rustin and Agnes Rötig for their comments and suggestions.

References

Adham IM, Tessmann D, Soliman KA, Murphy D, Kremling H, Szpirer C, Engel W (1996) Cloning, expression and chromosomal localization of the rat mitochondrial capsule selenoprotein gene (MCS): the reading frame does not contain potential UGA selenocysteine codons. DNA Cell Biol 15:159–166

Aitken J, Fisher H (1994) Reactive oxygen species generation and human spermatozoa: the balance of benefit and risk. BioEssays 16: 259–267

Alcivar AA, Hake LE, Millette CF, Trasler JM, Hecht NB (1989) Mitochondrial gene expression in male germ cells of the mouse. Dev Biol 135:263–271

Alexandre C, Bisson JP, David G (1978) Asthenospermie totale avec anomalie ultrastructurale du flagelle chez deux frères stériles. J Gynaecol Obstet Biol Reprod (Paris) 7:31–38

Allen JF (1996) Separate sexes and the mitochondrial theory of aging. J Theor Biol 180:135–140

Ankel-Simons F, Cummins JM (1996) Misconceptions about mitochondria and mammalian fertilization: implications for theories on human evolution. Proc Natl Acad Sci USA 93: 13859–13863

Armbrust EV, Ferris PJ, Goodenough UW (1993) A mating type-linked gene cluster expressed in *Chlamydomonas* zygotes participates in the uniparental inheritance of the chloroplast genome. Cell 74:801–811

Bartoov B, Eltes F, Pansky M, Langzam J, Reichart M, Soffer Y (1994) Improved diagnosis of male fertility potential via a combination of quantitative ultramorphology and routine semen analyses. Hum Reprod 9:2069–2075

Bernadini G, Belgiojoso P, Canatini M (1988) *Xenopus* spermatozoon: is there any correlation between motility and oxygen consumption? Gamete Res 21:403–408

Bleau G, Lemarbre J, Faucher G, Roberts KD, Chapdelaine A (1984) Semen, selenium and human fertility. Fertil Steril 42:890–894

Bourgeron T, Chretien D, Rötig A, Munnich A, Rustin P (1993) Fate and expression of the deleted mitochondrial DNA differ between heteroplasmic skin fibroblast and Epstein-Barr virus-transformed lymphocyte cultures. J Biol Chem 268:19369–19376

Bourgeron T, Rustin P, Chretien D, Birch-Machin M, Bourgeois M, Viegas-Péquignot E, Munnich A, Rötig A (1995) A mutation in the flavoprotein subunit gene of the succinate dehydrogenase: the first nuclear gene mutation in mitochondrial respiratory chain deficiency. Nat Genet 11:144–148

Brokaw CJ (1967) Adenosine triphosphate usage by flagella. Science 156:76–78
Brokaw CJ, Benedict B (1968) Mechanochemical coupling in flagella. II. Effects of viscosity and thiourea on metabolism and motility of *Ciona* spermatozoa. J Gen Physiol 52:283–299
Calvin HI, Cooper GW, Wallace E (1981) Evidence that selenium in rat sperm is associated with a cystein-rich structural protein of the mitochondrial capsule. Gamete Res 4:139–149
Cardullo RA, Baltz JM (1991) Metabolic regulation in mammalian sperm: mitochondrial volume determines sperm length and flagellar beat frequency. Cell Motil Cytoskelet 19:180–188
Cardullo RA, Cone RA (1986) Mechanical immobilization of rat sperm does not change their oxygen consumption rate. Biol Reprod 34:820–830
Cataldo L, Baig K, Oko R, Mastrangelo M-A, Kleene KC (1996) Developmental expression, intracellular localization, and selenium content of the cysteine-rich protein associated with the mitochondrial capsules of mouse sperm. Mol Reprod Dev 45:320–331
Chretien D, Gallego J, Barrientos A, Casademont J, Cardellach F, Munnich A, Rötig A, Rustin P (1998) The biochemical parameters for the diagnosis of respiratory chain deficiency in man and their lack of age-related changes. Biochem J 329: 249–254
Comporti M (1989) Three models of free radical-induced cell injury. Chem Biol Interact 72:1–56
Cummins JM, Jequier AM, Kan R (1994) Molecular biology of human male infertility: links with aging, mitochondrial genetics, and oxidative stress? Mol Reprod Dev 37:345–362
de Lamirande E, Gagnon C (1992a) Reactive oxygen species and human spermatozoa. I. Effects on the motility of intact spermatozoa and on sperm axonemes. J Androl 13:368–378
de Lamirande E, Gagnon C (1992b) Reactive oxygen species and human spermatozoa. II. Depletion of adenosine triphosphate plays an important role in the inhibition of sperm motility. J Androl 13:379–386
De Martino C, Floridi A, Marcante ML, Malorni W, Scorza Barcellona P, Bellocci M, Silvestrini B (1979) Morphological, histochemical and biochemical studies on germ cell mitochondria of normal rats. Cell Tissue Res 196:1–22
Folgero T, Bertheussen K, Lindal S, Torbergsen T, Oian P (1993) Mitochondrial disease and reduced sperm motility. Hum Reprod 8:1863–1868
Ford WCL, Harrison A (1981) The role of oxidative phosphorylation in the generation of ATP in human spermatozoa. J Reprod Fertil 63:271–278
Fraga CG, Motchnik PA, Shigenaga MK, Helbock HJ, Jacob RA, Ames BN (1991) Ascorbic acid protects against endogenous oxidative DNA damage in human sperm. Proc Natl Acad Sci USA 88:11003–11006
Frank SA, Hurst LD (1996) Mitochondrial and male disease. Nature 383:224
Gallina FG, Gerez de Burgos NM, Burgos C, Coronel CE, Blanco A (1994) The lactate/pyruvate shuttle in Spermatozoa: operation in vitro. Arch Biochem Biophys 308:515–519
Gopalkrishnan K, Padwal V, D'Souza S, Shah R (1995) Severe asthenozoospermia: a structural and functional study. Int J Androl 18:67–74
Gottlieb C, Svanborg K, Bygdeman M (1991) Adenosine triphosphate (ATP) in human spermatozoa. Andrologia 23:421–425
Gyllensten U, Wharton D, Josefsson A, Wilson AC (1991) Paternal inheritance of mitochondrial DNA in mice. Nature 352:255–257
Halangk W, Bohnensack R, Kunz W (1985) Interdependence of mitochondrial ATP production and extramitochondrial ATP utilization in intact spermatozoa. Biochim Biophys Acta 808: 316–22
Halangk W, Dietz H, Bohnensack R, Kunz W (1987) Regulation of oxidative phosphorylation in mitochondria of epididymal bull spermatozoa. Biochim Biophys Acta 893:100–108
Hammerstedt RH, Lardy HA (1983) The effect of substrate cycling on the ATP yield of sperm glycolysis. J Biol Chem 258:8759 8768
Hansen JC, Deguchi Y (1996) Selenium and fertility in animals and man – a review. Acta Vet Scand 37:19–30
Hanson MR (1991) Plant mitochondrial mutations and male sterility. Annu Rev Genet 25: 461–486

Hayashi JI, Ohta S, Kagawa Y, Kondo H, Kaneda H, Yonekawa H, Takai D, Miyabayashi S (1994) Nuclear but not mitochondrial genome involvement in human age-related mitochondrial dysfunction. J Biol Chem 269:6878–6883
Hecht NB, Liem H (1984) Mitochondrial DNA is synthesized during meiosis and spermiogenesis in the mouse. Exp Cell Res 154:293–298
Houshmand M, Holme E, Hanson C, Wennerholm UB, Hamberger L (1997) Paternal mitochondrial DNA transferred to the offspring following intracytoplasmic sperm injection. J Assist Reprod Genet 14:223–227
Hsu H-S, Wei Y-H, Li AF, Chen M-T, Chang LS (1986) Defective mitochondrial oxidative phosphorylation in varicocele-bearing testicles. Urology 46:545–549
Huang CC, Chen RS, Chen CM, Wang HS, Lee CC, Pang CY, Hsu HS, Lee HC, Wei YH (1994) MELAS syndrome with mitochondrial tRNA Leu (UUR) gene mutation in a chinese family. J Neurol Neurosurg Psychiatry 57:586–589
Hurt GS, Turner TT, Howard SS (1986) Repair of experimental varicocele in the rat: long-term effects on testicular blood flow and temperature and caudal epididymal sperm concentration and motility. J Androl 7:271–276
Huszar G, Vigue L, Corrales M (1990) Sperm creatine kinase activity in fertile and infertile oligospermic men. J Androl 11: 40–46
Huszar G, Vigue L, Morshedi M (1992) Sperm creatine phosphokinase M-isoform ratios and fertilizing potential of men: a blinded study of 84 couples treated with in vitro fertilization. Fertil Steril 57:882–888
Jacobus WE, Lehninger AL (1973) Creatine kinase of rat heart mitochondria. J Biol Chem 248: 4803–4810
Jones R, Mann T, Sherins R (1979) Peroxidative breakdown of phospholipids in human spermatozoa, spermicidal properties of fatty acid peroxides, and protective action of seminal plasma. Fertil Steril 31:531–537
Kaldis P, Stolz M, Wyss M, Zanolla E, Rothen-Rutishauser B, Vorherr T, Wallimann T (1996) Identification of two distinctly localized mitochondrial creatine kinase isoenzymes in spermatozoa. J Cell Sci 109: 2079–2088
Kamp G, Büsselmann G, Lauterwein J (1996) Spermatozoa: models for studying regulatory aspects of energy metabolism. Experientia 52:487–494
Kaneda H, Hayashi JI, Takahama S, Taya C, Lindahl KF, Yonekawa H (1995) Elimination of paternal mitochondrial DNA in intraspecific crosses during early mouse embryogenesis. Proc Natl Acad SCi USA 92:4542–4546
Kao S-H, Chao H-T, Wei Y-H (1995) Mitochondrial deoxyribonucleic acid 4977-bp deletion is associated with diminished fertility and motility of human sperm. Biol Reprod 52: 729–736
Killian GJ, Gelerinter E, Chapman DA (1985) Alteration of oxygen uptake and the redox state of ubiquinone in rabbit sperm exposed to a variety of physiologic treatments. Biol Reprod 33: 859–869
Kleene KC (1989) Poly (A) shortening accompanies the activation of translation of five mRNAs during spermatogenesis in the mouse. Development 106:367–373
Kondo R, Satta Y, Matsuura ET, Ishiwa H, Takahata N, Chigusa SI (1990) Incomplete maternal transmission of mitochondrial DNA in *Drosophila*. Genetics 126:657–63
Larsson N-G, Garman JD, Oldorfs A, Barsh GS, Clayton DA (1996) A single mouse gene encodes the mitochondrial transcription factor A and a testis-specific nuclear HMG-box protein. Nat Genet 13:296–302
Larsson N-G, Oldorfs A, Garman JD, Barsh GS, Clayton DA (1997) Down-regulation of mitochondrial transcription factor A during spermatogenesis in humans. Hum Mol Genet 6:185–191
Luft R, Ikkos D, Palmieri G (1962) Severe hypermetabolism of non-thyroid origin with a defect in the maintenance of mitochondrial respiratory control: a correlated clinical, biochemical and morphological study. J Clin Invest 41:1776–1804
McClure RD, Brawer J, Robaire B (1983) Ultrastructure of immotile spermatozoa in an infertile male: a spectrum of structural defects. Fertil Steril 40: 395–399

McLeod J (1941) The metabolism of human spermatozoa. Am J Physiol 132: 193–201
Mita M, Yasumasu I (1983) Metabolism of lipid and carbohydrate in sea urchin spermatozoa. Gamete Res 7: 133–144
Morshedi M (1990) Sperm ATP concentrations: the Norfolk experience. In: Acosta A, Swanson RJ, Ackerman SB et al. (eds) Human spermatozoa in assisted reproduction. Williams and Wilkins, Baltimore, pp 200–206
Mundy AJ, Ryder TA, Edmonds DK (1995) Asthenozoospermia and the human sperm midpiece. Hum Reprod 10: 116–119
Munnich A, Rötig A, Chretien D, Cormier V, Bourgeron T, Bonnefont JP, Saudubray JM, Rustin P (1996) Clinical presentation of mitochondrial disoders in chidhood. J Inherited Metab Dis 19: 521–527
Olds-Clarke P (1996) How does poor motility alter sperm fertilizing ability? J Androl 17: 183–186
Olson GE, Winfrey V (1990) Mitochondria-cytoskeleton interactions in the sperm midpiece. J Struct Biol 103: 13–22
Olson GE, Winfrey V (1992) Structural organization of surface domains of sperm mitochondria. Mol Reprod Dev 33: 89–98
Otani H, Tanaka O, Kasai K-I, Yoshioka T (1988) Development of mitochondrial helical sheath in the middle piece of the mouse spermatid tail: Regular dispositions and synchronized changes. Anat Rec 222: 26–33
Palermo G, Joris H, Devroey P, Van Steirteghem AC (1992) Pregnancies after intracytoplasmic sperm injection of a single spermatozoon into an oocyte. Lancet 340: 17
Pallini V, Baccetti B, Burrini AG (1979) A peculiar cysteine-rich polypeptide related to some unusual properties of mammalian sperm mitochondria. In: Fawcett DW, Bedfords JM (eds) The spermatozoon. Urban and Schwartzenberg, Baltimore
Pedersen H, Rebbe H, Hammen R (1971) Human sperm fine structure in a case of severe asthenospermia-necrospermia. Fertil Steril 22: 156–164
Peterson RN, Freund M (1970) ATP synthesis and oxidative metabolism in human spermatozoa. Biol Reprod 3: 47–54
Phillips DM (1977) Mitochondrial disposition in mammalian spermatozoa. J Ultrastruct Res 2: 144–154
Reynier P, Chrétien M-F, Penisson-Besnier I, Malthiéry Y, Rohmer V, Lestienne P (1997) Male infertility associated with multiple mitochondrial DNA rearrangements. C R Acad Sci 320: 629–636
Ross A, Christie S, Edmond P (1973) Ultrastructural tail defects in the spermatozoa from two men attending a subfertility clinic. J Reprod Fertil 32: 243–251
Rötig A, Cormier V, Blanche S, Bonnefont JP, Ledeist F, Romero N, Schmitz J, Rustin P, Fischer A, Saudubray JM, Munnich A (1990) Pearson's marrow-pancreas syndrome. A multisystem mitochondrial disorder in infancy. J Clin Invest 86: 1601–1608
Roveri A, Cassaco A, Maiorino M, Dalan P, Calligaro A, Ursini F (1992) Phospholipid hydroperoxide glutathione peroxidase of rat testis: gonadotropin dependence and immunocytochemical identification. J Biol Chem 267: 6142–6146
Rustin P, Bourgeron T, Parfait B, Chretien D, Munnich A, Rotig A (1997) Inborn errors of the Krebs cycle: a group of unusual mitochondrial diseases in humans. Biochim Biophys Acta 1361: 185–197
Saypol DC, Howards SS, Turner TT, Miller ED Jr (1981) Influence of surgically induced varicocele on testicular blood flow, temperature, and histology in adult rats and dogs. J Clin Invest 68: 39–45
Schoff PK, Cheetham J, Lardy HA (1989) Adenylate kinase activity in ejaculated bovine sperm flagella. J Biol Chem 264: 6086–6091
Selley ML, Lacey MJ, Bartlett MR, Copeland CM, Ardlie NG (1991) Content of significant amounts of a cytotoxic end-product of lipid peroxidation in human semen. J Reprod Fertil 92: 291–298
Shadel GS, Clayton DA (1997) Mitochondrial DNA maintenance in vertebrates. Annu Rev Biochem 66: 409–435

Shih DM, Kleene KC (1992) A study by in situ hybridization of the stage of appearance and disappearance of the transition protein 2 and mitochondrial capsule seleno-protein mRNA during spermatogenesis in the mouse. Mol Reprod Dev 33:222–227

Shoubridge EA, Karpati G, Hastings KE (1990) Deletion mutants are functionally dominant over wild-type mitochondrial genomes in skeletal muscle fiber segments in mitochondrial disease. Cell 62:43–49

Steeghs K, Oerlemans F, Wieringa B (1995) Mice deficient in ubiquitous mitochondrial creatine kinase are viable and fertile. Biochim Biophys Acta 1230:130–138

St Johns JC, Cooke ID, Barratt CLR (1997) Mitochondrial mutations and male infertility. Nat Med 3:124–125

Suomalainen A, Kaukonen J Amati P, Timonen R, Haltia M, Weissenbach J, Zeviani M, Somer H, Peltonen (1995) An autosomal locus predisposing to deletions of mitochondrial DNA. Nat Genet 9:146–151

Sutovsky P, Navara CS, Shatten G (1996) Fate of sperm mitochondria, and the incorporation, conversion, and disassembly of the sperm tail structures during bovine fertilization. Biol Reprod 55:1195–1205

Takahashi J, Higashi Y, La Nasa JA, Yoshida K-I, Winters SJ, Oshima H, Troen P (1983) Studies of the human testis. XVIII. Simultaneous measurement of nine intratesticular steroids: evidence for reduced mitochondrial function in testes of elderly men. J Endocrinol Metab 56: 1178–1187

Thyagarajan B, Padua RA, Campbell C (1996) Mammalian mitochondria pocess homologous DNA recombination activity. J Biol Chem 271:27536–27543

Tombes RM, Shapiro BM (1985) Metabolite channeling: a phosphorylcreatine shuttle to mediate high energy phosphate transport between sperm mitochondrion and tail. Cell 41:325–334

Tosic J, Walton A (1950) Metabolism of spermatozoa and its effect on motility and survival. Biochem J 47:199–212

Tyler DD (1992) The mitochondrion in health and disease 1st edn VCH Publishers, New York

Vigue C, Vigue L, Huszar G (1992) Adenosine triphosphate (ATP) concentrations and ATP/adenosine diphosphate ratios in human sperm of normospermic, oligospermic, and asthenospermic specimens and in their swim-up fractions: lack of correlation between ATP parameters and sperm creatine kinase concentrations. J Androl 13:305–311

Wallace DC (1986) Mitotic segregation of mitochondrial DNAs in human cell hybrids and expression of chloramphenicol resistance. Somatic Cell Mol Genet 12:41–49

Wallace DC (1992) Diseases of the mitochondrial DNA. Annu Rev Biochem 61:1175–1212

Wallace DC, Shoffner JM, Trounce I, Brown MD, Ballinger SW, Corral-Debrinski M, Horton T, Jun AS, Lott MT (1995) Mitochondrial DNA mutations in human degenerative diseases and aging. Biochim Biophys Acta 1271:141–151

Wallace E, Cooper GW, Calvin HI (1983) Effects of selenium deficiency on the shape and arrangement of rodent sperm mitochondria. Gamete Res 4:389–399

Wallimann T, Hemmer W (1994) Creatine kinase in non-muscle tissues and cells. Mol Cell Biochem 133/134:193–220

Wilton LJ, Temple-Smith PD, de Kretser DM (1992) Quantitative ultrastructural analysis of sperm tails reveals flagellar defects associated with persistent asthenozoospermia. Hum Reprod 7:510–516

Wu SH, Olfield JE, Shull LR, Cheeke PR (1979) Specific effect of selenium deficiency on rat sperm. Biol Reprod 20, 793–798

Zinni A, de Lamirande E, Gragnon C (1993) Reactive oxygen species in semen of infertile patients: levels of superoxide dismutase and catalase-like activities in seminal plasma and spermatozoa. Int J Androl 16:183–188

Zouros E. Oberhauser Ball A. Saavedra C. Freeman KR (1994) An unusual type of mitochondrial DNA inheritance in the blue mussel Mytilus. Proc Natl Acad Sci USA 91:7463–7467

The Human Y Chromosome and Male Infertility

Ken McElreavey[1], Csilla Krausz[2], and Colin E. Bishop[3]

1
Structure of the Human Y Chromosome

1.1
Pseudoautosomal Regions

Although most of the human Y does not normally recombine with the X chromosome, there are two limited regions of sequence identity with the X that permit pairing and recombination during male meiosis (see Rappold 1993). These are the pseudoautosomal regions located at the distal portions of the short and long arms of the Y chromosome. The Yp pseudoautosomal region consists of 2.6 Mb. Absence of this region is associated with short stature, and male infertility. During meiotic prophase, germ cells require the Y chromosome as a pairing partner for the X. In the absence of pairing, caused by deletions of the pseudoautosomal region, germ cells undergo meiotic arrest resulting in azoospermia. The Yq pseudoautosomal region is 0.4 Mb in size and contains the genes for the interleukin 9 receptor (*ILR9*) and synaptobevine (*SYBL1*). Both genes have X homologues. The gene for *ILR9* escapes X-inactivation and is expressed from the Y whereas *SYBL1* is subject to X-inactivation and is not expressed from the Y chromosome (Vermeesch et al. 1997). The boundary between the pseudoautosomal region and the non-recombining region is defined by an Alu element of 303 bp followed by 220 bp with 78% identity, after which the sequences diverge completely (Ellis et al. 1989). This Alu repeat element inserted at a pre-existing boundary sometime after the Old World monkey and great ape lineages diverged. During male meiosis there is an obligatory crossing-over event between Xp and Yp pseudoautosomal

[1] Immunogénétique Humaine, Institut Pasteur, 25 rue du Dr Roux, 75724 Paris Cedex 15, France
[2] Andrology Unit, Department of Clinical Physiopathology, University of Florence, Florence, Italy
[3] Baylor College of Medicine,
Department of Obstetrics & Gynecology and Molecular & Human Genetics,
6550 Fannin Street, Houston, Texas 77030, USA

Results and Problems in Cell Differentiation, Vol. 28
McElreavey (Ed.): The Genetic Basis of Male Infertility

regions which maintains X-Y identity. There is a gradient of recombination which decreases as one approaches the pseudoautosomal boundary (Rouyer et al. 1986).

Various other regions of sequence homology with the X chromosome are found along the non-recombining region of Yp and Yq but do not undergo pairing and recombination. For example, the region Yq11.21 shares similarities with Xp22.3, including the pseudogenes of *KALIG-1* and steroid sulfatase (*STSP*). The presence of this block of homology with Xp22.3 has been interpreted as a consequence of a large pseudoautosomal region that was disrupted by a pericentric inversion during early primate evolution. A second large block of homology is shared between a large portion of the non-recombining region of the Y short arm and the region Xq21 (Page et al. 1984). This block of homology arose by an X-Y transposition that occurred during human evolution after the divergence of humans from chimpanzee and gorilla lineages (Lambson et al. 1992).

1.2 Non-Recombining Region

It has been estimated that between 50–70% of the non-recombining region of the human Y chromosome is composed of a variety of highly repeated DNA elements, the majority of which appear to be unique to the human Y chromosome (Cooke 1976). Thus the alphoid sequences that are clustered near the centromere differ from other genomic alphoid sequences, both in sequence and repeat organization (Smith et al. 1987). Y chromosome *Alu*-elements also show unique sequence arrangements compared with non-Y sequences. Y chromosome short interspersed repetitive elements (SINE) and long interspersed repetitive elements (LINE) sequences are abundant throughout the chromosome and are similar to elements found elsewhere in the human genome (Smith et al. 1987). Other repetitive sequences are unique to the Y chromosome and are locus-specific. There are at least 40 copies of the Y-DNA sequence pY6H65 which extend more than 300 kb in distal Yq (Vogt et al. 1991). The Y chromosome also contains two major sets of tandem repeat sequences organized in large arrays, that are specific for the Y chromosome and are located in the long arm, mainly in the heterochromatin region (*DYZ1* and *DYZ2*; Cooke 1976). An average Y chromosome contains about 3000 copies of the *DYZ1* repeat, which can be further subdivided into different families (Nakagome et al. 1991). *DYZ1* consists of a 3.4 kb repeat unit which mainly consists of a tandem array of pentanucleotides (TTCCA; Nakahori et al. 1986). There are about 2000 copies of the *DYZ2* repeat family (Ludena et al. 1993). The human Y chromosome exhibits considerable variation in the size of the heterochromatin between different populations and it may be entirely lacking in some individuals without any phenotypic effect. The size variation correlates with the amount of *DYZ1* and *DYZ2* repeats (Schmid et al. 1990). This suggests

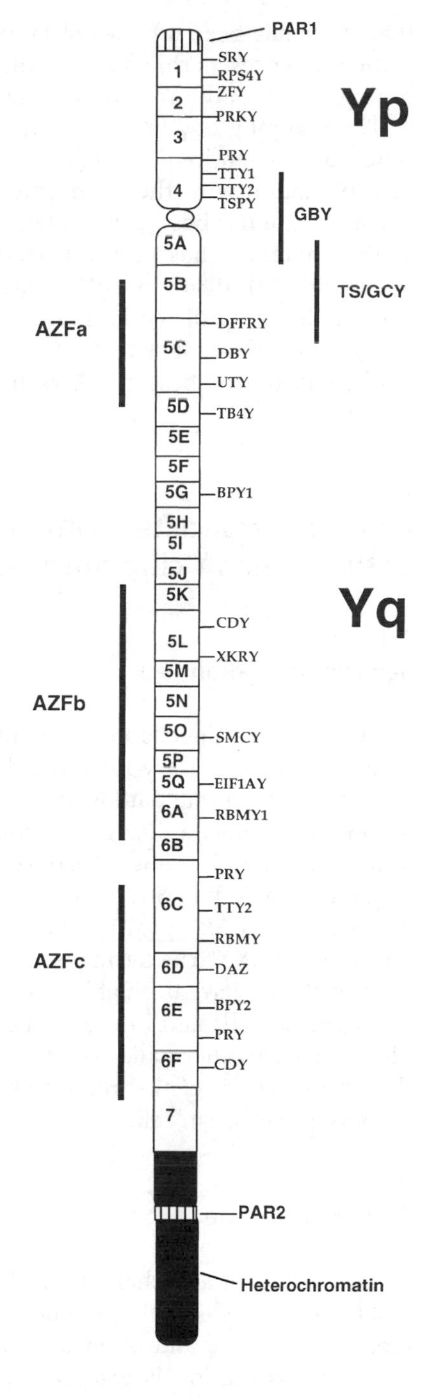

Fig. 1. Schematic representation of the human Y chromosome showing genes and gene families in the non-recombining region. Pseudoautosomal regions 1 and 2 are indicated. The deletion intervals are shown (Vergnaud et al. 1986), together with the positions of the three AZF regions associated with spermatogenic failure. The approximate positions of loci associated with gonadaoblastoma *GBY*, short stature *GCY* and the skeletal anomalies of the Turner syndrome are indicated by the *solid bar*

that there has been a series of amplification events of particular Y chromosome sequences, perhaps as a result of inherent instability of the non-recombining region of the Y chromosome.

The mapping of genes and phenotypes to the human Y chromosome has relied almost entirely on deletion maps because of the absence of genetic recombination with the X chromosome. The entire length of the human Y chromosome has been subdivided into seven deletion intervals (Fig. 1). Each of these intervals has been further subdivided into subintervals (A, B, C etc.; Vergnaud et al. 1986). In 1992, Vollrath and colleagues constructed a 43-interval deletion map of the human Y chromosome which contained an ordered array of sequence tagged sites (STS) that span the short and long arms of the Y chromosome. These markers have been widely used for microdeletion screens.

2 Functions associated with the Non-Recombining Region of the Human Y chromosome

2.1 Sex Determination

As described in the chapter by Smith and Sinclair, the Y-located gene that is essential and necessary for testis determination, *SRY* is located immediately proximal to the pseudoautosomal region in Yp. Mutations in this gene result in Sywers syndrome, a 46,XY individual with a completely female phenotype, although puberty is absent. Conversely, the accidental transfer of Y-specific sequences including *SRY*, onto the X chromosome during male meiosis can result in 46,XX *SRY*-positive individuals that may have a male phenotype or develop as 46,XX true hermaphrodites. *SRY* is expressed in both somatic and germ cells. We have been able to observe the presence of SRY protein in both adult Sertoli cells and in round spermatids (unpubl. observations). Although SRY may have other male-specific roles, these would be difficult to determine due to the nature of the mutant phenotype, which is completely female with the absence of germ cells.

2.2 Turner Syndrome

Turner's syndrome is characterized by gonadal dysgenesis, short stature, skeletal anomalies (shield thorax, cubitis valgus and short metacarpals) lymphodema, low hairline and other anomalies. Embryonic lethality is greater than 90%. This condition is usually associated with a 45,X karyotype. The syndrome is probably the consequence of haploinsufficiency of genes that are common to both the X and Y chromosome and escape X-inactivation. The genes have so far eluded identification, but there appear to be distinct loci on proximal Yq associated with short stature (growth control Y, GCY) and skeletal anomalies (TCY; Barbaux et al. 1995), whereas deletions of Yp are associated with short stature (short stature homeobox-containing gene (SHOX) gene in pseudoautosomal region; Rao et al. 1997) and lymphodema (deletions of the non-recombining region of Yp).

2.3 Histocompatibility Y antigen (H-Y)

Histocompatibility Y antigen is a minor histocompatibility antigen that can lead to rejection of male organ and bone marrow grafts by female recipients. It was orginally identified in female inbred mice which rejected isogenic male skin grafts (Eichwald and Silmser 1955). The H-Y antigen can be detected by cytotoxic T-lymphocytes and for years it eluded biochemical characterization. However, it was clear that there were several H-Y antigens because the H-Y phenotype depended on the method used for its detection. One antigen detected by the transplantation assay was termed H-Yt, another identified by cytotoxic T lymphocytes H-Yc and the serological one H-Ys. At one point H-Y antigens were considered to play a role in sex determination because they could also be detected in the heterogametic sex in nonmammalian vertebrates. One human H-Y antigen is presented by HLA-B7 and corresponds to an 11-residue peptide derived from the gene Selected mouse cDNA on the Y (*SMCY*). A second H-Y epitope is encoded by the gene *UTY* (ubiquitously transcribed tetraticopeptide repeat gene on the Y chromosome).

2.4 Gonadoblastoma

Gonadoblastoma is a rare neoplasm composed of germ cells intermixed with cells resembling immature Sertoli and granulosa cells. These tumors arise within dysgenetic gonads of individuals who possess Y-chromosomal material. Using sequence-tagged-sites, Tsuchiya et al. (1995) have narrowed the region associated with this tumor to a 1–2 Mb fragment immediately surrounding the centromere, in the region 4A–4B. It is possible that this gene may have a function in normal males such as spermatogenesis. The *TSPY* (testis-specific protein, Y-encoded) gene family consists of at least 20–40 copies located within the *DYZ5* tandem repeat array, immediately adjacent to the centromere on Yp (Arnemann et al. 1991; Manz et al. 1993). This localisation immediately makes *TSPY* a candidate gonadoblastoma gene. Expression of *TSPY* early in spermatogenesis immediately prior to the spermatogonia-to-spermatocyte transition is also consistent with this hypothesis (Schnieders et al. 1996).

2.5 Male Infertility

Although it has been established since the 1970s that deletions of the long arm of the Y chromosome are associated with spermatogenic failure, it is only in the last few years that these regions have been defined at a molecular level. In parallel, Y-linked genes and gene families have been identified that are candidates for deletion phenotypes. The first association between spermatogenic failure and an underlying genetic cause was demonstrated by Tiepolo and Zuffardi (1976) in a report of six azoospermic patients carrying microscopically detectable deletions of the distal portion of Yq in a screen of 1170 cases. In four cases the deletion was de novo (their fathers were tested and found to carry intact Y chromosome). On the basis of this finding they proposed the existence of a spermatogenesis factor, the "azoospermia factor" (AZF) encoded by a gene on distal Yq. In 1992 Vogt et al. described microdeletions in distal Yq in two of 19 infertile males with non-obstructive severe oligo- or azoospermia. Although both microdeletions were on the long arm of the Y, they were apparently not overlapping. In fact as it later turned out these two individuals "JOLAR" and "KLARD" had microdeletions of what would be later termed the AZFa and AZFc regions respectively. Further studies by Ma et al. 1993 revealed four men with non-overlapping deletions in the interval six region. These results suggested the presence of multiple loci on Yq associated with infertility. This was clearly established in 1996 by Vogt and colleagues who screened 370 men with idiopathic azoospermia or oligozoospermia for deletions of 76 DNA loci in Yq11 (Vogt et al. 1996). Twelve individuals harbored de novo microdeletions, whilst another patient had an inherited deletion. These deletions mapped to different subregions in Yq11. The three non-overlapping regions of the Y chromosome required for spermatogenesis were termed AZFa (or JOLAR), AZFb and AZFc (or ZLARD). The AZFa region is located in proximal Yq, AZFb in a region approximately in the center of the long arm and AZFc is immediately proximal to the heterochromatic region. The physical size of these regions has been difficult to establish because of blocks of highly repetitive sequences that have consistently led to underestimates of their size. However, AZFa and AZFb are estimated to be between 1–3 Mb and AZFc approximately 1.4 Mb.

In addition to the fact that there were several distinct AZF genes, it also became clear that microdeletions are not exclusively associated with azoospermia (Reijo et al. 1996a). As discussed below, subsequent studies have demonstrated deletions associated with a wide range of testis histological profiles ranging from Sertoli cell only syndrome (SCOS) to spermatogenic arrest and severe hypospermatogenesis.

3
Yq-Specific Genes and Gene Families

Several genes and gene families have been identified on the long arm of the Y chromosome (Lahn and Page 1997). Some of these genes fall within AZF deletion intervals and may therefore underlie the observed deletion phenotypes (Tables 1 & 2). These genes can be divided into those that may be involved in cellular "housekeeping" activities, and those that are expressed solely in the testis. The former group (See Table 1) includes Drosophila fats facets related Y(*DFFRY*), dead box Y (*DBY*), ubiquitous TPR motif Y (*UTY*), eukaryotic translation initiation factor 1A Y isoform (*eIF-1AY*), selected mouse cDNA on the Y (*SMCY*) and thymosin B4 Y isoform (*TB4Y*). These ubiquitously expressed genes each exist in a single copy on the Y chromosome, and each possesses a closely related X-linked homolog that escapes X inactivation. The testis-specific group (see Table 2) includes the RNA-binding motif Y chromosome (*RBMY*) gene and its relatives, deleted in azoospermia (*DAZ*), chromodomain Y (*CDY*), XK related Y (*XKRY*), PTP-BL related Y (*PRY*) and basic proteins Y1 and Y2 (*BPY1* and *BPY2*). These genes are present as in multiple copies on the Y and do not appear to have X homologs.

Table 1. Ubiquitously expressed "housekeeping" genes that map to the AZF deleted regions that have been implicated in male infertility. These genes are present in a single copy on the Y, but they all have X homologs that escape X inactivation. The degree of sequence identity between X- and Y-homologues is in all cases at least 84%

Gene symbol	Gene name	Comments	X homolog	Amino acid Identity
DFFRY	*Drosophila fats facets related Y*	Homologous to *Drosophila* deubiquinating enzyme (Brown et al. 1998)	*DFFRX*	91%
DBY	*Dead box Y*	Contains a DEAD box motif. May function as an RNA helicase (Linder et al. 1989)	*DBX*	91%
TB4Y	*Thymosin B4 Y*	May be involved in actin sequestration (Gondo et al. 1987)	*TB4X*	93%
UTY	*Ubiquitous TPR motif Y*	Contains 10 tandem TPR motifs that may be involved in protein-protein interactions (Greenfield et al. 1986)	*UTX*	85%
SMCY	*Selected mouse cDNA on the Y*	Encodes an H-Y antigen epitope (Agulnik et al. 1994a, b)	*SMCX*	84%
eIF-1AY	*Eukaryotic translation initiation factor 1A*	Eukaryotic translation intiation factor (Pestova et al. 1998)	*eIF1-1AX*	98%

Table 2. Genes and families with expression restricted to the testis that map to the AZF deleted regions of the Y chromosome

Gene symbol	Gene name	Comments	X or autosomal homolog
RBMY	*RNA-binding motif Y*	Several subfamilies including *RBMY1* and *RBMY2*. RBMY1 may be functional. Predicted to have RNA-binding activity	*RBMY* may be an ancestral *hnRNPG* gene
DAZ	*Deleted in azoospermia*	Predicted to have RNA-binding activity. *Xenopus Dazl* has RNA-binding activity in vitro	*DAZL1* chromosome 3p25
XKRY	*XK related Y*	Shows similarity to XK, a putative membrane transport protein (Ho et al. 1994)	Unknown
CDY	*Chromodomain Y*	Contains chromodomain (James and Elgin 1986). May be involved in chromatid modification during spermatogenesis?	Unknown
PRY	*PTP-BL related Y*	Shows similarity to PTP-BL a putative membrane transport protein (Hendriks et al.1995)	Unknown
BPY1	*Basic protein Y1*	Basic protein of unknown function	Unknown
BPY2	*Basic protein Y2*	Basic protein of unknown function	Unknown

DFFRY, *DBY* and *UTY* all fall within the AZFa deletion interval, so one or more of these genes may be implicated in SCOS or other male fertility disorders. AZFb includes copies of *CDY* and *XKRY* as well as *SMCY*, *eIF-1AY* and *RBMY*. Although sequences related to *RBMY* are found throughout the Y chromosome, functional copies appear to be restricted to AZFb, since deletions of distal AZFb (region sY142-sY145) lead to the absence of RBMY epitopes in testicular sections (Elliott et al. 1997). A number of transcripts are found within AZFc. At least six copies of *DAZ* are found in this region (Saxena et al. 1996; Yen et al. 1997), as are multiple copies of *PRY*, *BPY2*, *CDY* and *XKRY*. Any of these genes may contribute to the AZFc deletion phenotype, and most AZFc deletions probably remove all of these genes.

4
Function of Y-Specific Genes in spermatogenesis

4.1
RBMY

At the moment, we know very little about the biochemistry or biology of Y-encoded proteins. Only *RBMY* and *DAZ* have been extensively studied. *RBMY* was originally termed the *YRRM* gene family (Ma et al. 1993). More than 30 *RBMY* genes and pseudogenes occur over both arms of the Y chromosome (Ma et al. 1993; Prosser et al. 1996; Chai et al. 1997), and these sequences can be divided into several subfamilies (*RBMY1* to *RBMY6*). The *RBMY1* subfamily has at least seven members, which dffer from each other by one to seven bases (Prosser et al. 1996; Chai et al. 1997, 1998). Most if not all *RBMY1* gene family members are arrayed in tandem in the AZFb region, in proximal deletion interval 6 (Chai et al. 1998). These genes encode germ cell specific nuclear proteins that contain an RNA-binding motif (RBM) as well as four copies of an internal tandem repeat structure of a 37-aa peptide termed the SRGY box, the function of which is unknown. *RBMY2* genes share 88% homology with *RBMY1* and carry an RBM and a single SRGY repeat. *RBMY1* sequences are 67% similar to the autosomally expressed hnRNPG protein, a nuclear glycoprotein with RNA-binding activities, but with no known biological function (Soulard et al. 1993). *RBMY1* genes may derive from an *hnRNPG* gene that translocated to the Y chromosome and was subsequently amplified (Delbridge et al. 1997). In man, RBMY1 can be detected by immunostaining in pachytene spermatocytes, an interesting observation in the light of the spermatogenic arrest often see in association with AZFb deletions (Elliott et al. 1997, 1998). In spermatocytes, RBMY1 co-localizes with pre-mRNA splicing components in a discrete area of the nucleus but, by late meiosis, it is found diffusely throughout the nucleoplasm of round spermatids. Hence, RBMY1 may play distinct roles during different phases of spermatogenesis. The *RBMY2* subfamily may be nonfunctional since there is a frameshift in exon 11 resulting in a modified C-terminal portion of the molecule compared with RBMY1 and hnRNPG proteins (Chai et al. 1998). There is also very low expression of *RBMY2* in adult testis (Chai et al. 1997). Although seven *RBMY1* genes are present in interval 6, within the AZFb region, it has been estimated that there are at least 20 copies of *RBMY1* and five copies of *RBMY2*. Some of these copies may lie within other regions of the Y that are recurrently deleted in infertile males (Pasantes et al. 1997).

4.2
DAZ

Like *RBMY*, *DAZ* encodes a testis-specific protein that has a single RBM and a series of between 8 and 24 copies of a 24 amino acid unit termed the "DAZ repeat" (Reijo et al. 1995; Yen et al. 1997). Although *DAZ* was originally considered to be a single copy gene on the basis of a Southern blot of *Eco*RI-digested genomic DNA (Reijo et al. 1995), subsequent analysis by high resolution FISH and sequencing of genomic clones indicated that there are probably three to six copies of the gene all located within the AZFc region. *DAZ* is homologous to an autosomal gene with a single DAZ repeat named *DAZL1* (DAZ like-autosomal 1; Saxena et al. 1996; Yen et al. 1996), and the Y-linked *DAZ* probably originated from the translocation and amplification of this ancestral autosomal gene. Mice lack the Y-located *DAZ* gene, but they do carry a single autosomal *Dazl1* gene (Reijo et al. 1996b; Cooke et al. 1996). Immunostaining has revealed human DAZ in the innermost layer of male germ cell epithelium and in the tails of spermatozoa (Habermann et al. 1998). This observation is consistent with the expression of *DAZ* transcripts just inside the perimeter of seminiferous tubules in spermatogonia (Menke et al. 1997). However, some caution must be exercised in interpreting these results, as cross-hybridization with DAZL1 mRNA or protein cannot be excluded.

The biological function of the DAZ motif is unknown, however, since it is conserved as a single copy in the autosomal *DAZL1* genes it probably has a functional significance. It may play a role in RNA-binding or it may be involved in interaction with other proteins. The human *DAZ* genes differ substantially in the sequence and organization of these repeats (Yen et al. 1997). Five *DAZ* cDNA clones were isolated from a testis cDNA library made with RNA pooled from four individuals who had normal spermatogenesis. Each cDNA clone differed not only in the number of copies of the DAZ repeat but also in the order of the repeat units. These DAZ isoforms could potentially have different properties (for example their affinity for RNA). There is evidence to support a variation of *DAZ* gene copy number between normal men. Agulnik et al. 1998 described an *Mbo*I polymorphism in the 5' region of the *DAZ* gene. There were population differences for the presence or absence of this polymorphism. In a small percentage of males the *DAZ* probe (p49f) detects one *Sfi*I fragment instead of two, again suggesting variation in *DAZ* copy number (Jobling 1996). Since *DAZ* gene copy number and DAZ repeat vary from one individual to another, could this have an effect on sperm counts.

Insights into human *DAZ* function may come from the analysis of its autosomal homologues in other species. Targeted disruption of *Dazl1* in mice leads to a complete absence of gamete production in both testis and ovary, demonstrating that Dazl1 is essential for development or survival of germ cells (Ruggiu et al. 1997). In the mouse ovary Dazl1 is present in maturing follicles. In embryonic and prepubertal ovaries, Dazl1 is cytoplasmic in the oocyte

(Ruggiu et al. 1997). In *Drosophila*, mutation of the *boule* gene, another homologue of *DAZL*, results in spermatocyte arrest at the G2/M transition and complete azoospermia (Castrillon et al. 1993; Eberhart et al. 1996). In *Xenopus*, *Xdazl* is expressed in premeiotic germ cells in adult testis (Houston et al. 1998). Interestingly, the *Xenopus Xdazl* gene can rescue meiotic entry of spermatocytes in *Drosophila boule* mutants, suggesting that there may be functional conservation of the *DAZ* family over evolutionary time (Houston et al. 1998). Xdazl protein has RNA-binding properties in vitro, and exhibits a specificity for G- or U-rich RNA sequences (Houston et al. 1998). Perhaps other members of the *DAZ* family play a role in RNA metabolism during gamete development. In *Drosophila* the Boule protein occurs in the nucleus of primary spermatocytes until the end of the meiotic prophase, after which it is found in the cytoplasm (Cheng et al. 1998). The nuclear localisation of Boule is dependent on the presence of a single fertility factor on the short arm of the Y chromosome. The *Drosophila* Y chromosome encodes six fertility loci, each spanning four or more megabases of DNA. These repetitive sequences are transcribed stage-specifically in the nucleus of the primary spermatocyte as continuous long transcription units. During their transcription, the RNA accumulates specific nuclear proteins, which can be observed under light microscopy as distinct intranuclear structures (termed lampbrush structures or Y loops). Absence of the *ks-1* fertility locus results in the cytoplasm localisation of Boule in primary spermatocyes and does not seem to interfere with spermatid differentiation indicating that that the role of Boule in meiotic entry is independent of its nuclear localisation. As suggested by Wasserman and collegues, the Y chromosome loops may serve as a storage for Boule and other RNA-binding proteins (Cheng et al. 1998). The breakdown of the lampbrush Y chromosome loops which occurs at the end of the spermatocyte growth phase may result in the release of these factors in a synchronous fashion (Cheng et al. 1998).

Other genes on the long arm of the Y may also be involved in RNA metabolism. DBY is predicted to have RNA helicase activity (Lahn and Page 1997) and *eIF-1AY* encodes an essential translation initiation factor (Pestova et al. 1998). During the latter stages of spermatogenesis, transcription terminates and post-transcriptional regulation plays a primary role (see chapter by Braun). RNA synthesis peaks during the spermatocyte stage and is gradually reduced in subsequent stages and ceases as round spermatids differentiate into elongated spermatids. Numerous mRNA have been identified that are under post-translational control during spermatogenesis. It is tempting to speculate that many of the factors encoded by Y-linked genes play key roles in this process.

5 Which Genes Underlie the AZF Phenotypes?

No direct genetic evidence has confirmed that any of the published Y chromosome genes are responsible for infertility. By direct genetic evidence is meant point mutations or microdeletions of a part of any of these genes. A mutation screen of the, presumably functionally important, RBM domain in a group of 60 infertile males failed to find evidence of mutations associated with the infertile phenotype (Prosser et al. 1996). Although large deletions of Yq11 that include the *DAZ* gene are responsible for azoo- or oligospermia, no point mutations nor specific rearrangements of the *DAZ* gene family have been described in infertile males (Vereb et al. 1997). This suggests that we may be dealing with a deletion phenotype, mutations involving individual genes may be rare and even then difficult to detect in a multicopy gene family.

6 Frequency of Yq Microdeletions

In recent years several combined clinical and molecular studies have sought to (1) define recurrently deleted regions of Yq, (2) determine the incidence of microdeletions among azoo- and oligozoospermic men, and (3) try to correlate the size and position of the deletions with the infertile phenotype. The incidence of microdeletions in infertile men varies considerably between studies, from 1% to 55% (Reijo et al. 1995; Quereshi et al. 1996; Stuppia et al. 1996; 1997; Foresta et al. 1997; Pryor et al. 1997; Simoni et al. 1997; Van der Ven et al. 1998). The major factor influencing this frequency appears to be the study design. Study populations have included either azoospermic patients, azoospermic and oligozoospermic patients or azoo/oligozoospermic and infertile normospermic patients.

The majority of clinical studies "select" patients with idiopathic azoo- or oligozoospermia whilst others include "unselected" infertile men with known or unknown causes of infertility. However, one of the problems encountered is the difficulty in agreement over what constitutes "idiopathic infertility". Varicocele and history of cryptorchidism are considered as idiopathic by some studies and as non-idiopathic by others. The variation in deletion frequency is also caused by important differences in the number of patients because studies with low patient numbers report a higher deletion frequency. This could be due to a more stringent selection of the study population by the smaller studies. Another variable which may also affect Yq deletion frequency is marker density or the position of markers. However, detailed analysis by Gromel and colleagues suggest that a higher number of markers does not result in an increased deletion frequency. Some deletions may be deletion polymorphisms that are present in normal fertile males in the population. For example, the marker RBM1/C is absent in about 50% of the European population. Pryor et al. (1997) described a screen of 200 infertile and 200 fertile men. Of the 200 infertile men, 7% were found to have Y chromosome microdeletions. However, microdeletions of the Y chromosome were found in 4 of the 200 fertile men. The deleted markers were sY207 and sY272. Our own studies suggest that the absence of these markers are polymorphisms in the population and have no causal relationship to infertility.

However, it is possible that differences in deletion frequency and/or localisation between studies may reflect genuine geographic or ethnic differences, perhaps related to a particular Y chromosome haplogroup, the genetic background or environmental influences (see sect. 9).

There may be an underestimation of Y chromosome anomalies associated with male infertility. The majority of studies have used PCR to detect Y deletions. As well as the obvious problem of sample contamination leading to false positive results, many of the genes that are candidates for infertility are present as multicopies on the Y and PCR amplification may not be able to distinguish between these copies. As a result, deletion of the functionally important gene copy cannot be detected. Simple deletion analysis by PCR would not detect other rearrangements of the Y chromosome such as inversions, duplications and so forth. These types of rearrangement are relatively common events on the Y chromosome and they have the potential to alter gene expression. The third problem associated with using the PCR technique to detect deletions is that mosaicisms will not be detected. There is indirect, but strong evidence that this is the case, since there have been published reports of babies born after ICSI treatment who carried Y deletions, whilst DNA analysis of their infertile father's DNA was apparently normal (Kent-First et al. 1996).

7 Microdeletions and Genotype/Phenotype Relationships

Vogt and colleagues attempted to correlate the deletion intervals AZFa, AZFb and AZFc with the phase in which spermatogenesis was blocked (Vogt et al. 1996). Testis histology suggested disruption of spermatogenesis at the same phase when the microdeletion occurred in the same Yq11 subregion, but at a different phase when the microdeletion occurred in a different Yq subregion. Each loci would therefore act at a different stage of spermatogenesis. The testis histology of one patient with an AZFa deletion indicated type I Sertoli cell only (SCO) syndrome. Only Sertoli cells, but no germ cell were visible in all tubules, after analysing more than 100 testis tubules from different testis sections. Individuals with type I SCO have small testis volumes (5–10 ml). The primary event of spermatogenic disruption in patients with type I SCO probably occurs premeiotically before or during the proliferation phase of spermatogonia. Testis histology was also examined in three patients with AZFb deletions. In all three cases testicular histology revealed populations of spermatogonia and primary spermatocytes but no postmeiotic germ cells could be detected in any of the tubules analysed. The conclusion from this observation was that microdeletions of the AZFb region disrupt spermatogenesis before or during meiosis resulting in spermatogenic arrest at the spermatocyte stage. Five patients with AZFc deletions presented with a different testis histology. Most tubules did not contain germ cells but some tubules contained germ cells at all developmental stages. Four of the five AZFc deleted patients produced some motile sperm (0.1–2 million per ml), although the morphology of the sperm was abnormal. This profile of spermatogenetic defects suggests that as one proceeds from the Y chromosome centromere along the length of the long arm, the asbence of proximal AZFa sequences is responsible for a prepuberal phase of spermatogenic arrest, the absence of AZFb sequences is responsible for a maturation arrest of germ cells at puberty before or during meiosis, and finally a variable phenotype is associated with AZFc deletions. Since some mature sperm may still be produced in AZFc deleted patients, the AZFc gene product may be involved in the maturation process of postmeiotic germ cells or sperm.

Germ cells lacking all *DAZ* copies due to an AZFc deletion have been used successfully for intracytoplasmic sperm injection indicating that spermatozoa from men carrying Y deletions are functionally competent (Mulhall et al. 1997). Vogt et al. described a case of an AZFc deletion in a man with a sperm count of 0.1 million per ml that was also present in his father. Stuppia et al. (1997) described a case where the father carried an AZFc deletion that was smaller that the deletion present in his son. The latter had a sperm count of 2 million sperm per ml. These observations raise several important points. Transmission of AZFc deletions from father to son indicates that the *DAZ* gene family or any other genes within the AZFc deletion interval are not absolutely required for complete sperm cell production, fertilization and embryonic development.

A major problem in defining genotype/phenotype correlations is the change of the phenotype in the same man over time. There have been two published examples of this. In the first a patient with an AZFc deletion, including the *DAZ* gene copies had a progressive decrease in sperm concentration from severe oligozoospermia to azoospermia over 30 months (Girardi et al. 1997). In a second report by Simoni et al. (1997) an oligozoospermic man with an AZFc deletion was found to be azoospermic 9 months later.

From the published studies several conclusions can be drawn:

1. Microdeletions have been found almost exclusively in males affected by azoospermia or severe oligozoospermia.
2. Microdeletions have also been found also in patients with abnormal andrological findings.

3. A higher frequency of Yq deletions are found in azoospermic vs oligozoospermic patients and in well defined idiopathic infertility vs infertility with known aetiology.
4. Large deletions are, in general, associated with more a severe spermatogenic defect.
5. AZFa deletions are less common (1–5%) and are usually associated with SCOS type I.
6. AZFc and AZFc + b deletions are the most frequent deletions, and they may be associated with a variety of spermatogenic failure including oligozoospermia.

8 Mechanism of Y Chromosome Microdeletions

The relatively high frequency of Y deletions suggests that the Y chromosome is susceptible to the spontaneous loss of genetic material. However, it is not clear if the Y chromosome is a "hot spot" for deletions or if they survive better because the Y chromosome is not essential for viability. The instability of the Y chromosome may be related to the high frequency of repetitive elements clustered along the length of the chromosome. Deletion interval 6 for example is rich in both inverted and direct repeats, many of which are several hundred kilobases in length (Yen 1998). Deletions may be caused by aberrant recombination events (between areas of homologous or similar sequence repeats, between the X and Y chromosomes or Y chromosome unbalanced sister chromatid exchange) or by slippage during DNA replication. Large inverted repeats exhibit novel genome destabilizing features. In *E. coli* palindromic sequences are deleted at very high rates. The proposed mechanism of inverted repeat deletions includes the generation of a hairpin structure in single stranded DNA that arises during replication. A deletion could occur if the hairpin is cleaved by a structure-specific nuclease and the surrounding DNA is end-joined. If the inverted repeat is flanked by short direct repeats, replication slippage could occur near the base of the hairpin structure (Collins 1981).

As described above, the Y chromosome is particularly rich in a variety of *Alu*-like repetitive elements. Clusters of *Alu* repeats are also associated with genomic instability resulting in deletion and/or duplication events. For example, clusters of intragenic *Alu* repeats in the human C1 inhibitor locus predispose this region to deleterious rearrangements and are responsible for about 20% of hereditary angioedema (Stoppa-Lyonnet et al. 1990). Short direct repeats can also be a source of microdeletion formation. The best examples of this are the microdeletions in the mitochondrial genome. The 4977bp deletion is the most common and is found in some patients with mitochondrial myopathies as well as in old human tissues (Wallace 1992; and see chapt. by

Bourgeron). The majority of deleted sequences in the mitochondrial genome are flanked by two short repetitive elements. Since the mitochondrial genome does not undergo recombination, the most likely explanation for deletion formation is replication slippage between the direct repeats.

Y chromosome deletions may be facilitated by a combination of both organisation of Y chromosome genomic sequences and by environmental influences. One of the best studied environmental effects causing deletions is ionizing radiation. Most mutations induced by ionising radiation are large deletions. Studies on the chinese hamster ovary hypoxanthine phosphoribosyl transferase gene indicated that 70% of X-ray induced mutations were deletions, with most resulting in the total absence of the gene (Nelson et al. 1994). Several stages of spermatogenesis are particularly sensitive to the effects of radiation. In the mouse meiotic and early-post meiotic stages as well as mature spermatozoa are particularly sensitive to radiation (Sirlin and Edwards, 1957, 1958). As well as environmental influences, there could be a paternal age effect contributing to the loss of Y sequences. This effect could be determined by correlating deletion incidence with the age of the father at conception. Paternal age effects have been described in Marfan syndrome, neurofribosis and Apert syndrome (Crow 1997). It should be noted that in most of these cases the mutations are single base pair substitutions and not deletions.

9 Y Chromosome Susceptibility Haplotypes

There may be particular Y chromosome sequence organisations that facilitate deletion formation of the AZF regions. Consequently, some individuals may be more susceptible to deletion formation than others. A Y chromosome haplotype that protects against aberrant X/Y exchange during male meiosis, leading to Y-positive 46,XX maleness has recently been defined (Jobling et al. 1998b). XX males, althought they carry the testis determining gene *SRY*, lack the Y-specific genes necessary for spermatogenesis and hence are azoospermic. The commonest form of XX males result from a recombination between the homologous genes *PRKX* (on the X chromosome) and *PRKY* (on the Y chromosome). The reciprocal translocation accounts for about one third of *SRY*-deleted XY females (Schiebel et al. 1997). In the European population an inversion polymorphism lies on Yp. The *PRKY* gene lies within this inversion. *PRKY*/*PRKX* translocations may preferentially take place if the *PRKY* gene lies in one orientation or the other. If so, the Y chromosome haplotypes associated with the susceptible orientation would be found preferentially in the translocation XX males and XY females. This possibility was investigated using two biallelic base pair substitutional markers (DYS257 {G or A} and 92R7 {C or T}) and the hypervariable minisatellite MSY1. The Y specific locus MSY1 displays

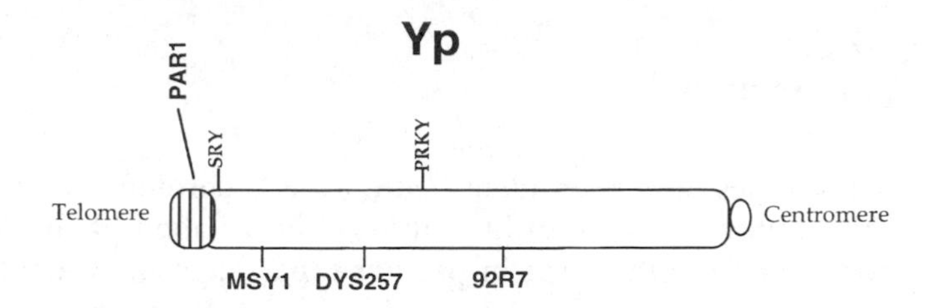

	MSY1 class 1	DYS257(A)	92R7(T)
European population	53/81	54/81	54/81
XX males	2/24	2/24	n.p.
XX males' fathers	n.t.	n.t.	0/2
XY females	n.p.	n.p.	0/5
XY female's father	0/1	0/1	0/1
Total independent chromosomes	2/25	2/25	0/7
P value	<0.001	<0.001	<0.005

Fig. 2. The frequency of Y alleles in the European population and in PRKX/PRKY translocation individuals. Thirty-two individual DNA samples that had undergone PRKX/PRKY translocations and three paternal Y chromosomes were studied. *PAR1* pseudoautosomal region 1, *n.p.*, not present. *n.t.*, not tested

an extremely high degree of structural diversity. This minisatellite is composed of 48–114 copies of a 25 bp repeat unit which is AT rich and predicted to form stable hairpin structures (Jobling et al. 1998a). MSY1 is highly polymorphic in both repeat number and in the sequence of repeat units. This can be assessed by minisatellite variant repeat PCR (MVR-PCR) in which PCR products are generated between a fixed flanking primer and discriminator primers designed to anneal specifically to one kind of repeat variant. This permits the fine structure of alleles to be defined. The results of this study are shown in Fig. 2. Of 81 chromosomes studied, 53 had an MSY1 code "MSY1 class 1" associated with DYS257A and 92R7T (class 1/A/T). The others (about one third) were defined as MSY1 non-class 1/G/C. However, in the XX males only 2/24 had the class1/A/T haplotype (Fig. 2). This demonstrates that, in one form of infertility, XX males a 4 chromosome haplotype protects against X-4 recombination. It is possible that AZF-deletion phenotypes may also occur on particular Y backgrounds, or that some haplotypes offer protection against deletion events.

10 Perspectives

Many genes have been identified on the Y chromosome each of which may contribute to male infertility. Little is known about their biological function. Two hybrid or other protein-partner experiments may identify proteins interacting with these molecules. In addition, RNA-binding studies could identify target sequence sites. Gene targeting in mice (of autosomal or X-linked homologs of Y genes) may provide insights into their function. Similarly, the transgenic rescue of these mice by human Y genes may be possible; for example the rescue of *Dazl1* knockout mice by human *DAZ*. The Y genes responsible for male infertility could be identified by (1) determining the minimal deletion size, and (2) screening infertile men with probes corresponding to the genes rather than anonymous markers.

In the long term, the identification of proteins responsible for male infertility could lead to novel, knowledge-based therapies to stimulate spermatogenesis in oligozoospermic males.

References

Agulnik AI, Mitchell MJ, Lerner JL, Woods DR, Bishop CE (1994a) A mouse Y chromosome gene encoded by a region essential for spermatogenesis and expression of male-specific histocompatibility antigens. Hum Mol Genet 3:873–878

Agulnik AI, Mitchell MJ, Mattei MG, Borsani G, Avner PA, Lerner JL, Bishop CE (1994b) A novel X gene with a widely transcribed Y homologue escapes X inactivation in mouse and human. Hum Mol Genet 3:879–884

Agulnik AI, Zharkikh A, Boettger-Tong H, Bourgeron T, McElreavey K, Bishop CE (1998) Evolution of the DAZ gene family suggests that Y-linked DAZ plays little, or a limited, role in spermatogenesis but underlines a recent African origin for human populations. Hum Mol Genet 7:1371–1377

Arnemann J, Jakubiczka S, Thuring S, Schmidtke J (1991) Cloning and sequence analysis of a human Y-chromosome-derived, testicular cDNA, *TSPY*. Genomics 11:108–114

Barbaux S, Vilain E, Raoul O, Gilgenkrantz S, Jeandidier E, Chadenas D, Souleyreau N, Fellous M, McElreavey K (1995) Proximal deletions of the long arm of the Y chromosome suggest a critical region associated with a specific subset of characteristic Turner stigmata. Hum Mol Genet 4: 1565–1568

Bardoni B, Zuffardi O, Guioli S, Ballabio A, Simi P, Cavalli P, Grimoldi MG, Fraccaro M, Camerino G (1991) A deletion map of the human Yq11 region: implications for the evolution of the Y chromosome and tentative mapping of a locus involved in spermatogenesis. Genomics 11: 443–451

Brown GM, Furlong RA, Sargent CA (1998) Characterisation of the coding sequence and fine mapping of the human DFFRY gene and comparative Dffry gene. Hum Mol Genet 7: 97–107

Castrillon DH, Gönczy P, Alexander S, Rawson R, Eberhart CG, Viswanathan S, Dinardo S, Wasserman SA (1993) Towards a molecular genetic analysis of spermatogenesis in *Drosophila melanogaster*: characterisation of male-sterile mutants generated by single P element mutagenesis. Genetics 135:489–505

Chai NN, Salido EC, Yen PH (1997) Multiple functional copies of the RBM gene family, a spermatogenesis candidate on the human Y chromosome. Genomics 45:355–361

Chai NN, Zhou H, Hernandez J, Najmabadi H, Bhasin S, Yen PH (1998) Structure and organisation of the RBMY genes on the human Y chromosome: transposition and amplification of an ancestral autosomal hnRNPG gene. Genomics 49:283–289

Cheng MH, Maines JZ, Wasserman SA (1998) Biphasic subcellular localisation of the DAZL-related protein Boule in *Drosophila* spermatogenesis. Dev Biol 204: 567–576

Collins J (1981) Instability of palindromic DNA in *Escherichia coli.* Cold Spring Harbor Symp Quant Biol 45:409–416

Cooke HJ (1976) Repeated sequence specific to human males. Nature 262:182–186

Cooke HJ, Lee M, Kerr S, Ruggiu M (1996) A murine homologue of the human DAZ gene is autosomal and expressed only in male and female gonads. Hum Mol Genet 5:513–516

Crow JF (1997) The high spontaneous mutation rate: is it a health risk? Proc Natl Acad Sci USA 94:8380–8386

Delbridge ML, Harry JL, Toder R, O'Neill RJ, Ma K, Chandley AC, Graves JA (1997) A human candidate spermatogenesis gene, RBM1, is conserved and amplified on the marsupial Y chromosome. Nat Genet 15:131–136

Eberhart CG, Maines JZ, Wasserman SA (1996) Meiotic cell requirement for a fly homologue of human Deleted in Azoospermia. Nature 381:783–785

Eichwald EJ and Silmser CR (1995) Communication. Transplant Bull 2:148–149

Elliott DJ, Millar MR, Oghene K, Ross A, Kiesewetter F, Pryor J, McIntyre M, Hargreave TB. Saunders PT. Vogt PH. Chandley AC. Cooke H (1997) Expression of RBM in the nuclei of human germ cells is dependent on a critical region of the Y chromosome long arm. Proc Natl Acad Sci USA. 94:3848–3853.

Elliott DJ, Oghene K, Makarov G, Makarova, Hargreave TB, Chandley AC, Eperon IC, Cooke HJ (1998) Dynamic changes in the subnuclear organisation of pre-mRNA splicing proteins and RBM during germ cell development. J Cell Sci 111:1255–1265

Ellis NA, Goodfellow PJ, Pym B, Smith M, Palmer M, Frischauf AM, Goodfellow PN, (1989) The pseudoautosomal boundary in man is defined by an Alu repeat sequence inserted on the Y chromosome. Nature 337:81–84

Foresta C, Ferlin A, Garolla A, Rossato M, Barbaux S, Bortoli A (1997) Y-chromosome deletions in idiopathic severe testiculopathies. J Clin Endocrinol Metabol 82:1075–1080

Foresta C, Ferlin A, Garolla A, Moro E, Pistorello M, Barbaux S, Rossato M (1998) High frequency of well-defined Y-chromosome deletions in idiopathic Sertoli cell-only syndrome. Hum Rep 13:302–307

Girardi SK, Mielnik A, Schlegel PN (1997) Submicroscopic deletions in the Y chromosome of infertile men. Hum Reprod 12:1635–1641

Gondo H, Kudo J, White JW, Barr C, Selvanayagam P, Saunders GF (1987) Differential expression of the human thymosin-beta 4 gene in lymphocytes, macrophages, and granulocytes. J Immunol 13:3840–3848

Greenfield A, Scott D, Pennisi D Ehrmann I, Ellis P Cooper L, Simpson E, Koopman P. (1996) An H-YDb epitope is encoded by a novel mouse Y chromosome gene. Nat Genet 14: 474–478

Habermann B. Mi HF. Edelmann A. Bohring C. Backert IT. Kiesewetter F. Aumuller G. Vogt PH (1998) DAZ (Deleted in AZoospermia) genes encode proteins located in human late spermatids and in sperm tails. Hum Reprod 13:363–369

Hecht NB (1998) Molecular mechanism of male germ cell differentiation. BioEssays 20: 555–561

Hendriks W, Schepens J, Bachner D, Rijss J, Zeeuwen P, Zechner U, Hameister H, Wieringa B (1995) Molecular cloning of a mouse epithelial protein-tyrosin phosphatase with similarities to submembranous proteins. J Cell Biochem 59:418–430

Ho M, Chelly J, Carter N, Danek A, Crocker P, Monaco AP (1994) Isolation of the gene for McLeod syndrome that encodes a novel membrane transport protein. Cell 77:869–880

Houston DW, Zhang J, Maines JZ, Wasserman SA, King ML (1998) A *Xenopus* DAZ-like gene encodes an RNA component of germ plasm and is a functional homologue of *Drosophila* boule. Development 125:171–180

James TC, Elgin SC (1986) Identification of a nonhistone chromosomal protein associated with heterochromatin in *Drosophila melanogaster* and its gene. Mol Cell Biol 11:3862–3872

Jobling MA, Samara V, Pandya A, Fretwell N, Bernasconi B, Mitchell RJ, Gerelsaikhan T, Dashnyam B, Sajantila A, Salo PJ, Nakahori Y, Disteche CM, Thangaraj K, Singh L, Crawford MH, Tyler-Smith C (1996) Recurrent duplication and deletion polymorphisms on the long arm of the Y chromosome in normal males. Hum Mol Genet 5:1767–1775

Jobling MA, Bouzekri N, Taylor PG (1998a) Hypervariable digital DNA codes for human paternal lineages: MVR-PCR at the Y-specific minisatellite, MSY1 (*DYF155S1*). Hum Mol Genet 7: 643–653

Jobling MA, Williams G, Schiebel K, Pandya A, McElreavey K, Salas L, Rappold GA, Affara NA, Tyler-Smith C. (1998b) A selective difference between human Y-chromosomal DNA haplotypes. Curr Biol 8:1391–1394

Kent-First MG, Kol S, Muallem A, Ofir R, Manor D, Blazer S, First N, Itskovitz-Eldor J (1996) The incidence and possible relevance of Y-linked microdeletions in babies born after intracytoplasmic sperm injection and their infertile fathers. Mol Hum Reprod 2:943–50

Kohler MR, Vogt PH (1994) Interstitial deletions of repetitive DNA blocks in dicentric human Y chromosomes. Chromosoma 103:324–330

Lahn BT, Page D (1997) Functional coherence of the human Y chromosome. Science 278:675–680

Lambson B, Affara NA, Mitchell M, Ferguson-Smith MA (1992) Evolution of DNA sequence homologies between the sex chromosomes in primate species. Genomics. 14:1032–1040

Linder P, Lasko PF, Ashburner M, Leroy P, Nielsen PJ, Nishi K, Schnier J, Slonimski PP (1989) Birth of the D-E-A-D box. Nature 337:121–122

Ludena P, Fernandez-Piqueras J, Sentis C (1993) Distribution of DYZ2 repetitive sequences on the human Y chromosome. Hum Genet 90:572–574

Ma K, Inglis JD, Sharkey A, Bickmore WA, Hill RE, Prosser EJ, Speed RM, Thomson EJ, Jobling M, Taylor K, Wolfe J, Cooke HJ, Hargreave TB, Chandley AC (1993) A Y chromosome gene family with RNA-binding protein homology: candidates for the azoospermia factor AZF controlling spermatogenesis. Cell 75:1287–1295

Manz E, Schnieders F, Brechlin AM, Schmidtke J (1993) *TSPY*-related sequences represent a microheterogeneous gene family organized as constitutive elements in DYZ5 tandem repeat units on the human Y chromosome. Genomics 17:726–731

Menke DB, Mutter GL, Page DC (1997) Expression of *DAZ*, an azoospermia factor candidate in human spermatogenesis. Am J Hum Genet 60:237–241

Mulhall JP, Reijo R, Alagappan R, Brown L, Page D, Carson R, Oates RD (1997) Azoospermic men with deletion of the DAZ gene cluster are capable of completing spermatogenesis: fertilization, normal embryonic development and pregnancy occur when retrieved testicular spermatozoa are used for intracytoplasmic sperm injection. Hum Reprod 12:503–508

Nakagome Y, Nagafuchi S, Seki S, Nakahori Y, Tamura T, Yamada M, Iwaya M (1991) A repeating unit of the DYZ1 family on the human Y chromosome consists of segments with partial male-specificity. Cytogenet Cell Genet 56:74–77

Nakahori Y, Mitani K, Yamada M, Nakagome Y (1986) A human Y-chromosome specific repeated DNA family (DYZ1) consists of a tandem array of pentanucleotides. Nucleic Acids Res 14 :7569–7580

Nelson SL, Giver CR, Grosovsky AJ (1994) Spectrum of X-ray-induced mutations in the human hprt gene Carcinogenesis 15:495–502

Page DC, Harper ME, Love J, Botstein D (1984).Occurrence of a transposition from the X-chromosome long arm to the Y-chromosome short arm during human evolution. Nature 311: 119–123

Pasantes JJ, Rottger S, Schempp W (1997) Part of the RBM gene cluster is located distally to the DAZ gene cluster in human Yq11.23. Chromosome Res 5:537–540

Pestova TV, Borukhov SI, Hellen CTU (1998) Eukaryotic ribosomes reuire initiation factors 1 and 1A to locate initiation codons. Nature 394:854–859

Prosser J, Inglis JD, Condie A, Ma K, Kerr S, Thakrar R, Taylor K, Cameron JM Cooke HJ (1996) Degeneracy in human multi-copy RBM (YRRM), a candidate spermatogenesis gene. Mammal Genome 7:835–842

Pryor JL, Kent-First M, Muallem A, Van Bergen AH, Nolten WE, Meisner L, Roberts KP (1997) Microdeletions in the Y chromosome of infertile men. New Eng J Med 336:534–539

Quereshi SJ, Ross AR, Ma K, Cooke HJ, Intyre MAM, Chandley AC, Hargreave TB (1996) PCR screening for Y chromosome microdeletions: a first step towards the diagnosis of genetically determinined spermatogenic failure in men. Mol Hum Reprod 2:775–779

Rao E, Weiss B, Fukami M, Rump A, Niesler B, Mertz A, Muroya K, Binder G, Kirsch S, Winkelmann M, Nordsiek G, Heinrich U, Breuning MH, Ranke MB, Rosenthal A, Ogata T, Rappold GA (1997) Pseudoautosomal deletions encompassing a novel homeobox gene cause growth failure in idiopathic short stature and Turner syndrome. Nature Genetics 16:54–63

Rappold GA (1993) The pseudoautosomal regions of the human sex chromosomes. Hum Genet 92:315–324

Reijo R, Lee TY, Salo P, Alagappan R, Brown LG, Rosenberg M, Rozen S, Jaffe T, Straus D, Hovatta O, de la Chapelle A, Silber S, Page DC. (1995) Diverse spermatogenic defects in humans caused by Y chromosome deletions encompassing a novel RNA-binding protein gene. Nat Genet 10:383–393

Reijo R, Alagappan RK, Patrizio P, Page D (1996a) Severe oligospermia resulting from deletions of azoospermia factor gene on Y chromosome. Lancet 347:1290–1293

Reijo R, Seligman J, Dinulos MB, Jaffe T, Brown LG, Disteche CM, Page DC (1996b) Mouse autosomal homolog of *DAZ*, a candidate male sterility gene in humans is expressed in male germ cells before and after puberty. Genomics 35:346–352

Rouyer F, Simmler MC, Johnsson C, Vergnaud G, Cooke HJ, Weissenbach J (1986) A gradient of sex linkage in the pseudoautosomal region of the human sex chromosomes. Nature 319: 291–295

Ruggiu M, Speed R, Taggart M, McKay SJ, Kilanowski F, Saunders P, Dorin J, Cooke HJ (1997) The mouse *Dazla* gene encodes a cytoplasmic protein essential for gametogenesis. Nature 389:73–77

Saxena R, Brown LG, Hawkins T, Alagappan RK, Skaletsky H, Reeve MP, Reijo R, Rozen S, Dinulos MB, Disteche CM, Page DC (1996) The DAZ gene cluster on the human Y chromosome arose from an autosomal gene that was transposed, repeatedly amplified and pruned. Nat Genet 14:292–299

Schiebel K, Winkelmann M, Mertz A, Xu X, Page DC, Weil D, Petit C Rappold GA (1997) Abnormal XY interchange between a novel isolated protein kinase gene, PRKY, and its homologue, PRKX, accounts for one third of all (Y+) XX males and (Y-)XY females. Hum Mol Genet 6:1985–1989

Schmid M, Guttenbach M, Nanda I, Studer R, Epplen JT (1990) Organisation of DYZ2 repetitive DNA on the human Y chromosome. Genomics 6:212–218

Schnieders F, Dork T, Arnemann J, Vogel T, Werner M, Schmidtke J (1996) Testis-specific protein, Y-encoded (TSPY) expression in testicular tissues. Hum Mol Genet. 5:1801–1807

Simoni M, Gromoll J, Dworniczak B, Rolf C, Abshagen K, Kamischke A, Carani C, Meschede D, Behre HM, Horst J, Nieschlag E (1997) Screening for deletions of the Y chromosome involving the DAZ (Deleted in AZoospermia) gene in azoospermia and severe oligozoospermia. Fertil Steril 67:542–547

Sirlin JL, Edwards RG (1957) Duration of spermatogenesis in the mouse Nature 180: 1137–1139

Sirlin JL, Edwards RG (1958) The labelling of mammalian spermatozoa with radioactive tracers J Exp Zool 137: 363–387

Smith KD, Young KE, Talbot CC, Schmeckpeper BJ (1987) Repeated DNA of the Y chromosome. Development 101s:77–92

Soulard M, Valle VD, Siomi M, Pinol-Roma S, Codogno P, Bauvy C, Bellini M, et al. (1993) HnRNPG: Sequence and characterisation of a glycosylated RNA-binding protein. Nucleic Acids Res 21:4210–4217

Stoppa-Lyonnet D, Carter PE, Meo T, Tosi M (1990) Clusters of intragenic Alu repeats predispose the human C1 inhibitor locus to deleterious rearrangements. Proc Natl Acad Sci USA 87: 1551–1555

Stuppia L, Mastroprimiano G, Calabrese G, Peila R, Tenaglia R, Palka G (1996) Microdeletions in interval 6 of the Y chromosome detected by STS-PCR in 6 of 33 patients with idiopathic oligo- and azoospermia. Cytogenet Cell Genet 72:155–158

Stuppia L, Gatta V, Mastroprimiano G, Pompetti F, Calabrese G, Guanciali Franchi P, Morizio E, Mingarelli R. Nicolai M. Tenaglia R. Improta L. Sforza V. Bisceglia S. Palka G (1997) Clustering of Y chromosome deletions in subinterval E of interval 6 supports the existence of an oligozoospermia critical region outside the DAZ gene. J Med Genet 34:881–883

Tiepolo L, Zuffardi O (1976) Localization of factors controlling spermatogenesis in the nonfluorescent portion of the human Y chromosome long arm. Hum Genet 34:119–124

Tsuchiya K, Reijo R, Page DC, Disteche CM (1995) Gonadoblastoma: molecular definition of the susceptibility region on the Y chromosome. Am J Hum Genet 57:1400–1407

van der Ven K, Montag M, Peshka B, Leygraaf J, Schwanitz G, Haidl G, Krebs D. van der Ven H (1997) Combined cytogenetic and Y chromosome microdeletion screening in males undergoing intracytoplasmic sperm injection. Mol Hum Reprod 3:699–704

Vereb M, Agulnik AI, Houston JT, Lipschultz LI, Lamb DJ, Bishop CE (1997) Absence of DAZ gene mutations in cases of non-obstructed azoospermia. Mol Hum Reprod 3:55–59

Vermeesch JR, Petit P, Kermouni A, Renauld JC, Van Den Berghe H, Marynen P (1997) The IL-9 receptor gene, located in the Xq/Yq pseudoautosomal region, has an autosomal origin, escapes X inactivation and is expressed from the Y. Hum Mol Genet 6:1–8

Vergnaud G, Page DC, Simmler MC, Brown L, Rouyer F, Noel B, Botstein D, de la Chapelle A, Weissenbach J (1986) A deletion map of the human Y chromosome based on DNA hybridization. Am J Hum Genet 38:109–124

Vogt P, Keil R, Kohler M, Lengauer C, Lewe D, Lewe G (1991) Selection of DNA sequences from interval 6 of the human Y chromosome with homology to a Y chromosomal fertility gene sequence of *Drosophila hydei*. Hum Genet 86:341–349

Vogt PH, Chandley AC, Hargreave TB, Keil R, Ma K Sharkey A (1992) Microdeletions in interval 6 of the Y chromosome of males with idiopathic sterility point to disruption of AZF, a human spermatogenesis gene. Hum Genet 89: 491–496

Vogt PH, Edelmann A, Kirsch S, Henegariu O, Hirschmann P, Kiesewetter F, Kohn FM, Schill WB, Farah S, Ram C, Hartmann M, Hartshuh W, Meschede D, Behre H, Castel A, Nieschlag E, Weidner W, Grone H-J, Jung P, Engel W, Haidl G (1996) Human Y chromosome azoospermia factors (AZF) mapped to different subregions in Yq11. Hum Mol Genet 5:933–943

Vollrath D, Foote S, Hilton A, Brown LG, Beer-Romero P, Bogan JS, Page DC (1992) The human Y chromosome: a 43-interval map based on naturally occurring deletions. Science 258:52–59

Wallace DC (1992) Mitochondrial genetics: a paradigm for ageing and degenerative diseases? Science 256:628–632

Yen PH (1998) A long range restriction map of deletion interval 6 of the human Y chromosome: a region frequently deleted in azoospermic males. Genomics 54:5–12

Yen PH, Chai NN, Salido EC (1996) The human autosomal gene DAZLA: testis specificity and a candidate for male infertility. Hum Mol Genet 5:2013–2017

Yen PH, Chai NN, Salido EC (1997) The human DAZ genes, a putative male infertility factor on the Y chromosome, are highly polymorphic in the DAZ repeat regions. Mammal Genome 8: 756–759

Spermatogenesis and the Mouse Y Chromosome: Specialisation Out of Decay

Michael J. Mitchell[1]

1
The Unique Y Chromosome

The Y chromosome has been extensively studied in humans and mice. In both species it is required for testis determination and spermatogenesis, but in the human it is also required for normal somatic development. Rich transcription maps are now available for the non-recombining portions of the human and mouse Y chromosome (Lahn and Page 1997; Mazeyrat et al. 1998), but so far the function of only one gene product has been elucidated: *SRY* is required for primary testis determination in both species. Compared with the human Y chromosome, the mouse Y chromosome is specialised in spermatogenesis, but the gene maps of the two chromosomes are very similar, suggesting that the functions of the genes on the Y chromosome are evolving much more rapidly than on other chromosomes. An understanding of the evolutionary forces that shape the genetic content of the mammalian Y chromosome may provide valuable insights into the potential functions of the Y chromosome genes, which in turn will help to correlate the genetic map with the gene map.

The Y chromosome differs from other mammalian chromosomes in two fundamental ways: it is the only chromosome which does not recombine along the majority of its length, and it is present only in one sex, the male. The evolution of the genetic functions, and therefore the gene content, of the Y chromosome are believed to be a reflection of these basic properties. Two principal theories for the evolution of genes on the Y chromosome have been formulated, each based on one of its two distinctive properties. Firstly, it is theorised that the absence of recombination prevents the segregation of deleterious mutations from advantageous mutations, thus leading to an inevitable deterioration in the genetic content of the non-recombining Y chromosome (NRY; Muller 1964; Charlesworth 1978). The lack of Y-linked genetic functions, which originally inspired this theory over 80 years ago (Muller 1914), is still the best known feature of the Y chromosome, after sex-determination. Secondly, the

[1] Inserm U.491, Génétique médicale et développement, Faculté de médecine, 27 bd. Jean Moulin, 13385 Marseille Cedex 05, France

Results and Problems in Cell Differentiation, Vol. 28
McElreavey (Ed.): The Genetic Basis of Male Infertility

male-specific nature of the Y chromosome may promote the accumulation of male-enhancing/female-damaging alleles, sexually antagonistic (SA) alleles (Fisher 1931), leading to the Y chromosome becoming a specialised male chromosome. These two theories are not mutually exclusive but are contradictory, in the sense that the first proposes a loss of genetic information while the second proposes a gain of genetic information.

In this review, I will discuss the role of the Y chromosome in spermatogenesis from an evolutionary perspective. The mouse Y chromosome will be presented in detail, as will the comparative studies which illustrate some of the possible steps in the evolution of a Y chromosome gene. There is now substantial evidence that, as the theory predicts, genes which have been conserved on the Y chromosome during mammalian evolution are eventually lost from the genome. Thus genes on the Y chromosome are decaying and, based on the mouse map, it seems clear that an intermediate step in this process is the elimination of expression from somatic tissues and its restriction to the germ cells.

2 The Functions of the Mouse Y Chromosome

There are many similarities between the genetic maps of the mouse and human Y chromosome. They both determine the male sex by the action of the homologous genes *SRY* and *Sry* (Berta et al. 1990; Koopman et al. 1991), and both are necessary for post-natal germ cell development. Unlike the human Y chromosome, however, the mouse Y chromosome is not essential to somatic development and XO mice which are effectively deleted for the entire Y chromosome develop as normal fertile females. Thus, it would appear that the mouse Y chromosome is a more male-specialised chromosome than its human counterpart.

2.1 Somatic Functions of the Mouse Y Chromosome

In the XO mouse, the absence of the Y chromosome is compatible with normal development and fertility. The in utero survival rate of XO embryos carrying a paternal X chromosome (X_P) is slightly reduced, at an estimated 65% (Searle 1990). No decrease in viabilty is seen, however, in XO embryos where the X chromosome is of maternal origin (X_M; Hunt 1991), as it is in normal XY embryos. This is in stark contrast to the situation in humans where in utero survival to term is estimated to be only 1% (Ogata and Matsuo 1995). It has been suggested that the shortened reproductive life of XO female mice is due to the presence of an unpaired X chromosome during oogenesis (Burgoyne and Baker 1985). The small litter size of XO female mice can be explained by the elimination of OY zygotes taken together with the reduced viability of X_PO

embryos (Searle 1990). It is therefore without doubt that the mouse Y chromosome is not critical for development, although it cannot be ruled out that it acts to improve the efficiency of the process.

Certain Y chromosomes have been shown to accelerate the preimplantation growth rate of the fetus (Burgoyne 1993; Burgoyne et al. 1995). This effect is not seen with all Y chromosomes, and the RIII Y chromosome, from which the *Sxr* deletion mutations are derived (see Sect. 3.3.3), has no detectable effect on the rate of preimplantation development. These developmental effects probably stem from the genes on the Y chromosome which have a distinct homologue on the X chromosome, as these are the only Y chromosome genes expressed in the soma. The combined expression of the X and Y gene pair may be required in order to achieve optimal levels of gene product at a particular stage in the development of the preimplantation embryo.

The other somatic effect of the mouse Y chromosome is the expression of the male-specific minor transplantation antigen H-Y. In inbred strains, this antigen is responsible for the rejection of tissue grafts from male donors by female recipients (Eichwald and Silmser 1955). The antigen is composed of several peptides (epitopes) which are each presented on the cell surface bound to a class I or class II major histocompatibility molecule (H-2). They are recognised by H-Y-specific H-2-restricted T-cells. All defined H-Y epitopes have been mapped to the Sxr^b deletion interval of the Y short arm (McLaren et al. 1988a, b; King et al. 1994). Three epitopes have so far been identified: H-YKk and H-YDb coded by *Smcy* (Scott et al. 1995; Markiewicz et al. 1998) and a distinct H-YDb coded by *Uty* (Greenfield et al. 1996). The functional significance of these epitopes and the genes that code them is unclear.

2.2
Germ Cell Functions

The Y chromosome is not necessary for somatic development in the mouse, but it is critical for transmission of the genetic material to the next generation. XO germ cells in the testis of XO/XY chimeric males can develop normally as far as the post-natal spermatogonial stage, but they proliferate poorly, and rarely enter meiosis (Levy and Burgoyne 1986). This shows that the Y chromosome produces a cell-autonomous factor which is necessary for the successful post-natal development of the mitotically dividing spermatogonial germ cells, and this factor has been named *Spy* (Spermatogenesis on the Y chromosome). This is the earliest stage at which the Y chromosome is seen to have an effect on germ cell development, after determination of the testis. Central to our current understanding of the location of germ cell functions on the mouse Y chromosome are five deletion mutations. These deletions have no noticeable effect on somatic development, but they do affect germ cell development. The associated phenotypes range from reduced fertility to an absence of germ cells in the adult testis. These deletions have provided a more precise localisation of

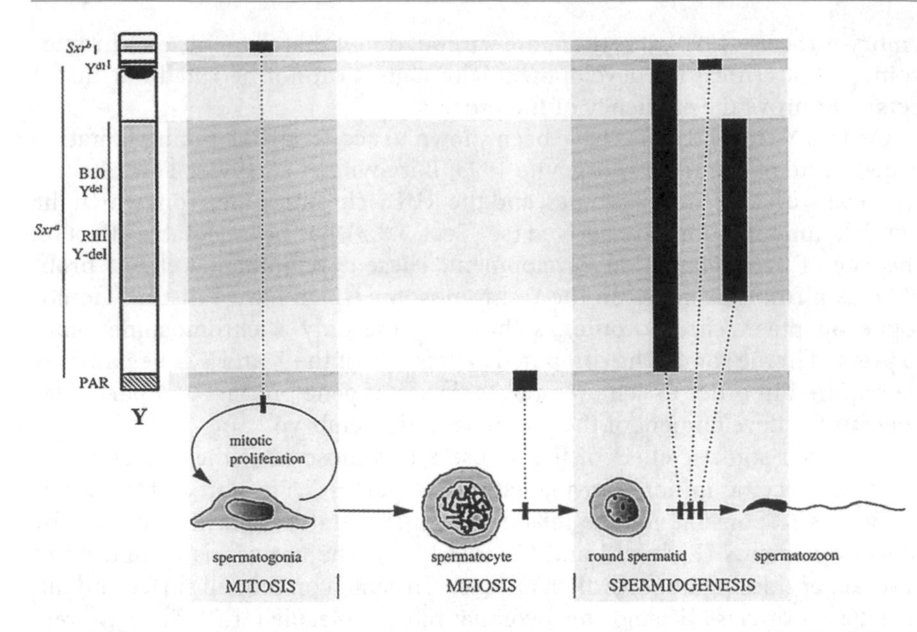

Fig. 1. Deletion mapping the role of the mouse Y chromosome in spermatogenesis. The different deletions are shown at the *left* of the diagram as is the pseuodautosomal region (*PAR*). Spermatogenesis is divided into three major stages; mitotic, meiotic and spermiogenic. The block in spermatogenesis associated with each deletion is indicated by the *dotted line*. Absence of the *PAR* causes a meiotic block, and factors on the minute short arm are required for mitotic stages, but most of the chromosome is required uniquely for spermiogenesis

Spy to the minute short arm, and have led to the identification of a spermiogenesis factor on the long arm, *Smy* (Sperm head morphology on the Y chromosome), which is necessary for the production of spermatozoa with functional heads. The role of the different intervals of the mouse Y chromosome in spermatogenesis is summarised in Fig. 1. The deletion variants will be presented in detail below, in the context of the current molecular and gene maps of the mouse Y chromosome.

3 The Molecular Genetics of the Mouse Y Chromosome

3.1 Overview

The mouse Y chromosome, like its human counterpart, is generally held to be approximately 60 Mb in length (Bishop 1992). A recent study, however, shows this to be a considerable underestimate, and using bivariate flow-sorting, the mouse Y chromosome has been calculated to contain 94.7 Mb of DNA (Bergstrom et al. 1998). It is mainly composed of a homogeneously staining long arm, but unlike the other mouse chromosomes is not acrocentric, as it has a minute short arm (Ford 1966). Pairing with the X chromosome during male meiosis occurs via a pseudoautosomal region (PAR) located at the distal end of the long arm. The steroid sulfatase gene (Sts) is the only functional mouse PAR gene identified to date (Salido et al. 1996). The short arm is barely visible cytogenetically, and on this basis is estimated to be less than 5 Mb. At the molecular level, the minute short arm is the best characterised segment of the mouse Y chromosome. Outside the short arm, the chromosome remains ill-characterised, despite considerable efforts to isolate chromosome-wide markers, using the same techniques which have proved so successful in the human (Baron et al. 1986; Bishop et al. 1987; Navin et al. 1996; Bergstrom et al. 1997, 1998). The human map is, however, highly resolved over 30 Mb (Foote et al. 1992; Vollrath et al. 1992), while well-defined maps of the mouse Y chromosome together cover less than 5 Mb (King et al. 1994; Mazeyrat et al. 1998). The difference in the states of the maps seems to lie in the fact that while in the human, single copy or low complexity markers are readily found, such markers are very rarely found in the mouse. The most common type of low copy marker derived for the mouse Y chromosome detects Y-specific repeat sequences which are present in several tens of copies (Bishop et al. 1987; Navin et al. 1996; Bergstrom et al. 1997, 1998). This indicates that the majority of sequences on the mouse Y chromosome have arisen by recent amplification events, and it is only on the short arm that single copy sequences have been identified (Mitchell and Bishop 1992; Mazeyrat et al. 1998). Here, I will present what is known about the molecular structure and the genes of the mouse Y chromosome. The phenotypes associated with deletions of different intervals of the chromosome will be discussed in this context.

3.2
The Long Arm

3.2.1
Molecular Structure

Even though the mouse Y chromosome is estimated to be more than 50% longer than its human counterpart, it does not have a discrete cytogenetically detectable region of heterochromatin. In the human, this region accounts for 30 Mb, or half the length of the entire chromosome. Nevertheless, a large proportion of the mouse Y-chromatin does consist of sequences that have been amplified specifically on the Y chromosome. There are an estimated 300–500 copies of a complex element on the chromosome which is composed of at least two Y-specific proviral genomes: a murine retrovirus, *MuRVY*, and an intracisternal A particle, *IAPE-Y* (Eicher et al. 1989; Fennelly et al. 1996). The full extent of the repeat unit has not been defined, but it is greater than 25 kb, and therefore accounts for a minimum of 9–12% of the Y-chromatin. FISH analysis shows that both the proviral components of this repeat are spread throughout the Y chromatin (Fennelly et al. 1996).

Attempts to isolate low-copy number sequences specifically from the Y chromosome have led to the isolation of Y-specific low-level repeat sequences which are present on the Y chromosome in 10 to 250 copies. Analysis of the distribution of these elements by FISH shows that most are dispersed throughout the long arm of the chromosome (Navin et al. 1996; Bergstrom et al. 1997). In the majority of cases, Southern analysis with these Y-fragments reveals a single intense male-specific band, suggesting that the amplification of these sequences occurred recently in the evolution of the Muridae, and that the sequences defined are part of a larger element. Thus, it is reasonable to conclude that most of the 94.7 Mb of the mouse Y chromosome have arisen recently through amplification events, and that the long arm of the mouse Y chromosome is predominantly composed of several hundred copies of one large composite element, interspersed with a variety of low to medium-copy elements.

One of the most interesting findings of the recent systematic cDNA selection performed in the human was that many human Y chromosome genes are in fact encoded by low-copy repeat elements (Lahn and Page 1997). In the mouse, one such gene, *Ssty*, has been isolated from the Y chromosome, where it is present as a heterogeneous multi-copy gene family dispersed along the long arm (Bishop and Hatat 1987; Prado et al. 1992; Conway et al. 1994). Two subclasses of transcript which have only 84% nucleotide identity have been defined and a related multi-copy family (DXF34) has been identified on the human and mouse X chromosome (Laval et al. 1997). *Ssty* shares homology (75% amino acid similarity) with the *Spindlin* gene, which represents a more distantly related autosomal homologue. The *Spindlin* gene encodes a protein that associates with the meiotic spindle in the unfertilised egg and persists in the embryo until some time between the two to eight cell stage (Oh et al. 1997). Unlike *Ssty* and its X homologue, *Spindlin* contains introns (B. Oh, pers. comm.), suggesting that the sex chromosome genes arose through retroposition of a *Spindlin* transcript. The expression of *Ssty* is limited to the spermatids, consistent with its having a role in spermiogenesis (Conway et al. 1994). Presently, the Ssty genes represent the only long arm sequences known to be transcribed, and so it remains to be determined whether any of the other repeats encode similar gene families.

3.2.2
Deletion of the Long Arm

Two partial deletions of long arm sequences have been reported: B10.BR-Y^{del} (Styrna et al. 1991b) and RIII Y-del (Conway et al. 1994), which are associated with a reduction in fertility. The B10.BR-Y^{del} deletion arose in the B10.BR strain of mice, and was detected by cytogenetic screening. In chromosome spreads, the Y^{del} is seen as a small chromosome, about 25% the size of the normal Y chromosome. Meiotic chromosome spreads revealed that the Y^{del} pairs normally with the X chromosome, indicating that Y^{del} has a pseudoautosomal region, and must therefore have arisen through interstitial deletion of most of the long arm (Styrna et al. 1991b).

B10.BR-Y^{del} males are fertile, but produce increased numbers of spermatozoa with abnormal head structures (Styrna et al. 1991a, b). Invitro tests further indicated that even the spermatozoa with normal morphology from B10.BR-Y^{del} mice fertilize with reduced efficiency, when compared with those from B10.BR (Styrna and Krzanowska 1995; Xian et al. 1992). In normal B10.BR ejaculate ~22% of sperm are abnormal, but this rises to ~64% in mice carrying the Y^{del}. This increase is largely due to the presence of spermatozoa with an almost normal head structure but a flattened acrosome (Styrna et al. 1991a, b). This type of spermatozoon is not found at all in the ejaculate of normal B10.BR mice, but represents ~28% of all the spermatozoa in the Y^{del} ejaculate. A slightly increased frequency is also seen in two types of more grossly deformed spermatozoa which are present in the normal ejaculate.

In the case of RIII Y-del (Conway et al. 1994), a similar threefold increase in the percentage of abnormal sperm was observed, although cytogenetically, the deletion was slightly smaller, removing about half the long arm. A slight distortion in the sex ratio in favor of females was observed in the offspring of the males carrying the Y-del chromosome. This was measured on two different genetic backgrounds, and the percentage of males dropped from 57% with the undeleted Y to 46% with the Y-del on a C57BL/10 background, and from 51% to 38% on an MF1 background. The sex-ratio distortion suggests that the deleted region contains factors that act during the haploid stages of spermatogenesis and favour transmission of the Y-bearing gamete. At the molecular level, the RIII Y-del deletion removes approximately half the *Ssty* copies from the Y chromosome, and this translates to a threefold decrease in the levels of *Ssty* transcripts. This reduction in *Ssty* expression could underlie the increase in abnormal spermatozoa, but the deletion intervals are very large and may contain other genes.

Thus, there are factors on the long arm of the Y chromosome which, although not essential for spermiogenesis, are required to maximise the production of normal spermatozoa. The expression of the multi-copy spermatid-specific gene *Ssty* is diminished from at least one of these deleted Y chromosomes, making it a candidate for the sperm-head morphology gene, *Smy*. Given the apparent multi-copy nature of the majority of sequences on the long arm of the mouse Y chromosome, it is tempting to conclude that the spermiogenesis functions of the Y chromosome are fulfilled by such amplified gene families. The role of the testis-specific amplified gene families that have been identified on the human Y chromosome may therefore also be in spermiogenesis.

3.3 The Pericentric Region

3.3.1 *Molecular Structure*

A composite repeat element has been partially defined, which is present in 20–50 copies in the pericentromeric region of the mouse Y chromosome (Navin et al. 1996; Mahadevaiah et al. 1998). It is known to comprise in part the *Rbmy* gene (Laval et al. 1995; Elliot et al. 1996), together with a sequence defined by anonymous marker pSx1 (Roberts et al. 1988) and these two sequences are located within the same 15 kb stretch of DNA (Mahadevaiah et al. 1998). The repeat block is partially characterised and the total size of the repeat unit has been estimated, from pulse field gel electrophoresis data, to be 50–70 kb (M. J. Mitchell, unpubl. data). Thus, copies of this element cover at least 1 Mb. It is not known whether the *Rbmy/Sx1* repeats are arranged as a single block or are split into multiple blocks. FISH analysis, using pSx1 as a probe, shows a hybridisation signal exclusively at the centromeric end of the chromosome (Navin et al. 1996). There is one copy of Sx1 that is known to be located outside the pericentric repeat, and it is localised on the short arm within the ΔSxr^b interval (Roberts et al. 1988; see Sect. 3.4.2). Screening a large contiguous stretch of DNA cloned from this interval (Mazeyrat et al. 1998), with an *Rbmy* cDNA probe, has not detected any homologous sequences, and so it is certain that this isolated copy of *Sx1* is not associated with a copy of *Rbmy* (M. J. Mitchell unpubl. data). The *Sry* gene is located close to the main *Rbmy/Sx1* repeat cluster (King et al. 1994; Laval et al. 1995), and the deletion of some copies of this repeat can lead to the transcriptional silencing of *Sry*, causing the development of sex-reversed XY females (Capel et al. 1993).

The mouse *Rbmy* gene is the homologue of the human Y chromosome gene *RBMY*(formerly *YRRM*; Ma et al. 1993). These genes are members of the heterogeneous nuclear ribonucleoprotein (hnRNP) family, which interact with mRNAs and appear to be involved in pre-mRNA processing and transport (Weighardt et al. 1996). The closest relative of Rbmy is hnRNP G (60–67% amino acid similarity), which contains an RNA binding domain of the RRM type (Chai et al. 1998; Delbridge et al. 1998). No specific functions are known for hnRNP G. The *Rbmy* gene is transcribed specifically in the testis, and its expression is limited to the germ cells (Elliot et al. 1996; Mahadevaiah et al. 1998). The mouse and human RBMY protein are found predominately in the nucleus (Elliott et al. 1997; Mahadevaiah et al. 1998). Both are present in spermatogonia and early spermatocytes, but while human RBMY protein persists until the round spermatid stage, the mouse protein disappears from the primary spermatocytes and does not reappear until the elongating spermatid stages. This suggests that the two proteins have distinct roles during the meiotic and spermiogenesis stages of spermatogenesis.

3.3.2
Deletion of the Pericentric Region

The Y^{d1} mutation arose in an X*Sxr*^{a}Y mouse (see Sect. 3.3.3) which carries a duplicated segment of the short arm, the *Sxr*^{a} region, on the X chromosome. The deletion is probably the result of unequal crossing-over between the *Sxr*^{a} region and the short arm of the Y chromosome (Capel et al. 1993; Laval et al. 1995; Mahadevaiah et al. 1998). The unequal crossing-over is believed to have occurred within the *Rbmy/Sx1* tandem repeats situated on the short arm, between the centromere and *Sry*. The resulting Y^{d1} carries an estimated two to four copies of this repeat. If the repeat is arranged in a single block on the Y chromosome then it is only copies of this repeat that have been deleted, but if it is split between two blocks then interstitial sequences will also have been deleted. Y^{d1} carries only about 10% of the normal number of copies of *Rbmy*, and there is a drastic reduction in the level of *Rbmy* transcripts.

The primary phenotype of the Y^{d1} deletion is sex-reversal, caused by the transcriptional silencing of *Sry* which remains intact on Y^{d1} (Capel et al. 1993; Laval et al. 1995). To see the effect of the deletion on spermatogenesis, an *Sry* transgene was introduced, to allow determination of the testis, and the resulting XY^{d1}*Sry* male mice were analysed (Mahadevaiah et al. 1998). This analysis showed that like the B10.BR-Y^{del} and the RIII Y-del mice, XY^{d1}*Sry* mice are fertile, but produce increased numbers of abnormal spermatozoa. The profile of spermatozoa morphology is different from that associated with the interstitial long arm deletions, with slightly abnormal forms being only marginally increased and grossly abnormal forms greatly increased. The converse being observed in B10.BR-Y^{del} and RIII Y-del, this suggests that the Y chromosome may act at at least two different stages in the spermiogenic process.

3.3.3
Deletion of the Long Arm and the Pericentric Region

The *Sxr*^{a} mutation has allowed an evaluation of the spermatogenesis in male mice lacking the long arm and the pericentric region. The *Sxr*^{a} (sex-reversed) mutation was originally described as an autosomal dominant sex-reversal mutation which caused the development of testis in embryos lacking a Y chromosome (Cattanach et al. 1971). The molecular basis for this mutation is a translocated fragment of the Y chromosome. XY*Sxr*^{a} carrier males have a duplication of most of the Y short arm, translocated to the distal end of the Y chromosome pseudoautosomal region (Cattanach et al. 1982; Evans et al. 1982; McLaren and Monk 1982; Singh and Jones 1982). The duplicated segment is called the *Sxr*^{a} region, and during male meiosis it is transferred to the X chromosome by pseudoautosomal recombination, giving rise to XX*Sxr*^{a} progeny (Evans et al. 1982). Like the Y^{d1} chromosome, the *Sxr*^{a} region has a reduced number of copies of the *Rbmy/Sx1* repeat and so it is likely that the *Sxr*^{a} translocation breakpoint lies within the *Rbmy/Sx1* cluster. (Roberts et al. 1988; Mahadevaiah et al. 1998). Thus, the *Sxr*^{a} region consists of an estimated three to seven copies of the *Rbmy/Sx1* repeat, together with all distally located short arm sequences.

XX*Sxr*^{a} mice develop as males because of the expression of *Sry* from the *Sxr*^{a} region, but have a limited post-natal germ-cell development due to the presence of two X chromosomes (Burgoyne et al. 1986). This latter effect is not present in X*Sxr*^{a}O mice, where all stages of spermatogenesis are present (Cattanach et al. 1971). XY/XO chimeric studies had shown that XO germ cells rarely progressed beyond spermatogonial stages, and established that a cell-autonomous spermatogenesis factor, *Spy*, mapped to the Y chromosome (Levy and Burgoyne 1986). The development of X*Sxr*^{a}O germ cells into spermatozoa showed that the gene, or genes, for *Spy* must be located in the *Sxr*^{a} region. Despite the presence of *Spy*, X*Sxr*^{a}O mice are nevertheless sterile, producing only a few malformed sperm. The reduced number of spermatozoa produced by X*Sxr*^{a}O mice was shown to result from the absence of a meiotic pairing partner when genetic crosses were used to introduce a rearranged chromosome, Y^{*X} (Burgoyne et al. 1992). The Y^{*X} is composed of a pseudoautosomal region attached to an unknown X or autosomal centromere (Eicher et al. 1991). X*Sxr*^{a}Y^{*X} mice produce almost normal numbers of spermatozoa, but all are grossly deformed and the mice are infertile (Burgoyne et al. 1992). The X*Sxr*^{a}Y^{*X} spermatozoa are equivalent to the more grossly deformed types produced with increased frequency in mice carrying long arm or pericentric deletions of the Y chromosome. Thus, the deletion of long arm and pericentric sequences provokes a more severe phenotype than either deletion individually. It can therefore be concluded that Y-borne spermiogenesis factors, necessary for normal sperm head morphogenesis, are located outside the *Sxr*^{a} region, but are not limited to a single locus. It can also be concluded that the tiny *Sxr*^{a} region contains sufficient Y chromosome genetic material for the production of spermatozoa in normal numbers.

3.4 The Short Arm

3.4.1 *Molecular Structure*

The most fully characterised region of the mouse Y chromosome is the short arm. The size of the short arm is estimated to be ≤5 Mb, from cytogenetic observations. The region carries a large number of the BKM tetranucleotide microsatellite repeats (Singh and Jones 1982), which are mostly located towards the Yp telomere (King et al. 1994). Relative to the rest of the chromosome, however, it appears to have a high density of single copy segments, and almost all the genes so far identified on the Y chromosome are located on the short arm. There are several segments that have been amplified from two to six copies and dispersed on the short arm, representing region-specific repeats (Mitchell and Bishop 1992; King et al. 1994). These segments include the X-Y homologous genes *Zfy* (two copies; Mardon et al. 1989) and *Ube1y* (five to six copies; Mitchell et al. 1991). Thus, much of the DNA content of the short arm may have arisen by duplication events during murine evolution. *Zfy1* and *Ube1y* are separated by less than 20 kb and lie adjacent to a segment of >600 kb of single copy sequences which encodes five functional X-Y homologous genes (Mazeyrat et al. 1998). No other X-Y homologous genes have yet been identifed on any other part of the mouse NRY.

3.4.2
Deletion of the Short Arm

The study of the *Sxr*^{a} mutation established that a factor necessary for mitotic germ cell survival is located on the short arm of the chromosome. This factor has been further localised by a deletion variant of the *Sxr*^{a} region which has been termed *Sxr*^{b}. The search for the *Spy* gene has made this deletion interval the best characterised segment of the mouse Y chromosome at the molecular level.

The *Sxr*^{b} deletion arose in a stock of mice carrying the *Sxr*^{a} mutation (McLaren et al. 1984, 1988b; Roberts et al. 1988), and the deleted interval has been termed Δ*Sxr*^{b}. Unlike X*Sxr*^{a}O mice, X*Sxr*^{b}O mice have a severe block in spermatogenesis (Burgoyne et al. 1986), affecting the proliferation of the differentiating A spermatogonia shortly after the exit of the germ cells from mitotic arrest, at birth. The fact that this deletion is a transmissible mutation, being complemented by the Y short arm in XY*Sxr*^{b} carrier males, has allowed post-natal X*Sxr*^{b}O germ cell development to be precisely characterised (Sutcliffe and Burgoyne 1989). The earliest effect is evident 3 days after birth as reduced numbers of A spermatogonia precursors, the T2 prospermatogonia. Decreased mitotic activity among the differentiating A spermatogonia results in the effective absence of B spermatogonia, and only very rare germ cells are seen to enter meiosis. The abnormal proliferation of spermatogonia in X*Sxr*^{b}O mice appears to activate a control system which eliminates the aberrantly dividing germ cells, leaving the adult testis tubules essentially empty of germ cells. The Δ*Sxr*^{b} gene or genes encoding the spermatogenesis factor necessary for normal spermatogonial development have been equated with *Spy*.

3.4.3
Genes in the Sxr^{b} *Deletion Interval*

Eight of the 11 genes isolated from the mouse NRY map into the Δ*Sxr*^{b} interval. The *Sxr*^{b} deletion has been shown to be the result of unequal crossing-over between duplicated *Sxr*^{a} region genes, *Zfy1* and *Zfy2*, which therefore define the limits of the deleted interval (Simpson and Page 1991). Chromosome walking has been initiated from these two genes, and approximately 1.2 Mb of ΔSxr^b genomic DNA isolated. The *Zfy1* and *Zfy2* contigs do not overlap, and so do not cover the entire interval. The ΔSxr^b interval is therefore large, and must be greater than 1.2 Mb. Seven functional ΔSxr^b genes, in addition to the *Zfy* genes, have now been identified by a variety of approaches (Kay et al. 1991; Mitchell et al. 1991; Agulnik et al. 1994; Greenfield et al. 1996; Brown et al. 1998; Ehrmann et al. 1998; Mazeyrat and Mitchell 1998; Mazeyrat et al. 1998), and they have all been localised to the 750 kb contig which includes *Zfy1* (Mazeyrat et al. 1998). No functional gene has to date been found in the *Zfy2* contig. One of the seven genes isolated is mouse *Tspy* which is a decayed, non-functional gene and this will be discussed later in Section 5.2.3.1. Like the *Zfy* genes, the other six genes appear to be functional and have a homologue on the X chromosome. These genes are arranged in the order *Zfy1-Ube1y-Smcy-Eif2γy-Tspyps-Uty-Dby-Dffry* (Fig. 2).

The functions of the proteins coded by some of these genes can be extrapolated from the known functions of their X homologues: activation of ubiquitin (Handley et al. 1991; *Ube1y*), initiation of translation (Gaspar et al. 1994; *Eif2γy*), deubiquitination of proteins (Huang et al. 1995; *Dffry*). In other cases, clues are provided by the functional analyses of homologous genes in yeast: transcriptional repression (Keleher et al. 1992; *Uty*) and initiation of translation (Chuang et al. 1997; *Dby*).

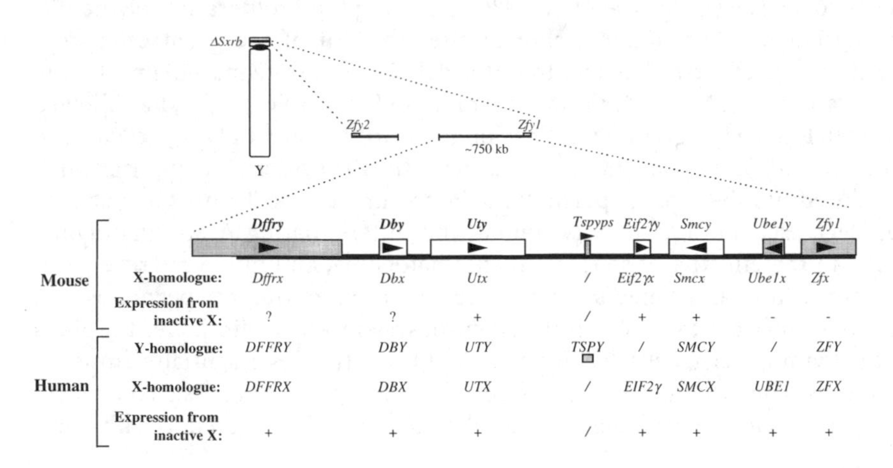

Fig. 2. The transcription map of the ΔSxr^b interval of the mouse Y chromosome. This interval is required for spermatogonial proliferation and it is dense in X-Y homologous genes. The genes that have been mapped to this interval are represented by the *boxes* and the *arrowhead* points towards their 3' end. The mouse and human X-homologues and the human Y homologues are shown *below* the map. The *grey boxes* indicate genes whose transcription is restricted to germ cells. It can be noted that, in the mouse, only the X genes which have a ubiquitously expressed Y homologue escape from X-inactivation

With respect to *Dby*, a closely related mouse gene has been used to complement a loss of function mutation in yeast. The X homologue of *Dby* is *Dbx* (also known as *ERH*, *mDEAD2* and *mDEAD3*), an RNA helicase containing a DEAD box motif (Gee and Conboy 1994; Sowden et al. 1995), and on mouse chromosome 1 there is a *Dbx* retroposon (N. Saut and M.J. Mitchell unpubl. data). This retroposon, called *PL10*, shares 89% nucleotide identity with *Dbx* and has maintained an open reading frame but, unlike *Dbx* which is ubiquitously expressed, it is only expressed in the testis (Leroy et al. 1989). In the yeast, *Saccharomyces cerivisiae*, it has been shown that the *DED1* gene, which encodes a polypeptide which has 53% amino acid identity to PL10, is required for the initiation of translation at the ribosome. A deletion of the *DED1* gene can be complemented by PL10, suggesting that the mouse genes are involved in translation initiation (Chuang et al. 1997). The X homologues of all these genes are widely transcribed, but not all the Y genes are expressed like their X homologues; *Zfy1*, *Zfy2*, *Ube1y*, and *Dffry* are only expressed in the testis. Thus, it would appear that one step in the evolution of a gene on the Y chromosome is the limitation of its expression to the germ cells, through the loss of expression in somatic tissues.

The interval of the mouse Y chromosome required for the proliferation and survival of the mitotic spermatogonia is dense in X-Y homologous genes. The transcription map of the Sxr^b deletion remains incomplete, but it nevertheless seems likely that X-Y homologous genes are required for normal spermatogenesis in the mouse. The primary cause of the phenotype associated with the Sxr^b deletion seems to be inefficient mitotic proliferation. Four of the genes in the ΔSxr^b interval are homologues of confirmed or probable housekeeping genes – *Ube1y*, *Eif2γy*, *Dffry* and *Dby* – and thus, it could be envisaged that the expression of these genes is critical for spermatogenesis, because they improve the efficiency of the mitotic cell-cycle in spermatogonia. The ΔSxr^b genes may function by augmenting the dose of their X homologue to achieve the critical level of product for the normal proliferation of the spermatogonial germ cells. The genes could each be critical at a separate point in the process, or may all act together, or in various combinations, to allow the proliferation, differentiation and survival of the spermatogonia.

3.5 The Mouse Y Chromosome in Spermatogenesis – Conclusions

A clear picture of the involvement of the mouse NRY in spermatogenesis is emerging (Fig. 1). The entire long arm and the pericentromeric region are required exclusively for the development of normal sperm heads, suggesting that most of the mouse Y chromosome is dedicated to genes that encode spermiogenesis factors. The region of the short arm defined by the ΔSxr^b interval is necessary for at least the mitotic stages of germ cell development. Thus, at its current resolution, the genetic map of the mouse Y chromosome can be divided into two very unequal parts, with the *Sry* gene or the *Sxr[a]* breakpoint defining the boundary. The preliminary transcription map of the mouse Y chromosome shows a striking correlation with the genetic map. X-Y homologous genes co-localise with the spermatogonial proliferation factor, *Spy*, while the multi-copy Y-specific genes co-localise with the spermiogenesis factors *Smy*.

4 Comparison of the Mouse and Human Y Chromosome Maps

The recent publication of an extensive transcription map of the human Y chromosome stressed that the majority of human Y chromosome genes fall into one of two categories (Lahn and Page 1997). Firstly, there are ubiquitously expressed, single copy genes with a readily detectable homologue on the X chromosome. These genes escape X-inactivation, suggesting that the X-Y homologous pair function as non-recombining autosomal genes to produce two doses of the appropriate gene product in males and females. Secondly, there are genes that do not have a readily detectable homologue on the X chromosome, and whose expression is limited to the testis. These latter genes are present in multiple copies on the human Y chromosome. Given that the human Y chromosome is required for both somatic and germ cell development, it is tempting to conclude that the X-Y homologous genes are somatic factors, while the Y-specific genes are spermatogenesis factors. A comparison with the transcriptional and genetic maps of the mouse Y chromosome suggest, however, that this view may be too reductionist, and may hinder our understanding of the relationship between the genes of the Y chromosome and spermatogenesis. The gene map of the mouse Y chromosome is very similar to the human Y chromosome, particularly in terms of its complement of X-Y homologous genes, but it is more specialised than its human counterpart, having a critical role in spermatogenesis and only a very minor role in somatic development. Furthermore, some of the mouse X-Y homologous genes are not expressed ubiquitously like their human counterparts, but are expressed only in the testis. Deletion mapping suggests that, in the mouse, X-Y homologous genes are required for the proliferation of the germ cells, while the multi-copy Y-specific genes are necessary for spermiogenesis.

4.1 Distinct Gene Organisation

Of the eleven genes that have been mapped to the mouse NRY, eight have homologues on the human Y chromosome. A comparison of the mouse and human Y chromosome transcription maps stresses just how different the organisation of genes is between the two species (Fig. 3). In the mouse, all such

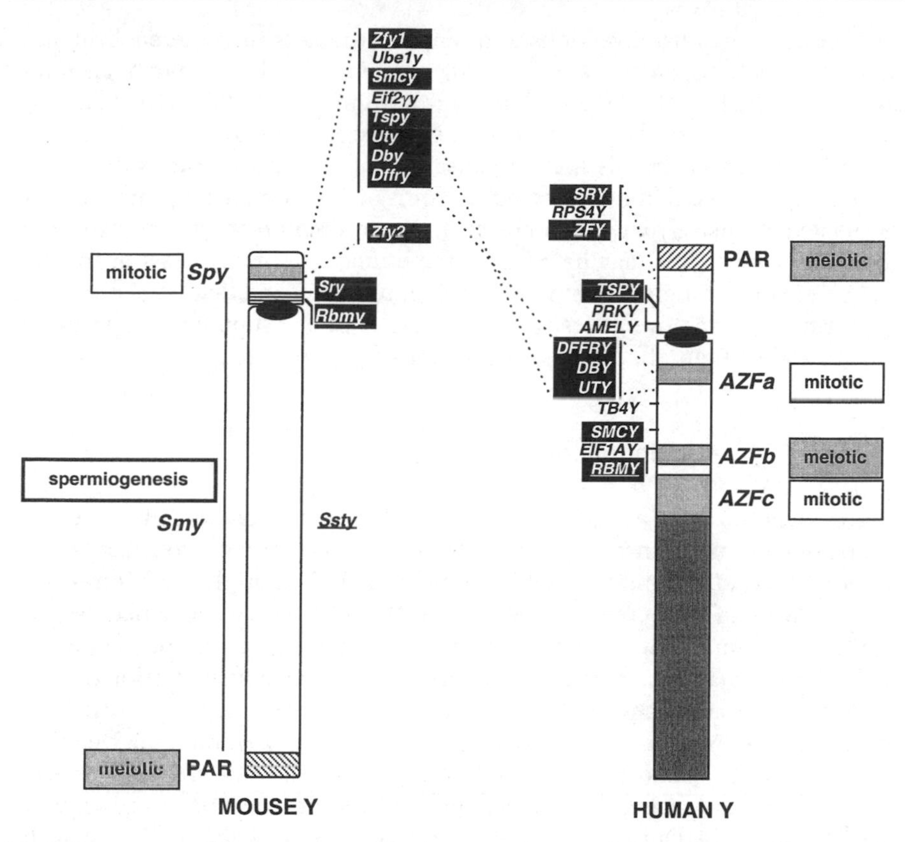

Fig. 3. Comparison of the gene maps of the human and mouse Y chromosomes. All genes with Y homologues in both species are shown (*boxed in grey or black*). All X-Y homologous genes are shown. Multi-copy genes are *underlined*. The genes boxed in *black* represent a syntenic homology between the two species. The genetic factors involved in spermatogenesis together with the stage at which they affect spermatogenesis are shown on the *opposite* side of each chromosome from the genes

genes are clustered within the ΔSxr^b interval, but in the human, they are spread out along 30 Mb (Lahn and Page 1997; Mazeyrat et al. 1998). This difference probably results from both a compression of the X-Y component during rodent evolution, and from a dispersal of the X-Y component during primate evolution. The X homologues of the ΔSxr^b X-Y homologous genes do not co-localise on the mouse or human X chromosome as they do on the mouse Y chromosome, and comparative analyses show that these X-Y homologous genes are of at least two distinct evolutionary origins (see Sect. 5.2.1). This, taken together with the presence of *Tspy*, a Y-specific gene at the centre of the cluster, indicates that these genes have been condensed in the ΔSxr^b interval. There is also convincing evidence that, recently in primate evolution, there have been insertions, inversions and amplifications within the euchromatin that now lies between the X-Y homologous genes on the human Y chromosome (Page et al. 1984; Wolfe et al. 1985; Tyler-Smith et al. 1988; Yen et al. 1988; Lambson et al. 1992; Glaser et al. 1998; Schwartz et al. 1998).

With respect to the distinct organisation of the two chromosomes, it is interesting to note that no reported deletion of the human Y chromosome is associated exclusively with an increase in the production of spermatozoa with abnormal heads. This may be because the human spermiogenesis or fertility factors are intermingled with X-Y homologous genes, and deletions removing the former also remove genes which act earlier in development or spermatogenesis and so their effects on spermiogenesis are obscured.

4.2 A Block of Syntenic Homology

There is a noteworthy exception to the lack of syntenic homology between the mouse and human Y chromosome, and that is the X-Y homologous genes coding for *DFFRY*, *DBY* and *UTY* which are closely linked on the Y chromosome in both species (Mazeyrat et al. 1998). *DFFRY* and *DBY* have also been shown to have the same relative orientation in both species, but the orientation of *UTY* in the human has not yet been determined. In the mouse deletion of these genes in *Sxr*^{b} is associated with a block in spermatogenesis at the mitotic stage. In the human, these genes map close to or within the *AZFa* interval. Deletions of *AZFa* have been shown to be associated with small testis size, and where a testicular biopsy has been available, the testis tubules have been shown to be free of germ cells (Qureshi et al. 1996; Vogt et al. 1996). This suggests that the *AZFa* phenotype arose through a prepubertal block in spermatogonial development. Thus, the *Sxr*^{b} and *AZFa* deletions are associated with similar early blocks in spermatogenesis and share an exceptional block of synteny. This suggests that the *DFFRY-DBY-UTY* genes, or closely linked genes that have yet to be identified, are required for the early stages of spermatogonial proliferation in the mouse and man.

4.3 Implications for Spermatogenesis

The outstanding feature of the current Δ*Sxr*^{b} interval transcription map is its high density of X-Y homologous genes. With the exception of *Zfy* and *Dffry*, these genes have been isolated independently of the human gene map. Therefore, the strong correlation with the transcription map of the human Y chromosome indicates that most, if not all, such genes map into this interval of the mouse Y chromosome. This then raises the possibility that the deletion of X-Y homologous genes underlies impaired spermatogenesis not only in the X*Sxr*^{b}O mouse, but also in certain cases of human infertility where a drastically reduced sperm count, caused by reduced numbers of mitotic germ cells, is associated with a deletion of the Y chromosome. The unique syntenic homology of *DFFRY-DBY-UTY* between the human and mouse Y chromosomes strongly suggests that this is the case with deletions of the *AZFa* interval. This in turn raises the question of whether the deletion of X-Y homologous genes is likely to underlie the *AZFb* and *AZFc* spermatogenesis blocks

The *AZFb* deletion is associated with a very precise block which takes place during meiosis that does not seem to affect the numbers of spermatogonia or primary spermatocytes in the testis tubule, although no post-meiotic germ cells can be detected (Vogt et al. 1996). *AZFc* deletions are, however, frequently associated with a severe reduction in the population of mitotic germ cells (Reijo et al. 1995; Vogt et al. 1996). No X-Y homologous genes have yet been identified in the *AZFc* interval. Four genes, *DAZ*, *BPY2*, *PRY* and *CDY*, have now been mapped into *AZFc*, and it is implied that all four genes have been recently inserted into the Y chromosome from an autosome (Lahn and Page 1997), although this has only been shown to be the case for *DAZ* (Saxena et al. 1996). The mouse model would suggest that these genes are more likely to be spermiogenesis factors, as opposed to being genes required for efficient mitotic germ cell proliferation, and would, moreover, indicate the presence of an X-Y homologous gene in the *AZFc* interval. The transcription maps of the mouse ΔSxr^b interval and the human *AZF* intervals are perhaps not definitive, and these ideas will remain hypothetical until the final maps are available, or until functional analyses reveal the contribution of these genes to spermatogenesis.

The vision of the role of the Y chromosome, based on the mouse map, appears to be in direct conflict with the view of Y gene function that has been proposed on the basis of the human map, which sees X-Y homologous genes as somatic factors, and testis-specific multi-copy genes as germ cell factors. It should, however, be remembered that a ubiquitously expressed house-keeping gene will also be expressed in, and have a role in, the germ cells. The data from the mouse suggests that a restriction of the expression and function to the germ line of the X-Y component can occur during evolution. This is manifested in the mouse by the testis-specific expression of *Dffry*, *Ube1y* and *Zfy*, which is in sharp contrast to the situation in the human, where all X-Y homologous genes are expressed in a wide range of tissues, like their X-homologue. This difference between mouse and man is all the more striking because *Dffry* and *Zfy* have a ubiquitously expressed homologue on the human Y chromosome. Thus, the mouse Y chromosome may be at a later evolutionary stage than the human Y chromosome, a stage where the functions of some genes have become limited to spermatogenesis. This interpretation is consistent with the respective contributions of the two Y chromosomes to the development of the soma. In the next section I will discuss what is understood about the evolution of the sex chromosomes in relation to germ cell development and the transmission of the gametes.

5 Evolution of the Y Chromosome

5.1 Sex Chromosome Evolution Theory

5.1.1 *The Origin of the Non-Recombining Y Chromosome (NRY)*

Sex chromosomes are generally believed to be descended from a homologous pair of autosomes on which a testis-determining allele arose, defining the Y chromosome (Charlesworth 1978; Graves 1995). Once the testis determining allele had arisen on the Y chromosome, this allelic variant would only ever be present in males, but the proto-X and -Y would continue to recombine. The genes closely linked to the testis-determining allele would, however, be effectively male-specific and thus be evolving in a male environment. This provides the conditions necessary for the accumulation of male-enhancing/female detrimental mutations or sexually antagonistic (SA) mutations (Fisher 1931). To prevent these alleles being transferred to females, there would be selection for the suppression of recombination in this region. The truly male-specific portion of the Y chromosome would have enlarged, and this would then spread in the same way to the adjacent genes until recombination had been suppressed along the whole chromosome (Charlesworth 1991). Only a minimal region, the pseudoautosomal region, would be maintained in order to fulfil the needs of X-Y pairing during male meiosis. In *Drosophila*, genetic evidence indicates that SA alleles do arise in sex-limited portions of the genome (Rice 1992). Implicit to this model is that the Y chromosome accumulates male-enhancing alleles but, as will be discussed, their effect may be expected to be short-lived as a direct consequence of the suppression of recombination.

5.1.2 *The Decay of Genes on the Non-Recombining Y Chromosome (NRY)*

The Y chromosome is renowned for its lack of function and a theoretical explanation for the decay of its genetic content has been developed over the past 80 years (Muller 1914; Muller 1964; Charlesworth 1978; Rice 1987; Kondrashov 1988). The basis for this decay is believed to be the absence of recombination. All the genes on a non-recombining chromosome form a single linkage group and selection will be exerted on the whole chromosome haplotype. Optimal haplotypes cannot be constructed by the segregation of deleterious from advantageous mutations. In finite populations, deleterious mutations can, therefore, become fixed by chance and, as back-mutations are extremely rare, this will represent an irreversible deterioration in the genetic content of the Y chromosome. In this way there is presumed to be a steady accumulation of deleterious mutations on the NRY (Charlesworth 1978; Muller 1964). This process is known as "Muller's ratchet" (Felsenstein 1974). Given that Y genes are descended from autosomal genes, it is probable that the loss of the majority of degenerating NRY genes can be complemented by their healthy homologue on the recombining X chromosome. This will not be the case if the dose of the gene product is critical for its function and the transcription of the X and the Y alleles are required to achieve this dose. Haplotypes will initially be selected because they bear functional alleles of such dose-critical genes, resulting in the relatively rapid loss of any non-essential genes.

The critical-dose genes can only be lost from the Y chromosome once the expression of the X-allele is increased, and can therefore independently provide the necessary levels of gene product. It has been proposed that, in such cases, the reduction in the effective dose from the decaying Y-allele will select for X-alleles that are transcribed at a higher rate (Charlesworth 1978). It could also be reasoned that increased levels of expression from the X-allele will be a prerequisite for the selection of haplotypes carrying less functional Y-alleles and thus increased levels of expression of the X-allele may be followed by decay of the Y-allele. Ultimately, the critical dose will be produced in males from one X chromosome, but twice as much X-product as is necessary will be produced in females, from two X chromosomes. This inefficient over-expression in females may be reduced by the inactivation of the gene on the inactive X chromosome. Thus, even if the expression of a Y gene is essential, the disadvantage resulting from the absence of no recombination may be so great that there will be strong selection for the X-allele to adapt, and it is only during this adaptation period that the Y gene will be maintained. Thus, genes which are active on both sex chromosomes are undergoing a dynamic process of co-evolution.

5.1.3
Accumulation of Male-Enhancing Mutations

It is predicted that since the Y chromosome is male-limited it will accumulate male-enhancing mutations. This is based on the theory of Fisher, that the sexually antagonistic allele of a gene can only be conserved in the heterogametic sex if it is linked to the sex-determining locus, and so is never transferred to the homogametic sex (Fisher 1931). It has been shown in *Drosophila* that, initially at least, the transmission of a large segment of the genome exclusively by the male germ line results in an increase of male fitness (Rice 1998). This suggests that there are many genes in the genome whose evolution in males is constrained by counter-selection in females. If these genes are inserted into the NRY chromosome they could evolve to enhance male fitness, irrespective of their effect on female fitness. Male-enhancing mutations could be either mutations in genes already on the Y chromosome, or entire genes that are translocated directly into the NRY. It has also been proposed that the occurrence of such mutations might speed up "Muller's ratchet" by allowing the "genetic hitchhiking" of deleterious mutations: a haplotype carrying a suboptimal set of alleles is selected, and becomes fixed in the population, because it also carries a new beneficial mutation (Rice 1987). The male-enhancing mutations will initially confer a non-essential advantage, and so might not be expected to survive a long time on an undifferentiated Y chromosome, where the primary concern is the transmission of essential Y alleles of X-Y homologous genes. As the X-Y homologous component degenerates, however, the relative contribution of these mutations to the fitness of Y haplotypes will gradually increase.

Theoretical considerations then lead to the view of the NRY gene content as a combination of decaying autosomal genes which were once part of the recombining sex chromosomes, and genes which enhance male fitness, as defined by improving the transmission frequency of the Y-bearing gamete. These male-enhancing genes will also be decaying. The model is dominated by decay. Paradoxically, even the transient boosts to male fitness afforded by male-enhancing alleleic variants in genes freed from the constraints of maintaining female fitness, only accelerate the decay. The genes which have been identified to date on the mouse and human Y chromosome fit well with this model.

5.2 Evolution of Y Genes and Spermatogenesis

NRY genes can be more or less divided into three different groups based on the map position of their closest homologue in mouse or human and the extent to which homologues can be detected on the Y chromosome in different mammalian orders (Vogt et al. 1997). There are two categories for which the evolutionary origins are reasonably clear, and which may represent two extreme types of gene on the Y chromosome. These are the X-Y homologous genes, which are found on the Y chromosome in a wide range of mammalian orders and so have probably been on the Y chromosome since before the radiation of eutherian mammals, and the *DAZ* and *RBMY* genes which have arrived on the NRY from an autosome. The third group are the so-called Y-specific genes, *TSPY* and *Ssty*, whose origins remain obscure because a closest homologue has not yet been defined. Y-specific is a relative term, and it seems likely that this third class will eventually be shown to be an example of one of the other two classes. The evolution of these different genes on the Y chromsome is outlined in Fig. 4. Seven new testis-specific gene families have recently been described on the human Y chromosome. No data is yet available to indicate whether they represent recent insertions into the Y chromosome or genes that have been conserved on the Y chromosome during mammalian evolution, and so they will not be discussed further in this review. The comparative analysis of these

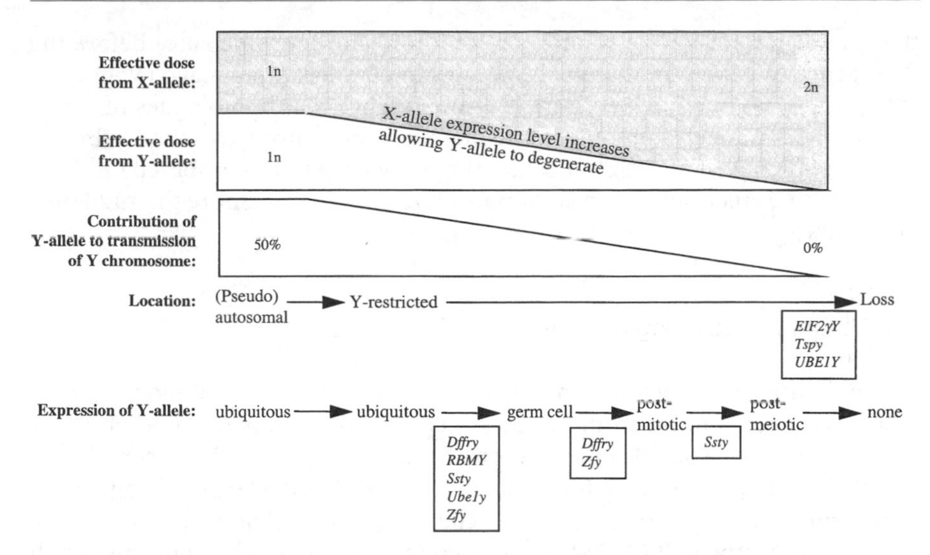

Fig. 4. The evolution of genes on the Y chromosome. The location and transcription pattern of Y genes is shown in relation to the hypothesised changes in dose of the X and Y gene pair, and the contribution of the Y allele to transmission of the Y-bearing gamete. The genes that provide evidence for the different evolutionary steps are shown in the *boxes*

genes and their non-Y homologues promises an important boost to our understanding of the forces that shape the gene content of the Y chromosome.

5.2.1
X-Y Homologous Genes

Of the three classes of genes, the X-Y homologous genes provide the most obvious support for the classical model of sex chromosome evolution and the view of the Y chromosome as a degenerating X chromosome. These genes are believed to have been part of the original autosomal pair that became the sex chromosomes, or to have been later additions to the differentiated sex chromosomes via the pseudoautosomal region (Graves 1995). The X-Y homologous genes can be defined as those NRY genes whose closest homologue is on the X chromosome: *AMELY, DFFRY/Dffry, DBY/Dby, EIF1AY, Eif2γy, PRKY, RPS4Y, SMCY/Smcy, SRY/Sry, Ube1y, UTY/Uty* and *ZFY/Zfy* (Page et al. 1987; Mardon et al. 1989; Fisher et al. 1990; Mitchell et al. 1991; Nakahori et al. 1991; Agulnik et al. 1994; Collignon et al. 1996; Greenfield et al. 1996; Lahn and Page 1997; Schiebel et al. 1997; Brown et al. 1998; Ehrmann et al. 1998; Mazeyrat et al. 1998). In a given species these heterologous gene pairs show approximately 80–90% nucleotide identity. This high sequence identity has made them ideal tools for comparative studies by Ark blot Southern analysis. In general, male-specific bands are detected in a wide range of eutherian mammals, showing that these genes have been conserved on the Y chromosome since before the divergence of eutherian mammals more than 80 million years ago (Vogt et al. 1997; Mazeyrat et al. 1998). There is also evidence that homologues of *Smcy*, *Sry* and *Ube1y* are also present on the metatherian Y chromosome (Agulnik et al. 1994; Mitchell et al. 1992), showing that they have been conserved on the Y chromosome since before the eutherian-metatherian split, more than 120 million years ago.

5.2.1.1 The Ubiquitin Activating Enzyme

The ubiquitin activating enzyme gene (*UBE1*) is the X-Y homologous gene which has been studied in the greatest depth at the comparative level. *UBE1* may represent an example of a gene which was on the original autosomal pair, since a homologue has been found at the distal end of the large pairing region of the monotreme X chromosome (Mitchell et al. 1998). Evidence supports the existence of distinct *UBE1* X and Y homologues on the sex chromosomes of all orders of mammals tested (Mitchell et al. 1991, 1992, 1998). *UBE1* Y-homologues have been isolated and sequenced from kangaroo, mouse, lemur and new world monkeys. A faster rate of nucleotide substitution in intron sequences, compared with exon sequences, shows that the gene has been conserved on the Y chromosome. Despite this broad conservation, the *UBE1* Y homologue appears to have been lost from certain lineages during primate evolution, as no trace of *UBE1* homologues has been found on the Y chromosome of human, chimpanzee or macacque (Mitchell et al. 1991, 1998). Thus *UBE1* illustrates that, as the theory predicts, a pseudoautosomal gene can evolve into non-recombining X and Y restricted genes and then finally be lost from the Y chromosome to become an X-specific gene.

5.2.1.2 Dosage Compensation

A comparison of the Y chromosome gene map with that of the X chromosome makes it evident that the X-Y homologous genes that have remained on the Y chromosome are certainly only a subset of the genes that were on the original autosomal pair. Thus, the majority of genes are relatively rapidly lost from the Y chromosome when recombination ceases. A functional X-Y homologous gene pair is probably conserved, because a critical dose of the gene product can only be achieved by transcription from two alleles. Loss of the Y allele will, therefore, cause a haploinsufficiency (Charlesworth 1978). Examples of the evolution of genes that are not dose-dependent may be provided by the human Y-homologues of the X-linked arylsulfatase gene, *STS*, and the *KAL* gene. Males deleted for *STS* or *KAL* suffer from X-linked ichthyosis or Kallmann syndrome, respectively. Both genes map to human Xp22.3, just outside the pseudoautosomal region, and they are expressed from both the active and inactive human X chromosome (Ballabio et al. 1987; Yen et al. 1987; Franco et al. 1991; Legouis et al. 1991). Sufficient transcript levels can nevertheless be produced from a single copy of the gene, as females carrying an X chromosome deleted for these genes are asymptomatic, showing that these are not dose-critical genes. Homologues of *STS* and *KAL* are present on the human Y chromosome, but they are decayed non-transcribed genomic copies (del Castillo et al. 1992; Incerti et al. 1992; Yen et al. 1988). A similar situation exists for other Xp22.3 genes: *ARSE*, *ARSD*, and *APLX* (Schiaffino et al. 1995; Meroni et al. 1996). The rate of nucleotide substitution in introns is the same as in exons, suggesting that they have never been functionally conserved on the Y chromosome.

If an X-Y homologous gene pair continues to behave like an autosomal gene whose expression is required from both the X-allele and the Y-allele in males, then it would be expected that the X-homologue of a conserved Y gene will be transcribed from both the active and inactive X chromosomes in females. In both mouse and human there is a correlation between transcription from the inactive X chromosome and the expression pattern of the Y-gene (see Fig. 2). Y-genes which are ubiquitously expressed have an X-homologue which escapes X-inactivation in the same species, while Y-genes whose expression is limited to the testis have an X-homologue which is subject to X-inactivation (Disteche 1995). The *ZFY/ZFX* genes illustrate this well, as *ZFY* is ubiquitously expressed in humans and its X-homologue *ZFX* escapes X-inactivation (Page et al. 1987; Schneider-Gädicke et al. 1989), whereas in the mouse, *Zfy* shows testis-specific expression and *Zfx* is not expressed from the inactive X chromosome (Mardon and Page 1989; Adler et al. 1991).

Ube1y and *Eif2γy* are two X-Y homologous genes which have been identified on the mouse Y chromosome (Mitchell et al. 1991; Ehrmann et al. 1998). The expression of *Ube1y* is limited to the testis and its X-homologue *Ube1x* does not escape X-inactivation. In contrast, the expression of *Eif2γy* is ubiquitous and its X-homologue *Eif2γx* does escape X-activation. Homologues of these two genes have not been found on the human Y chromosome and they have apparently been lost from the Y chromosome during primate evolution. Nevertheless, in the human, the X-homologues, *UBE1* and *EIF2γX*, escape X-inactivation (Brown and Willard 1989; Carrel et al. 1996; Ehrmann et al. 1998). This strongly suggests that incorporation into the X-inactivation system is one of the final steps in the evolution of these genes.

5.2.1.3
Restriction of Expression to the Germ Line

All the NRY X-Y homologous genes have been mapped together on the short arm of the mouse Y chromosome. Although their deletion does not prevent normal development, it cannot be ruled out that the expression of these genes confers a significant, although non-essential, advantage at the somatic level. The deletion of these genes is, however, incompatible with germ cell survival past the spermatogonial stage. It therefore seems likely that the critical functions of these genes have become restricted to germ cell development. This is further suggested by the fact that the expression of *Ube1y*, *Zfy* and *Dffry* is more or less limited to the germ cells. Thus, an intermediate step in the evolution of X-Y homologous genes may be the restriction of gene expression to the germ-line.

Expression studies with mouse Y chromosome genes indicate the occurrence of further intermediate steps. *Ube1y* is expressed in germ cells from before birth (Odorisio et al. 1996) but *Zfy* and *Dffry* are not expressed postnatally until 7–10 days after birth (Nagamine et al. 1990; Brown et al. 1998), around the time when germ cells are beginning to enter meiosis. Thus the expression of genes from the Y chromosome may become limited not only to germ cells but to particular stages of spermatogenesis. The testes are believed to be the site of much promiscuous transcription, resulting from a relaxation of transcriptional control in post-meiotic germ cells, brought about by the remodelling of chromatin during spermiogenesis. This does not seem to be the explanation for the testis-specific expression of these three genes as they are transcribed in testis in which the germ cells have not yet entered spermiogenesis.

This limitation of expression to the germ line could be presumed to occur for two separate, but not necessarily mutually exclusive reasons. Firstly, the X-Y homologous genes, although initially selected to avoid haploinsufficiency of the gene product, will be evolving in a male environment, and therefore may accumulate male-enhancing mutations which act during gametogenesis. After the dosage compensation of the X gene removes the need for the Y allele, it may be selectively maintained on the basis of its acquired male-enhancing functions. In the mouse, the expression of *Dffry*, *Ube1y* and *Zfy* is effectively limited to the germ cells. This would therefore suggest that the majority of male-enhancing mutations act during gametogenesis. The second explanation is that the restricted expression of the X-Y homologous genes is a direct consequence of the decay process (see Fig. 4). As the dose of the X allele increases, the Y allele will decay, and thus its selective advantage will diminish to a point where it can be lost without affecting fitness. It would be expected that this point will be reached in somatic tissues before germ cells. For Y genes, fitness translates to the efficiency with which the Y-bearing gamete is transmitted, and so the advantage conferred by continued expression of the Y allele will most likely be greatest in the gametogenic cells themselves. An allele with only a slightly advantageous effect on development may retain a significant selective advantage if it acts directly in gametogenesis to increase the number or fecundity of spermatozoa. Thus, the germ cell specific expression of X-Y homologous genes may reflect the diminishing contribution of the Y allele to cellular development.

5.2.2
Y-Autosomal Genes

5.2.2.1
DAZ

The *DAZ* gene (formerly also known as *SPGY*) is the gene which defines this class of Y chromosome gene (Reijo et al. 1995; Vogt et al. 1997). It is the homologue of the germ cell specific gene *DAZL1* (formerly also known as *DAZH*, *DAZLA*, or *SPGLA*), which has been mapped to chromosome 3 in humans and chromosome 17 in mice (Maiwald et al. 1996; Reijo et al. 1996; Saxena et al. 1996; Shan et al. 1996; Yen et al. 1996; Seboun et al. 1997). *DAZ* has only been found on the Y chromosome in primates and is believed to have been transposed from an autosome during primate evolution, after the divergence of the new world monkey lineage from other primates, about 35–55 million years ago (Shan et al. 1996; Agulnik et al. 1998). *DAZ* is present in multiple copies on the Y chromosome of humans and other primates. *DAZ* was initially isolated as *DYS1*, and was believed to be a pseudogene (Seboun et al. 1986). Its functionality remains controversial, however. Comparison of *DAZ* and *DAZL1* cDNA sequences indicated that the UTRs had diverged slightly further than the coding sequences, suggesting that *DAZ* has been conserved on the Y chromosome (Saxena et al. 1996). Recent sequence comparisons of *DAZ* genes from chimpanzee, rhesus monkey and baboon show, however, that there is no significant difference between the substitution rates at synonymous or non-synonymous positions, suggesting that there is at best only weak selection pressure to maintain the coding sequence of *DAZ* (Agulnik et al. 1998). Nevertheless, the *DAZ* gene has maintained its ORF and so it seems probable that *DAZ* has been selectively maintained on the NRY.

Despite the difficulty in proving that *DAZ* is functional, it has good credentials for being a fertility factor: most *DAZ* copies map to the *AZFc* deletion interval of the Y chromosome (Saxena et al. 1996), the knock-out of the mouse autosomal homologue *Dazl1* blocked male and female germ cell development (Ruggiu et al. 1997), and it is a homologue of *boule*, a gene required for germ cell development in *Drosophila* (Eberhart et al. 1996). Thus, *DAZ* has become the accepted evolutionary model for the testis-specific multi-copy Y chromosome genes which are assumed to have arisen on the Y chromosome by transposition and amplification of an autosomal gene (Saxena et al. 1996). These genes are presumed to be selectively maintained because they improve male reproductive fitness and therefore favour the transmission of the Y chromosome haplotype into which they have integrated.

The integration of *DAZ* into the Y chromosome most likely benefits spermatogenesis by allowing a function of *DAZL1* to be performed with greater efficiency. The neutral evolution of *DAZ* suggests that *DAZ* is unlikely to represent a "super-*DAZL1*". It is therefore reasonable to assume that the effects of the deletion of *DAZ* on spermatogenesis will not surpass the deletion of one copy of the autosomal *DAZL1*. The phenotype of the mouse heterozygous for a null allele of *Dazl1* may then provide an insight into the maximum effect expected if *DAZ* were deleted from the Y chromosome. The heterozygous *Dazl1* mice are fertile, but produce increased numbers of sperm with abnormal heads (Ruggiu et al. 1997). On this basis, it appears unlikely that the absence of *DAZ* alone is responsible for the severe early block in germ cell development associated with deletions that affect *AZFc*.

5.2.2.2 RBMY

RBMY homologues have been found in several orders of eutherian mammals and in metatherians, suggesting that, like *Smcy* and *Ube1y*, the origin of *RBMY* on the Y chromosome pre-dates the metatherian-eutherian split of 120 million years ago (Ma et al. 1993; Delbridge et al. 1997). Like *DAZ*, *RBMY* is present on the human Y chromosome as a testis-specific multi-copy gene family, and the mouse Y chromosome homologue *Rbmy* is similarly arranged (Laval et al. 1995; Elliot et al. 1996; Mahadevaiah et al. 1998). *RBMY* is homologous to *HNRPG* (60% nucleotide identity) which has been mapped to chromosome 6p12 in humans (Le Coniat et al. 1992).

Unlike *DAZL1*, whose expression is limited to the germ cells, *HNRPG* is expressed in a wide range of tissues (Soulard et al. 1993). This is almost certainly the closest homologue of *RBM*, as screening of sheep and wallaby (a marsupial) libraries with *RBMY* probes resulted in the isolation of one type of cDNA from each species, which has >93% amino acid identity to human *HNRPG* (Delbridge et al. 1998). It is proposed that *RBMY* originated on the Y chromosome through the transposition of an ancestral autosomal *HNRPG* gene (Chai et al. 1998; Delbridge et al. 1998). Thus, *RBMY* is concluded to be the same class of gene as *DAZ*. The only reservation regarding this conclusion is that *RBMY* has been conserved on the Y for more than 120 million years, while *DAZ* is a recent addition to the primate Y chromosome. Theoretical considerations suggest that such fertility factors will be short-lived on the degenerating Y chromosome, as they provide only non-essential benefits. It is therefore surprising to find that a fertility factor is one of the most ancient extant Y chromosome genes. Our recent mapping studies, however, indicate that in mouse and human an *HNRPG* gene containing introns maps to the X chromosome (M.J. Mitchell and S. Mazeyrat, unpubl. data). If this is confirmed then it suggests that *RBMY* is really an X-Y homologous gene.

5.2.3
Genes of Unknown Origin

5.2.3.1
TSPY

TSPY homologues have been found in several orders of eutherian mammal, suggesting that these genes have been conserved on the Y chromosome for more than 80 million years. Like *DAZ* and *RBMY*, *TSPY* is present on the human Y chromosome as a testis-specific, multi-copy gene family, and it is similarly arranged on the bovine Y chromosome (Arnemann et al. 1991; Schnieders et al. 1996; Vogel et al. 1997). *TSPY* genes on the mouse and rat Y chromosome are also transcribed exclusively in the testis, but are present in three or fewer copies and one copy, respectively (Mazeyrat and Mitchell 1998; Vogel et al. 1998). A comparison of the rat and mouse *Tspy* genes has provided the first definitive proof that genes which have been conserved on the Y chromosome can be lost from the genome (Mazeyrat and Mitchell 1998). Spread over six exons, the rat *Tspy* produces a spliced testis-specific transcript which codes a putative protein which is homologous to that coded by the human *TSPY*. In contrast, the mouse *Tspy* produces low levels of large unspliced or small "over-spliced" transcripts. Sequence comparison of the genomic sequence of the rat and mouse *Tspy* genes showed that although the mouse gene is very similar to the rat gene (92% nucleotide identity in the coding and 83% in the non-coding regions), the mouse gene contains mutations that destroy three splice sites, and two small deletions which cause frameshifts in the ORF. *Tspy* can therefore be concluded to have lost its function in the mouse lineage after its divergence from the rat lineage.

The origin of *Tspy* is difficult to discern. Its closest defined homologue is the proto-oncogene *SET*, with which it shares 53% amino-acid similarity (Schnieders et al. 1996). *SET*, like *DAZL1* and *HNRPG*, maps to an autosome (chromosome 9q34 in humans; von Lindern et al. 1992). The limited information available on the intron-exon structure of human *SET* indicates that it is distinct from that of *TSPY* (von Lindern et al. 1992). In addition, the genomic sequence of the *SET* homologue in the fish, *Tetraodon fluviatilis*, has been determined (Genbank acc. no. AF007219) and the intron-exon structure of this gene is very different from *TSPY*. It therefore seems unlikely that *SET* is the gene from which *TSPY* was derived, and the route by which *TSPY* found its way to the Y chromosome is not yet known.

5.2.3.2
Ssty

The mouse *Ssty* gene has a homologue on the X chromosome in mouse (*DXHXF34*) and human (*DXF34*; Laval et al. 1997). Both X genes are present as multiple copies close to the centromere, and *DXF34*-identical ESTs have been derived from a number of tissues, showing that its expression is not limited to the testis (M.J. Mitchell pers. observations). It is only in the mouse that copies of *Ssty* have been found on the Y chromosome (Bishop and Hatat 1987; Prado et al. 1992). Thus, it has been concluded that *Ssty* has been acquired from the X chromosome during rodent evolution by transposition directly into the NRY (Laval et al. 1997). The sequence of *DXHXF34* has unfortunately not yet been determined, but *Ssty* shares only 70% amino acid identity with the human *DXF34*. This low level of sequence identity indicates that *Ssty* has been conserved on the Y chromosome for a long time, and represents a strong counter-argument to the idea of a recent rodent origin for this gene. Given the low level of sequence identity between *Ssty* and *DXF34*, it is possible that the failure to detect Y homologues in other mammals is due to the fact that the *Ssty* Y-homologues do not cross-hybridise strongly enough to be detected by Southern analysis.

A complication in our understanding of the evolution of *Ssty* is the existence of the mouse *Spindlin* gene (Oh et al. 1997) which shows 85% nucleotide identity to the human *DXF34* gene. The *Spindlin* gene, however, contains introns within the coding part of the gene which are missing from the genomic copies of *DXF34* and *Ssty* (B. Oh, pers. comm.), and so it is likely that the sex chromosome genes were derived by retroposition of a *Spindlin* transcript. The fact that *DXF34* has only 85% nucleotide identity to *Spindlin*, but has maintained an ORF which codes for a putative protein with 90% amino acid similarity to *Spindlin*, indicates that this retrogene has been conserved on the X chromosome. The X localisation in mouse and human further suggests that the original retroposition occurred prior to the divergence of eutherian lineages over 80 million years ago.

There are therefore three possible origins for *Ssty*. Firstly, that it was within a recombining segment of the X and Y chromosomes, in which case it is a diverged X-Y homologous gene. Secondly, that it has transposed to the Y chromosome during rodent evolution from the X chromosome or other *Spindlin* retroposons present in the genome. Thirdly, that it arose as an independent retroposition event from *Spindlin*.

5.2.4 *Y Chromosome Gene Evolution and Function*

The majority of genes on the mammalian Y chromosome, perhaps including *RBMY*, are probably descendants of genes which were originally recombining gene pairs on an ancestral X and Y chromosome. The *DAZ* gene clearly represents another type of Y chromosome gene, which has translocated directly into the NRY from an autosome. The *DAZ* gene and its autosomal homologue *DAZL1* are expressed specifically in germ cells, but only some of the X-Y homologous genes are expressed exclusively in germ cells and the others

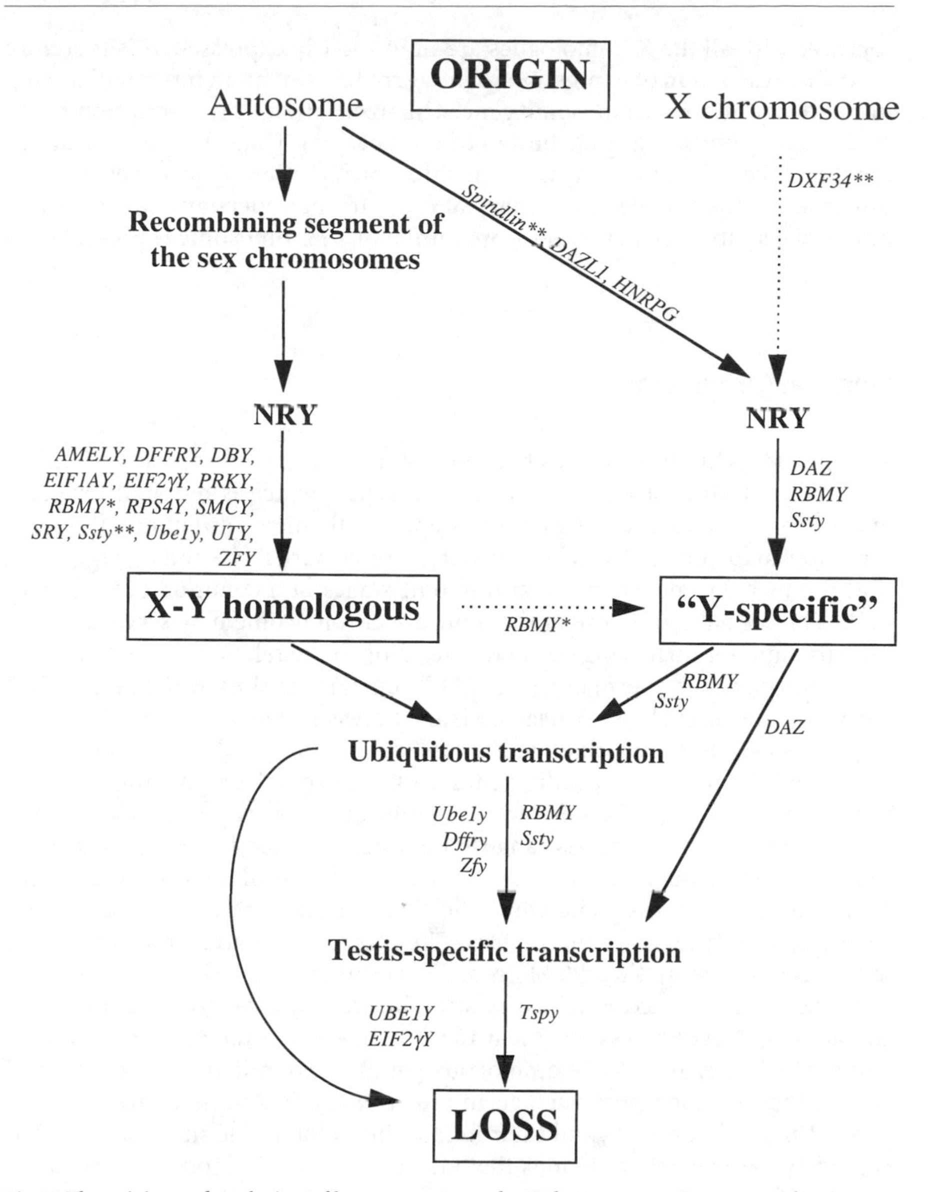

Fig. 5. The origins and evolution of known genes on the Y chromosome. Genes are either inserted directly into the NRY or indirectly via the pairing segment of the sex chromosomes. X-Y homologous genes may evolve into "Y-specific" genes. Steps for which the evidence is tentative are indicated by *dotted lines*. The genes that illustrate the different steps are shown. It is not known whether the loss of *UBE1Y* and *EIF2γY* in the primate lineage was preceded by restriction of transcription to the germ line. *RBMY** indicates the alternative possibility that *RBMY* originated as an X-Y homologous gene. ** indicates the three possible origins of the *Ssty* gene

together with all the X homologues are ubiquitously expressed. It is therefore clear that restriction of expression to the germ line can be an intermediate step in the decay of a Y chromosome gene. This restriction of transcription could reflect the diminishing contribution of the decaying Y allele to cellular development. It could also reflect the acquisition or retention of germ cell-specific functions by the Y allele. A summary of our current understanding of the origins and evolution of genes on the mammalian Y chromosome is presented in Fig. 5.

6 General Conclusions

The mouse Y chromosome has become specialised in germ cell development, and in this it differs from its human counterpart which is required for germ and somatic development. Despite this apparently major difference, the transcription maps of the two chromosomes are remarkably similar, suggesting that the two chromosomes are at different stages of a common evolutionary path. The specialised nature of the mouse Y chromosome may serve as a useful paradigm for furthering our understanding of the relationship between the Y chromosome and spermatogenesis. At its current resolution, the mouse NRY chromosome map shows a neat division between two types of genes: X-Y homologous genes are located in the *Sxr*^{b} deletion interval on the short arm, while multi-copy, testis-specific genes with autosomal or undefined homologues are located in the pericentric region and the long arm. The genetic functions in spermatogenesis can also be roughly divided along similar lines: the *Sxr*^{b} deletion interval is required for the mitotic proliferation and differentiation of the spermatogonia while the long arm and pericentric region are required for efficient spermiogenesis. The molecular characterisation of this latter region of more than 90 Mb remains rudimentary, and so caution needs to be exercised in making the correlation between the genetic and transcriptional maps. Nevertheless, it is clear that the interval of the mouse Y chromosome which is rich in X-Y homologous genes is essential for early stages of spermatogenesis and post-natal germ cell survival. In addition, three of these genes *Dffry*, *Dby* and *Uty* and their human homologues define a unique conserved syntenic block. Deletions that affect this block in mouse and man are associated with a severe early spermatogenic block. Thus, this syntenic homology suggests that X-Y homologous NRY genes are required for spermatogenesis, not only in mouse but also in human.

Molecular evidence now supports the prediction that the genes on the Y chromosome are decaying and will inevitably be lost from the genome. Thus all Y genes lie somewhere between being functionally interchangeable with their X homologues, and being non-functional. An intermediate step in this descent is that a gene can become non-functional in the soma while remaining functional in the germ line. As this decay process may occur relatively rapidly it is difficult to draw firm conclusions about the critical functions of a Y gene based on its expression pattern: a testis specific gene may no longer be functional, a ubiquitously expressed gene may only be functional in the testis. The Y chromosome is evolving rapidly towards its disappearance. This evolutionary end has been achieved in certain species of rodents (Just et al. 1995). The mouse Y chromosome is not yet gone but is diminished, being reduced to a role in germ cell development. The apparent specialisation of the mouse Y chromosome may only be a part of its decay.

7
The Future

Until now, the study of the Y chromosome has been dominated by the definition and comparative mapping of its gene content. Much has been learned about the evolution of the Y chromosome and indirectly about how it might fulfil its role in male development. The human gene map may be more or less complete but there is an urgent need to extend the mouse Y chromosome map to the large repeat-ridden long arm. This may provide the necessary tools for the elucidation of the impact of the Y chromosome on spermiogenesis and the transmission of the Y-bearing gamete. The glaring gap in our knowledge of the role of the NRY during spermatogenesis is the absence of functional data. With the exception of *SRY* we have no idea what any single NRY gene does during male development. In the human, no point mutations or single gene deletions have been described in candidate azoospermia genes. In the mouse, no transgenic experiments have been described which complement the effects of a deletion or knock out a gene. The decaying nature of the Y chromosome may complicate the functional analysis, because the genes with diminished doses may not be required individually for specific stages of spermatogenesis but may instead function in synergy with other Y genes to improve the efficiency of spermatogenesis. The design and successful implementation of informative experiments aimed at elucidating the role of NRY genes in male development is the current challenge for those working in the field. It is to be hoped that this will be achieved prior to the complete extinction of this singular mammalian chromosome.

References

Adler DA, Bressler SL, Chapman VM, Page DC, Disteche CM (1991) Inactivation of the mouse *Zfx* gene on the mouse X chromosome. Proc Natl Acad Sci USA 88:4592–4595

Agulnik AI, Mitchell MJ, Lerner JL, Woods DR, Bishop CE (1994) A mouse Y chromosome gene encoded by a region essential for spermatogenesis and expression of male-specific minor histocompatibility antigens. Hum Mol Genet 3:873–878

Agulnik AI, Zharkikh A, Boettger-Tong H, Bourgeron T, McElreavey K, Bishop CE (1998) Evolution of the DAZ gene family suggests that Y-linked DAZ plays little, or a limited, role in spermatogenesis but underlines a recent African origin for human populations. Hum Mol Genet 7:1371–1377

Arnemann J, Jakubiczka S, Thuring S, Schmidtke J (1991) Cloning and sequence-analysis of a human Y-chromosome-derived, testicular cDNA, *TSPY*. Genomics 11:108–114

Ballabio A, Parenti G, Carrozzo R, Sebastio G, Andria G, Buckle V, Fraser N, Craig I, Rocchi M, Romeo G, Jobsis AC, Persico MG (1987) Isolation and characterization of a steroid sulfatase cDNA clone: genomic deletions in patients with X-chromosome-linked ichthyosis. Proc Natl Acad Sci USA 84:4519–4523

Baron B, Metezeau P, Hatat D, Roberts C, Goldberg ME, Bishop C (1986) Cloning of DNA libraries from mouse Y chromosomes purified by flow cytometry. Somat Cell Mol Genet 12: 289–295

Bergstrom DE, Yan H, Sonti MM, Narayanswami S, Bayleran JK, Simpson EM (1997) An expanded collection of mouse Y chromosome RDA clones. Mamm Genome 8:510–512

Bergstrom DE, Grieco DA, Sonti MM, Fawcett JJ, Bell-Prince C, Cram LS, Narayanswami S, Simpson EM (1998) The mouse Y chromosome: enrichment, sizing, and cloning by bivariate flow cytometry. Genomics 48:304–313

Berta P, Hawkins JR, Sinclair AH, Taylor A, Griffiths BL, Goodfellow PN, Fellous M (1990) Genetic evidence equating SRY and the testis-determining factor. Nature 348:448–450

Bishop CE (1992) The mouse Y chromosome. Mamm Genome 3:S289–S293

Bishop CE, Hatat D (1987) Molecular cloning and sequence analysis of a mouse Y chromosome transcript expressed in the testis. Nucleic Acids Res 15:2959–2969

Bishop CE, Roberts C, Michot JL, Nagamine C, Winking H, Guenet JL, Weith A (1987) The use of specific DNA probes to analyse the *Sxr* mutation in the mouse. Development 101 Suppl: 167–175

Brown CJ, Willard HF (1989) Noninactivation of a selectable human X-linked gene that complements a murine temperature-sensitive cell cycle defect. Am J Med Genet 45:592–598

Brown GM, Furlong RA, Sargent CA, Erickson RP, Longepied G, Mitchell M, Jones MH, Hargreave TB, Cooke HJ, Affara NA (1998) Characterisation of the coding sequence and fine mapping of the human DFFRY gene and comparative expression analysis and mapping to the Sxr^b interval of the mouse Y chromosome of the Dffry gene. Hum Mol Genet 7:97–107

Burgoyne PS (1993) A Y-chromosomal effect on blastocyst cell number in mice. Development 117:341–345

Burgoyne PS, Baker TG (1985) Perinatal oocyte loss in XO mice and its implications for the aetiology of gonadal dysgenesis in XO women. J Reprod Fertil 75:633–645

Burgoyne PS, Levy ER, McLaren A (1986) Spermatogenetic failure in mice lacking H-Y antigen. Nature 320:170–172

Burgoyne PS, Mahadevaiah SK, Sutcliffe MJ, Palmer SJ (1992) Fertility in mice requires X-Y pairing and a Y-chromosomal "spermiogenesis" gene mapping to the long arm. Cell 71:391–398

Burgoyne PS, Thornhill AR, Boudrean SK, Darling SM, Bishop CE, Evans EP (1995) The genetic basis of XX-XY differences present before gonadal sex differentiation in the mouse. Philos Trans R Soc Lond B Biol Sci 350:253–261

Capel B, Rasberry C, Dyson J, Bishop CE, Simpson E, Vivian N, Lovell-Badge R, Rastan S, Cattanach BM (1993) Deletion of Y chromosome sequences located outside the testis determining region can cause XY female sex reversal. Nat Genet 5:301–307

Carrel L, Clemson CM, Dunn JM, Miller AP, Hunt PA, Lawrence JB, Willard HF (1996) X inactivation analysis and DNA methylation studies of the ubiquitin activating enzyme E1 and PCTAIRE-1 genes in human and mouse. Hum Mol Genet 5:391–401

Cattanach BM, Pollard CE, Hawkes SG (1971) Sex-reversed mice: XX and XO males. Cytogenetics 10:318–337

Cattanach BM, Evans EP, Burtenshaw MD, Barlow J (1982) Male, female and intersex development in mice of identical chromosome constitution. Nature 300:445–446

Chai NN, Zhou H, Hernandez J, Najmabadi H, Bhasin S, Yen PH (1998) Structure and organization of the RBMY genes on the human Y chromosome: transposition and amplification of an ancestral autosomal hnRNPG gene. Genomics 49:283–289

Charlesworth B (1978) Model for evolution of Y chromosomes and dosage compensation. Proc Natl Acad Sci USA 75:5618–5622

Charlesworth B (1991) The evolution of sex chromosomes. Science 251:1030–1033

Chuang RY, Weaver PL, Liu Z, Chang TH (1997) Requirement of the DEAD-Box protein ded1p for messenger RNA translation. Science 275:1468–1471

Collignon J, Sockanathan S, Hacker A, Cohen-Tannoudji M, Norris D, Rastan S, Stevanovic M, Goodfellow PN, Lovell-Badge R (1996) A comparison of the properties of *Sox-3* with *Sry* and two related genes, *Sox-1* and *Sox-2*. Development 122:509–520

Conway SJ, Mahadevaiah SK, Darling SM, Capel B, Rattigan AM, Burgoyne PS (1994) *Y353/B*: a candidate multiple-copy spermiogenesis gene on the mouse Y chromosome. Mamm Genome 5:203–210

Delbridge ML, Harry JL, Toder R, Oneill RJW, Ma K, Chandley AC, Graves JAM (1997) A human candidate spermatogenesis gene, *RBM1*, is conserved and amplified on the marsupial Y chromosome. Nat Genet 15:131–136

Delbridge ML, Ma K, Subbarao MN, Cooke HJ, Bhasin S, Graves JA (1998) Evolution of mammalian HNRPG and its relationship with the putative azoospermia factor RBM. Mamm Genome 9:168–170

del Castillo I, Cohen-Salmon M, Blanchard S, Lutfalla G, Petit C (1992) Structure of the X-linked Kallmann syndrome gene and its homologous pseudogene on the Y chromosome. Nat Genet 2:305–310

Disteche CM (1995) Escape from X inactivation in human and mouse. Trends Genet 11:17–22

Eberhart CG, Maines JZ, Wasserman SA (1996) Meiotic cell cycle requirement for a fly homologue of human *Deleted in Azoospermia*. Nature 381:783–785

Ehrmann I, Ellis P, Mazeyrat S, Duthie S, Brockdorf N, Mattei M-G, Gavin M, Simpson E, Mitchell MJ, Scott D (1998) Characterisation of genes encoding translation initiation factor eIF-2γ in mouse and human: sex chromosome localization, escape from X-inactivation and evolution. Hum Mol Genet 7:1725–1737

Eicher EM, Hutchison KW, Phillips SJ, Tucker PK, Lee BK (1989) A repeated segment on the mouse Y chromosome is composed of retroviral- related, Y-enriched and Y-specific sequences. Genetics 122:181–192

Eicher EM, Hale DW, Hunt PA, Lee BK, Tucker PK, King TR, Eppig JT, Washburn LL (1991) The mouse Y* chromosome involves a complex rearrangement, including interstitial positioning of the pseudoautosomal region. Cytogenet Cell Genet 57:221–230

Eichwald EJ, Silmser CR (1955) Untitled communication. Transplant Bull 2:148–149

Elliot D, Ma K, Kerr S, Thakrar R, Speed R, Chandley A, Cooke H (1996) An *RBM* homologue maps to the mouse Y chromosome and is expressed in germ cells. Hum Mol Genet 5:869–874

Elliott DJ, Millar MR, Oghene K, Ross A, Kiesewetter F, Pryor J, McIntyre M, Hargreave TB, Saunders PT, Vogt PH, Chandley AC, Cooke H (1997) Expression of RBM in the nuclei of human germ cells is dependent on a critical region of the Y chromosome long arm. Proc Natl Acad Sci USA 94:3848–3853

Evans EP, Burtenshaw MD, Cattanach BM (1982) Meiotic crossing-over between the X and Y chromosomes of male mice carrying the sex-reversing (Sxr) factor. Nature 300:443–445

Felsenstein J (1974) The evolutionary advantage of recombination. Genetics 78:737–756

Fennelly J, Harper K, Laval S, Wright E, Plumb M (1996) Co-amplification of tail-to-tail copies of *MuRVY* and *IAPE* retroviral genomes on the Mus musculus Y chromosome. Mamm Genome 7:31–36

Fisher EMC, Beer-Romero P, Brown LG, Ridley A, McNeil JA, Bentley Lawrence J, Willard HF, Bieber FR, Page DC (1990) Homologous ribosomal protein genes on the human X and Y chromosomes: escape from X inactivation and possible implications for Turner syndrome. Cell 63:1205–1218

Fisher R (1931) The evolution of dominance. Biol Rev 6:345–368
Foote S, Vollrath D, Hilton A, Page DC (1992) The human Y chromosome: overlapping DNA clones spanning the euchromatic region. Science 258:60–66
Ford CE (1966) The murine Y chromosome as a marker. Transplantation 4:333–335
Franco B, Guioli S, Pragliola A, Incerti B, Bardoni B, Tonlorenzi R, Carrozzo R, Maestrini E, Pieretti M, Taillon-Miller P, Brown CJ, Willard HF, Lawrence C, Persico MG, Camerino G, Ballabio A (1991) A gene deleted in Kallmann's syndrome shares homology with neural cell adhesion and axonal path-finding molecules. Nature 353:529–536
Gaspar NJ, Kinzy TG, Scherer BJ, Humbelin M, Hershey JW, Merrick WC (1994) Translation initiation factor eIF-2. Cloning and expression of the human cDNA encoding the gamma-subunit. J Biol Chem 269:3415–3422
Gee SL, Conboy JG (1994) Mouse erythroid cells express multiple putative RNA helicase genes exhibiting high sequence conservation from yeast to mammals. Gene 140:171–177
Glaser B, Grutzner F, Willmann U, Stanyon R, Arnold N, Taylor K, Rietschel W, Zeitler S, Toder R, Schempp W (1998) Simian Y chromosomes: species-specific rearrangements of *DAZ*, *RBM*, and *TSPY* versus contiguity of *PAR* and *SRY*. Mamm Genome 9:226–231
Graves J (1995) The origin and function of the mammalian Y chromosome and Y-borne genes – an evolving understanding. Bioessays 17:311–321
Greenfield A, Scott D, Penniis D, Ehrmann I, Ellis P, Cooper L, Simpson E, Koopman P (1996) An H-YDb epitope is encoded by a novel mouse Y chromosome gene. Nat Genet 14: 474–478
Handley PM, Mueckler M, Siegel NR, Ciechanover A, Schwartz AL (1991) Molecular cloning, sequence, and tissue distribution of the human ubiquitin-activating enzyme E1. Proc Natl Acad Sci USA 88:258–262
Huang Y, Baker RT, Fischer-Vize JA (1995) Control of cell fate by a deubiquitinating enzyme encoded by the fat facets gene. Science 270:1828–1831
Hunt PA (1991) Survival of XO mouse fetuses: effect of parental origin of the X chromosome or uterine environment? Development 111:1137–1141
Incerti B, Guioli S, Pragliola A, Zanaria E, Borsani G, Tonlorenzi R, Bardoni B, Franco B, Wheeler D, Ballabio A, Camerino G (1992) Kallmann syndrome gene on the X and Y chromosomes: implications for evolutionary divergence of human sex chromosomes. Nat Genet 2:311–314
Just W, Rau W, Vogel W, Akhverdian M, Fredga K, Graves JAM, Lyapunova E (1995) Absence of *Sry* in species of the vole *Ellobius*. Nat Genet 11:117–118
Kay GF, Ashworth A, Penny GD, Dunlop M, Swift S, Brockdorf N, Rastan S (1991) A candidate spermatogenesis gene on the mouse Y chromosome is homologous to ubiquitin activating enzyme E1. Nature 354:486–489
Keleher CA, Redd MJ, Schultz J, Carlson M, Johnson AD (1992) Ssn6-Tup1 is a general repressor of transcription in yeast. Cell 68:709–719
King TR, Christianson GJ, Mitchell MJ, Bishop CE, Scott D, Ehrmann I, Simpson E, Eicher EM, Roopenian DC (1994) Deletion mapping by immunoselection against the H-Y histocompatibility antigen further resolves the Sxr^a region of the mouse Y chromosome and reveals complexity of the *Hya* locus. Genomics 24:159–168
Kondrashov AS (1988) Deleterious mutations and the evolution of sexual reproduction. Nature 336:435–440
Koopman P, Gubbay J, Vivian N, Goodfellow P, Lovell-Badge R (1991) Male development of chromosomally female mice transgenic for *Sry*. Nature 351:117–121
Lahn BT, Page DC (1997) Functional coherence of the human Y chromosome. Science 278: 675–680
Lambson B, Affara NA, Mitchell M, Ferguson-Smith MA (1992) Evolution of DNA sequence homologies between the sex chromosomes in primate species. Genomics 14:1032–1040
Laval S, Glenister P, Rasberry C, Thornton C, Mahadevaiah S, Cooke H, Burgoyne P, Cattanach B (1995) Y chromosome short arm-*Sxr* recombination in X*Sxr*/Y males causes deletion of *Rbm* and XY female sex reversal. Proc Natl Acad Sci USA 92:10403–10407

Laval SH, Reed V, Blair HJ, Boyd Y (1997) The structure of *DXF34*, a human X-linked sequence family with homology to a transcribed mouse Y-linked repeat. Mamm Genome 8:689–691
Le Coniat M, Soulard M, Della Valle V, Larsen CJ, Berger R (1992) Localization of the human gene encoding heterogeneous nuclear RNA ribonucleoprotein G (hnRNP-G) to chromosome 6p12. Hum Genet 88:593–595
Legouis R, Hardelin J-P, Levilliers J, Claverie J-M, Compain S, Wunderle V, Millasseau P, Le Paslier D, Cohen D, Caterina D, Bougueleret L, Delemarre-van de Waal D, Lutfalla G, Weissenbach J, Petit C (1991) The candidate gene for the X-linked Kallmann syndrome encodes a protein related to adhesion molecules. Cell 67:423–435
Leroy P, Alzari P, Sassoon D, Wolgemuth D, Fellous M (1989) The protein encoded by a murine male germ cell-specific transcript is a putative ATP-dependent RNA helicase. Cell 57: 549–559
Levy ER, Burgoyne PS (1986) The fate of XO germ cells in the testes of XO/XY and XO/XY/XYY mouse mosaics: evidence for a spermatogenesis gene on the mouse Y chromosome. Cytogenet Cell Genet 42:208–213
Ma K, Inglis JD, Sharkey A, Bickmore WA, Hill RE, Prosser EJ, Speed RM, Thomson EJ, Jobling M, Taylor K, Wolfe J, Cooke HJ, Hargreave TB, Chandley AC (1993) A Y chromosome gene family with RNA-binding protein homology: candidates for the azoospermia factor AZF controlling human spermatogenesis. Cell 75:1287–1295
Mahadevaiah SK, Odorisio T, Elliott DJ, Rattigan A, Szot M, Laval S, Washburn LL, McCarrey JR, Cattanach BM, Lovell-Badge R, Burgoyne PS (1998) Mouse homologues of the human *AZF* candidate gene *RBM* are expressed in spermatogonia and spermatids, and map to a Y chromosome deletion interval associated with a high incidence of sperm abnormalities. Hum Mol Genet 7:715–727
Maiwald R, Luche RM, Epstein CJ (1996) Isolation of a mouse homolog of the human *DAZ* (*Deleted in Azoospermia*) gene. Mamm Genome 7:628
Mardon G, Page D (1989) The sex-determining region of the mouse Y chromosome encodes a protein with a highly acidic domain and 13 zinc fingers. Cell 56:765–770
Mardon G, Mosher R, Disteche CM, Nishioka Y, McLaren A, Page DC (1989) Duplication, deletion and polymorphism in the sex-determining region of the mouse Y chromosome. Science 243:78–80
Markiewicz MA, Girao C, Opferman JT, Sun J, Hu Q, Agulnik AA, Bishop CE, Thompson CB, Ashton-Rickardt PG (1998) Long-term T cell memory requires the surface expression of self- peptide/major histocompatibility complex molecules. Proc Natl Acad Sci USA 95: 3065–3070
Mazeyrat S, Mitchell M (1998) Rodent Y chromosome TSPY gene is functional in rat and nonfunctional in mouse. Hum Mol Genet 7:557–562
Mazeyrat S, Saut N, Sargent C, Grimmond S, Longepied G, Ehrmann I, Ellis P, Greenfield A, Affara N, Mitchell M (1998) The mouse Y chromosome interval necessary for spermatogonial proliferation is gene dense with syntenic homology to the human *AZFa* region. Hum Mol Genet 7:1713–1724
McLaren A, Hunt R, Simpson E (1988a) Absence of any male-specific antigen recognized by T lymphocytes in X/X*Sxr*' male mice. Immunology 63:447–449
McLaren A, Monk M (1982) Fertile females produced by inactivation of an X chromosome of "*sex- reversed*" mice. Nature 300:446–448
McLaren A, Simpson E, Epplen JT, Studer R, Koopman P, Evans EP, Burgoyne PS (1988b) Location of the genes controlling H-Y antigen expression and testis determination on the mouse Y chromosome. Proc Natl Acad Sci USA 85:6442–6445
McLaren A, Simpson E, Tomonari K, Chandler P, Hogg H (1984) Male sexual differentiation in mice lacking H-Y antigen. Nature 312:552–555
Meroni G, Franco B, Archidiacono N, Messali S, Andolfi G, Rocchi M, Ballabio A (1996) Characterization of a cluster of sulfatase genes on Xp22.3 suggests gene duplications in an ancestral pseudoautosomal region. Hum Mol Genet 5:423–431

Mitchell MJ, Bishop CE (1992) A structural analysis of the *Sxr* region of the mouse Y chromosome. Genomics 12:26–34

Mitchell MJ, Wilcox SA, Watson JM, Lerner JL, Woods DR, Scheffler J, Hearn JP, Bishop CE, Graves JAM (1998) The origin and loss of the ubiquitin-activating enzyme gene on the mammalian Y chromosome. Hum Mol Genet 7:429–434

Mitchell MJ, Woods DR, Tucker PK, Opp JS, Bishop CE (1991) Homology of a candidate spermatogenic gene from the mouse Y chromosome to the ubiquitin-activating enzyme E1. Nature 354:483–486

Mitchell MJ, Woods DR, Wilcox SA, Graves JAM, Bishop CE (1992) Marsupial Y chromosome encodes a homologue of the mouse Y-linked candidate spermatogenesis gene *Ube1y*. Nature 359:528–531

Muller H (1914) A gene for the fourth chromosome of *Drosophila*. J Exp Zool 17:325–336

Muller H (1964) The relation of recombination to mutational advance. Mutat Res 1:2–9

Nagamine CM, Chan K, Hake LE, Lau YF (1990) The two candidate testis-determining Y genes (*Zfy-1* and *Zfy-2*) are differentially expressed in fetal and adult mouse tissues. Genes Dev 4: 63–74

Nakahori Y, Takenaka O, Nakagome Y (1991) A human X-Y homologous region encodes "amelogenin". Genomics 9:264–269

Navin A, Prekeris R, Lisitsyn NA, Sonti MM, Grieco DA, Narayanswami S, Lander ES, Simpson EM (1996) Mouse Y-specific repeats isolated by whole chromosome representational difference analysis. Genomics 36:349–353

Odorisio T, Mahadevaiah SK, McCarrey JR, Burgoyne PS (1996) Transcriptional analysis of the candidate spermatogenesis gene *Ube1y* and of the closely related *Ube1x* shows that they are coexpressed in spermatogonia and spermatids but are repressed in pachytene spermatocytes. Dev Biol 180:336–343

Ogata T, Matsuo N (1995) Turner syndrome and female sex chromosome aberrations: deduction of the principal factors involved in the development of clinical features. Hum Genet 95: 607–629

Oh B, Hwang SY, Solter D, Knowles BB (1997) *Spindlin*, a major maternal transcript expressed in the mouse during the transition from oocyte to embryo. Development 124:493–503

Page DC, Harper ME, Love J, Botstein D (1984) Occurrence of a transposition from the X-chromosome long arm to the Y-chromosome short arm during human evolution. Nature 311: 119–123

Page DC, Mosher R, Simpson EM, Fisher EMC, Mardon G, Pollack J, McGillivray B, de la Chapelle A, Brown LG (1987) The sex-determining region of the human Y chromosome encodes a finger protein. Cell 51:1091–1104

Prado VF, Lee CH, Zahed L, Vekemans M, Nishioka Y (1992) Molecular characterization of a mouse Y chromosomal repetitive sequence that detects transcripts in the testis. Cytogenet Cell Genet 61:87–90

Qureshi SJ, Ross AR, Ma K, Cooke HJ, Intyre MA, Chandley AC, Hargreave TB (1996) Polymerase chain reaction screening for Y chromosome microdeletions: a first step towards the diagnosis of genetically-determined spermatogenic failure in men. Mol Hum Reprod 2: 775–779

Reijo R, Lee TY, Salo P, Alagappan R, Brown LG, Rosenberg M, Rozen S, Jaffe T, Straus D, Hovatta O, Delachapelle A, Silber S, Page DC (1995) Diverse spermatogenic defects in humans caused by Y chromosome deletions encompassing a novel RNA-binding protein gene. Nat Genet 10:383–393

Reijo R, Seligman J, Dinulos MB, Jaffe T, Brown LG, Disteche CM, Page DC (1996) Mouse autosomal homolog of *DAZ*, a candidate male sterility gene in humans, is expressed in male germ cells before and after puberty. Genomics 35:346–352

Rice WR (1987) Genetic hitchhiking and the evolution of reduced genetic activity of the Y sex chromosome. Genetics 116:161–167

Rice WR (1992) Sexually antagonistic genes: experimental evidence. Science 256:1436–1439

Rice WR (1998) Male fitness increases when females are eliminated from gene pool: implications for the Y chromosome. Proc Natl Acad Sci USA 95:6217–6221

Roberts C, Weith A, Passage E, Michot JL, Mattei M, Bishop C (1988) Molecular and cytogenetic evidence for location of *Tdy* and *Hya* on the mouse Y chromosome short arm. Proc Natl Acad Sci USA 85:6446–6449

Ruggiu M, Speed R, Taggart M, McKay SJ, Kilanowski F, Saunders P, Dorin J, Cooke HJ (1997) The mouse Dazla gene encodes a cytoplasmic protein essential for gametogenesis. Nature 389:73–77

Salido EC, Li XM, Yen PH, Martin N, Mohandas TK, Shapiro LJ (1996) Cloning and expression of the mouse pseudoautosomal steroid sulphatase gene (*Sts*). Nat Genet 13:83–86

Saxena R, Brown LG, Hawkins T, Alagappan RK, Skaletsky H, Reeve MP, Reijo R, Rozen S, Dinulos MB, Disteche CM, Page DC (1996) The *DAZ* gene-cluster on the human Y-chromosome arose from an autosomal gene that was transposed, repeatedly amplified and pruned. Nat Genet 14:292–299

Schiaffino MV, Bassi MT, Rugarli EI, Renieri A, Galli L, Ballabio A (1995) Cloning of a human homologue of the *Xenopus laevis* APX gene from the ocular albinism type 1 critical region. Hum Mol Genet 4:373–382

Schiebel K, Mertz A, Winkelmann M, Glaser B, Schempp W, Rappold G (1997) FISH localization of the human Y-homolog of protein kinase *PRKX* (*PRKY*) to Yp11.2 and two pseudogenes to 15q26 and Xq12→q13. Cytogenet Cell Genet 76:49–52

Schneider-Gädicke A, Beer-Romero P, Brown LG, Nussbaum R, Page DC (1989) *ZFX* has a gene structure similar to *ZFY*, the putative human sex determinant, and escapes X inactivation. Cell 57:1247–1258

Schnieders F, Dork T, Arnemann J, Vogel T, Werner M, Schmidtke J (1996) Testis-specific protein, Y-encoded (TSPY) expression in testicular tissues. Hum Mol Genet 5:1801–1807

Schwartz A, Chan DC, Brown LG, Alagappan R, Pettay D, Disteche C, McGillivray B, de la Chapelle A, Page DC (1998) Reconstructing hominid Y evolution: X-homologous block, created by X-Y transposition, was disrupted by Yp inversion through LINE-LINE recombination. Hum Mol Genet 7:1–11

Scott DM, Ehrmann IE, Ellis PS, Bishop CE, Agulnik AI, Simpson E, Mitchell MJ (1995) Identification of a mouse male-specific transplantation antigen, H-Y. Nature 376:695–698

Searle AG (1990) Sex-chromosomal aneuploidy. In: Lyon MF, Searle AG (eds) Genetic variants and strains of the laboratory mouse. Oxford University Press, Oxford, pp 583–586

Seboun E, Leroy P, Casanova M, Magenis E, Boucekkine C, Disteche C, Bishop C, Fellous M (1986) A molecular approach to the study of the human Y chromosome and anomalies of sex determination in man. Cold Spring Harb Symp Quant Biol 51:237–248

Seboun E, Barbaux S, Bourgeron T, Nishi S, Agulnik A, Egashira M, Nikkawa N, Bishop C, Fellous M, McElreavey K, Kasahara M, Algonik A (1997) Gene sequence, localization, and evolutionary conservation of *DAZLA*, a candidate male sterility gene. Genomics 41: 227–235

Shan Z, Hirschmann P, Seebacher T, Edelmann A, Jauch A, Morell J, Urbitsch P, Vogt PH (1996) A *SPGY* copy homologous to the mouse gene *Dazla* and the *Drosophila* gene *boule* is autosomal and expressed only in the human male gonad. Hum Mol Genet 5:2005–2011

Simpson EM, Page DC (1991) An interstitial deletion in mouse Y chromosomal DNA created a transcribed *Zfy* fusion gene. Genomics 11:601–608

Singh L, Jones KW (1982) Sex reversal in the mouse (*Mus musculus*) is caused by a recurrent nonreciprocal crossover involving the X and an aberrant Y chromosome. Cell 28:205–216

Soulard M, Della Valle V, Siomi MC, Pinol-Roma S, Codogno P, Bauvy C, Bellini M, Lacroix J-C, Monod G, Dreyfuss G, Larsen C-J (1993) hnRNP G: sequence and characterization of a glycosylated RNA-binding protein. Nucleic Acids Res 21:4210–4217

Sowden J, Putt W, Morrison K, Beddington R, Edwards Y (1995) The embryonic RNA helicase gene (*ERH*): a new member of the DEAD box family of RNA helicases. Biochem J 308: 839–846

Styrna J, Krzanowska H (1995) Sperm select penetration test reveals differences in sperm quality in strains with different Y chromosome genotype in mice. Arch Androl 35:111–118

Styrna J, Imai HT, Moriwaki K (1991a) An increased level of sperm abnormalities in mice with a partial deletion of the Y chromosome. Genet Res 57:195–199

Styrna J, Klag J, Moriwaki K (1991b) Influence of partial deletion of the Y chromosome on mouse sperm phenotype. J Reprod Fertil 92:187–195

Sutcliffe MJ, Burgoyne PS (1989) Analysis of the testis of H-Y negative XO*Sxr*^{b} mice suggests that the spermatogenesis gene (*Spy*) acts during the differentiation of the A spermatogonia. Development 107:373–380

Tyler-Smith C, Taylor L, Muller U (1988) Structure of a hypervariable tandemly repeated DNA sequence on the short arm of the human Y chromosome. J Mol Biol 203:837–848

Vogel T, Dechend F, Manz E, Jung C, Jakubiczka S, Fehr S, Schmidtke J, Schnieders F (1997) Organization and expression of bovine *TSPY*. Mamm Genome 8:491–496

Vogel T, Boettger-Tong H, Nanda I, Dechend F, Agulnik AI, Bishop CE, Schmid M, Schmidtke J (1998) A murine *TSPY*. Chromosome Res 6:35–40

Vogt PH, Edelmann A, Kirsch S, Henegariu O, Hirschmann P, Kiesewetter F, Kohn FM, Schill WB, Farah S, Ramos C, Hartmann M, Hartschuh W, Meschede D, Behre HM, Castel A, Nieschlag E, Weidner W, Grone HJ, Jung A, Engel W, Haidl G (1996) Human Y chromosome azoospermia factors (AZF) mapped to different subregions in Yq11. Hum Mol Genet 5: 933–943

Vogt PH, Affara NA, Davey P, Hammer M, Jobling M, Lau Y-FC, Mitchell MJ, Schempp W, Tyler-Smith C, Williams G, Yen P, Rappold GA (1997) Report of the third international workshop on Y chromosome mapping 1997. Cytogenet Cell Genet 79:1–20

Vollrath D, Foote S, Hilton A, Brown LG, Beer-Romero P, Bogan JS, Page DC (1992) The human Y chromosome: a 43-interval map based on naturally occurring deletions. Science 258:52–59

von Lindern M, van Baal S, Wiegant J, Raap A, Hagemeijer A, Grosveld G (1992) Can, a putative oncogene associated with myeloid leukemogenesis, may be activated by fusion of its 3' half to different genes: characterization of the set gene. Mol Cell Biol 12:3346–3355

Weighardt F, Biamonti G, Riva S (1996) The roles of heterogeneous nuclear ribonucleoproteins (hnRNP) in RNA metabolism. Bioessays 18:747–756

Wolfe J, Darling SM, Erickson RP, Craig IW, Buckle VJ, Rigby PW, Willard HF, Goodfellow PN (1985) Isolation and characterization of an alphoid centromeric repeat family from the human Y chromosome. J Mol Biol 182:477–485

Xian M, Azuma S, Naito K, Kunieda T, Moriwaki K, Toyoda Y (1992) Effect of a partial deletion of Y chromosome on in vitro fertilizing ability of mouse spermatozoa. Biol Reprod 47: 549–553

Yen P, Marsh B, Allen E, Tsai S, Ellison J, Connolly L, Neiswanger K, Shapiro L (1988) The human X-linked steroid sulfatase gene and a Y-encoded pseudogene: evidence for an inversion of the Y chromosome during primate evolution. Cell 55:1123–1135

Yen PH, Allen E, Marsh B, Mohandas T, Wang N, Taggart RT, Shapiro LJ (1987) Cloning and expression of steroid sulfatase cDNA and the frequent occurrence of deletions in STS deficiency: implications for X-Y interchange. Cell 49:443–454

Yen PH, Chai NN, Salido EC (1996) The human autosomal gene *DAZLA* – testis specificity and a candidate for male infertility. Hum Mol Genet 5:2013–2017

The Comparative Genetics of Human Spermatogenesis: Clues from Flies and Other Model Organisms

Ron Hochstenbach[1] and Johannes H. P. Hackstein[2]

1 Introduction

Subfertility affects about 8–10% of healthy males (Bhasin et al. 1994; Skakkebaek et al. 1994; de Kretser 1997). Until recently, environmental factors or infectious diseases have been regarded as the principal causes, but there is ample evidence that genetic defects might be the main cause for male subfertility. It has been known for a long time that about 10% of subfertile males possess an abnormal karyotype; half of these males exhibit Klinefelter syndrome due to a 47,XXY karyotype (DeBraekeleer and Dao 1991; Van Assche et al. 1996). Moreover, translocations can cause subfertility in their male carriers, and deletions in the q11.23 region on the Y chromosome have been detected in up to 18% of the infertile males that were selected for study (Reijo et al. 1995; Vogt et al. 1996; Pryor et al. 1997). Lastly, the online Mendelian inheritance in man database (OMIM 1997) contains approximately 60 more or less defined heritable disorders that are associated with male subfertility.

Nevertheless, we are still rather ignorant about the genetic causes of male subfertility, even in those cases where a clinical diagnosis is possible. The reasons for the impairment of spermatogenesis remain obscure in 40% of the patients attending male infertility clinics (Bhasin et al. 1994; Aitken et al. 1995; de Kretser 1997). Such patients can be normospermic, oligospermic, or azoospermic, they may possess immotile or abnormally shaped sperm, or they may even lack germ cells. They have a normal karyotype and normal levels of testosterone and gonadotropin, and there are no indications that they suffer from sperm autoimmunity or abnormalities of the urogenital tract. Such males had neither undergone vasectomy, nor did they suffer from testicular cancer, infections, or damage from drugs or radiation.

There are indications that heritable defects rather than environmental hazards are the causes of these unexplained forms of non-obstructive male sub-

[1] Department of Medical Genetics, Huispostnr. KC.04.084.2, University Medical Center, P.O. Box 85090, NL-3508 AB Utrecht, The Netherlands

[2] Department of Microbiology and Evolutionary Biology, Faculty of Sciences, Catholic University of Nijmegen, Toernooiveld 1, NL-6525 ED Nijmegen, The Netherlands

Results and Problems in Cell Differentiation, Vol. 28
McElreavey (Ed.): The Genetic Basis of Male Infertility

fertility. Czyglik et al. (1986) observed low sperm counts and a decreased sperm motility in the brothers of subfertile males. In a case-control study of males that are involuntarily childless, Lilford et al. (1994) have shown that, in more than half the cases, male subfertility has a familial component with an autosomal recessive mode of inheritance. However, the genes that are involved have not been identified, and we do not have the faintest notion as to how many genes in the human genome can mutate to male sterile alleles. In addition, we do not have an estimate for the genetic load by recessive autosomal mutations that cause male subfertility. In this chapter we will discuss how a genetic dissection of spermatogenesis in *Drosophila* and other eukaryotic model organisms can help to answer these questions.

2 Model Organisms for Studying the Genetic Causes of Subfertility in Man

2.1 Mendelian Genetics and Male Subfertility

Male sterile mutations on the human Y chromosome cannot be transmitted to the next generation because – by definition – they cause dysfunction of the spermatozoa. Moreover, all Y chromosomal male sterile mutations must have arisen de novo, since the presence of male sex determining genes on the Y chromosome precludes a transmission to the next generation by females. In contrast, dominant male sterile mutations can be transmitted by female carriers if female reproduction is not affected. Also, X chromosomal male sterile mutations can be transmitted by heterozygous female carriers, and recessive autosomal mutations can be transmitted by both heterozygous males and females. Thus, with the exception of Y chromosomal male sterile mutations, all mutations that impair male fertility or fecundity contribute to the genetic load of human populations. However, the study of heritable male sterile mutations is highly dependent on the fortuitous identification of families with several affected, subfertile brothers. Therefore, any systematic approach to studying male infertility has to rely on model organisms such as the fruit fly and the mouse, that are accessible to mutation induction and genome-wide screening for male sterile mutations.

2.2 Spermatogenesis: An Ancient, Conserved Process of Cellular and Subcellular Differentiation

The use of model organisms for the genetic dissection of spermatogenesis has its justification in the extraordinary degree of conservation of this process

(Fig. 1). Most eukaryotes form flagellated or ciliated cells, and with a few exceptions, all male animals produce flagellated sperm – highly specialized, mobile cells (Baccetti and Afzelius 1976). Male germ cell development begins when the germ line separates from the somatic cell lineages during early embryogenesis. The primordial germ cells originate at an extragonadal location and subsequently migrate to the developing gonads (for mammals, see Denis and Lacroix 1993, McCarrey 1993, Kierszenbaum 1994; for *Drosophila*, see Foe et al. 1993). In both flies and mammals the male germ line accommodates stem cells that generate spermatogonia. The spermatogonia proliferate by a limited number of mitotic divisions to give rise to diploid, primary spermatocytes. Every primary spermatocyte undergoes two meiotic divisions that result in four haploid, round spermatids. Postmeiotic sperm cell development is characterized by a sequence of remarkably conserved changes in the morphology and the structural organization of the spermatids. The sperm nucle-

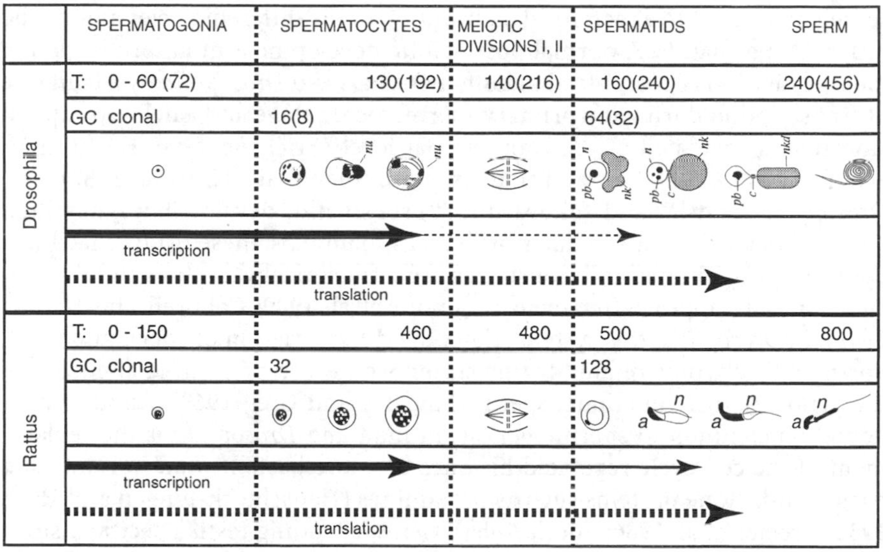

		SPERMATOGONIA	SPERMATOCYTES	MEIOTIC DIVISIONS I, II	SPERMATIDS	SPERM
Drosophila	T:	0 - 60 (72)	130(192)	140(216)	160(240)	240(456)
	GC clonal		16(8)		64(32)	
		transcription		translation		
Rattus	T:	0 - 150	460	480	500	800
	GC clonal		32		128	
		transcription		translation		

Fig. 1. Key events during spermatogenesis of flies (*Drosophila*) and mammals (*Rattus*). For both experimental systems, the *top line* in each diagram indicates the timing of germ cell development (in hours) for *D. melanogaster* and *R. norvegicus*. The *numbers in brackets* refer to *D. hydei*. The *second line* indicates the maximum number of cells in a germ cell clone (*GC*), again with *numbers in brackets* referring to *D. hydei*. In the *third line*, the major morphogenetic events are represented by schematic drawings (*nu* nucleolus, *n* nucleus, *pb* protein body, *c* centriolar adjunct, *nk* nebenkern, *nkd* nebenkern derivative, *a* acrosome). In the *bottom line*, the duration of translation and transcription are indicated by *arrows*. In both systems the bulk of RNA is synthetized premeiotically. Many genes are transcribed after meiosis during mammalian spermatogenesis (Erickson 1990), whereas in *Drosophila*, only a few genes are known to be transcribed postmeiotically (indicated by the *breaks*; see Fuller (1993) for review). Modified from Erickson (1990) and Hackstein (1991)

us condenses and changes in shape, the haploid DNA becomes tightly packed, an acrosomic system develops from the Golgi apparatus, the mitochondria become rearranged to form the so-called mitochrondrial derivative(s), the axoneme and the other tail structures are assembled, and most of the cytoplasm becomes excluded from the mature sperm cell. The formation of the axoneme, with its conserved structure of 9+2 microtubule doublets and associated proteins, highlights the conservation of spermatogenesis. Thus, despite certain interspecific differences in size and shape, the spermatozoa of the different animal phyla are basically identical (Baccetti and Afzelius 1976, Baccetti 1991; for mammals, see Fawcett 1975, Guraya 1987; for *Drosophila*, see Lindsley and Tokuyasu 1980, Hennig 1985, Hackstein 1991, Fuller 1993).

The similarities between spermatogenesis in *Drosophila* and mammals are not restricted to the cellular and subcellular differentiation of the germ cells. Male germ cells have to cope with a fundamental problem: postmeiotically, the spermatids are genetically different. Meiosis does not only lead to the distribution of the sex chromosomes into different spermatids; the autosomes also segregate, leading to a reciprocal distribution of the different alleles among the different spermatids. A normal postmeiotic development of all products of a meiotic division is guaranteed in both mammals and *Drosophila* by a high rate of transcription during the primary spermatocyte stage and a delayed expression that is regulated at the translational level (Erickson 1990; Hecht 1993; Schäfer et al. 1995). In addition, in both mammals (Guraya 1987) and *Drosophila* (Lindsley and Tokuyasu 1980) spermatids develop in a syncytium, being connected by intercellular bridges. In mammals, these bridges facilitate the sharing of cytoplasmic constituents postmeiotically, such as mRNA species that only appear after meiosis (Braun et al. 1989; Caldwell and Handel 1991). In *Drosophila*, this system allows the differentiation of spermatids that, postmeiotically, only retain the tiny chromosome 4 (representing about 2% of the genome), into functional sperm (Lindsley and Grell 1969). Another phenomenon common to spermiogenesis in man and *Drosophila* is the replacement of the cell-cycle regulated histones by more basic histone variants and basic non-histone proteins such as protamines (Hauschteck-Jungen and Hartl 1982; Kremer et al. 1986; Poccia 1986). Genes encoding testis-specific histone isoforms, which are not regulated by the cell cycle, are highly transcribed in the testis of both rat (Meistrich et al. 1985) and *Drosophila* (Akhmanova et al. 1995, 1996, 1997). Finally, a spermatozoon is not merely a vector that carries the paternal genome to the egg. In mammals, with the exception of rodents, and in *Drosophila*, the sperm cell also provides the egg with a centrosome that nucleates microtubule assembly in the course of the fertilization process (Foe et al. 1993; Simerly et al. 1995).

Spermatogenesis appears to be a highly conserved process of cellular and subcellular differentiation. Therefore, it is reasonable to assume that many of the genes and their products required for spermatogenesis in the fruit fly have also been conserved in mammals. A striking example is provided by the sperm tail-specific cysteine-glycine-proline-rich protein that is found in the outer dense fibers of mammalian sperm (Burfeind and Hoyer-Fender 1991). These fibers are thought to confer tensile strength, protecting sperm cells from shear forces during ejaculation (Baltz et al. 1990). In *Drosophila*, these sperm tail-specific proteins are encoded by the *Mst(3)CGP* gene family (Schäfer et al. 1993, 1995). Rat and *Drosophila* homologs share 42% identity in the carboxy-terminal 100 amino acids; the amino acid identity for the entire protein is 22% (Schäfer et al. 1993).

2.3 Model Organisms

In addition to the fruit fly species *Drosophila melanogaster* and *D. hydei*, several other organisms have proved to be of outstanding value for the understanding of the genetic basis of spermatogenesis in man. The yeast *Saccharomyces cerevisiae* has served as an invaluable model system to study genes that are required for meiosis, and the green alga *Chlamydomonas reinhardtii* has been of great importance for studying the genetics of flagellar assembly and function. The nematode *Caenorhabditis elegans* and the laboratory mouse *Mus musculus* have been successfully used for the identification of genes that function in spermatogenesis.

2.3.1 *Yeast*

The formation of flagellated spermatozoa is an ancient process that is intimately associated with meiosis. The budding yeast *S. cerevisiae* is uniquely suited for the molecular genetic, biochemical and cytological dissection of meiosis. When depleted of carbon and nitrogen, diploid yeast cells in a culture enter meiosis, and they perform the different stages of meiosis almost in synchrony. Yeast mutants defective in meiosis have allowed the identification of genes that control the meiotic cell cycle (such as cyclins and their associated kinases), the pairing of homologous chromosomes, the formation of the synaptonemal complex, genetic recombination, DNA (mismatch) repair and chromosome segregation (Roeder 1995; Page and Orr-Weaver 1997).

Abnormalities during meiosis have been diagnosed as the probable cause for male sterility in about 5% of oligospermic males and in about 14% of azoospermic patients (De Braekeleer and Dao 1991; Van Assche et al. 1996). The spectrum of meiotic anomalies that has been described for subfertile males (Koulischer et al. 1982; Lange et al. 1997) strongly resembles phenotypes of well known yeast meiotic mutants. The straightforward assumption of a genetic basis for the meiotic anomalies in man is supported by the observations of Chaganti et al. (1980): defects in chromosome pairing were shared by severely oligospermic brothers whose parents were first cousins. The consanguinity of the unaffected parents suggests an autosomal recessive mode of inheritance. Since several of the genes that control meiosis in yeast are highly conserved among eukaryotes, it is likely that searches for DNA sequence homologies will facilitate the identification of genes that have similar functions during meiosis in male mammals (Sassone-Corsi 1997).

2.3.2 *Chlamydomonas*

The unicellular, biflagellate, green alga *Chlamydomonas reinhardtii* is a model organism for studies on the assembly and function of the axoneme. With its 9+2 doublet microtubule structure and microtubule-associated proteins, the flagellar axoneme of *Chlamydomonas* is very similar to the axoneme of mammalian sperm tails (Fawcett 1975; Dutcher 1995). The flagellum can be detached by pH shock, and its membrane can be stripped off, permitting the recovery of isolated axonemes for biochemical analysis. Because the flagellum is not required for viability, mutants with abnormal flagella, or even without flagella, can be isolated and cultured.

How many genes are required to assemble an axoneme? Biochemical analysis has revealed that the axoneme of *Chlamydomonas* is composed of over 250 polypeptides (Luck 1984), and about 100 polypeptides have been assigned to specific structures in the axoneme (Dutcher 1995). It is not known to what extent the use of alternative promotors, alternative splicing, and post-translational modification leads to an overestimation of the number of genes that encode these polypeptides. Piperno et al. (1981) observed that 6 out of 17 polypeptides from the radial spokes show microheterogeneities in charge and apparent molecular weight in the two-dimensional gel system used for separating axonemal proteins, but this did not increase the number of distinct polypeptide spots. Over 25 genes have been identified by mutations that cause absence of flagella or defects in flagellar assembly, and over 52 genes have been identified by mutations that affect flagellar function by causing abnormal motility (Dutcher 1995). The synthesis of about half of the axonemal polypepetides is not affected in any of the 52 flagellar mutants. The genes encoding these polypeptides may have been missed in the mutant screens, either because the gene product is not needed for flagellar motility, or because it is required for a vital cellular function. Thus, as a rough approximation, the number of genes required for assembling and operating an axoneme is at least 77, but this number may be much higher.

It is very likely that many of the genes required for flagellar function in *Chlamydomonas* have been conserved in humans. Using electron microscopy, it could be shown that specific axonemal structures, such as inner and outer dynein arms, radial spokes, or the central microtubules, are missing in many *Chlamydomonas* mutants with impaired motility (Luck 1984; Dutcher 1995). Similar axonemal defects have been described in sterile male patients suffering from the immotile cilia syndrome; they have immotile sperm that lack inner dynein arms, outer dynein arms, or both (Afzelius et al. 1975). They also present with respiratory tract diseases because of slow mucociliary clearance due to immotility of bronchial cilia. The syndrome follows an autosomal recessive mode of inheritance: it has been described repeatedly in subfertile sons of consanguineous parents (Rott 1979). In other patients with immotile sperm, the various axonemal structures (heads of radial spokes, central microtubules) can be lacking (Afzelius and Eliasson 1979; Baccetti et al. 1981; Escalier and David 1984). In conclusion, mutations in a number of non-allelic genes affect the function of the axoneme of human sperm. Since the "human" phenotypes are very similar to those of flagellar mutants of *Chlamydomonas*, one can predict that the genes controlling axoneme formation are conserved between *Chlamydomonas* and man. Therefore, it is likely that at least some 80 genes – and probably many more – are involved in the formation of the axoneme of human spermatozoa.

2.3.3 *Caenorhabditis elegans*

The nematode *C. elegans* is exceptional among the eukaryotic model organisms, because most individuals in natural populations are hermaphrodites. During the larval development of the hermaphrodite a few hundred spermatozoa differentiate inside an ovotestis (L'Hernault 1997). These spermatozoa are crawling, amoeboid cells without flagella (Theriot 1996). About 60 genes specific for spermatogenesis have been recovered from genetic screens for hermaphrodites that are incapable of generating progeny by self-fertilization (L'Hernault 1997). This experimental design is biased in favour of the recovery of mutations that only affect the differentiation of male gametes, such as, for example, mutations that are specific for male meiosis. The value of *C. elegans* as a general model for human spermatogenesis seems rather limited because of the lack of flagellated sperm cells and the unusual differentiation of male germ cells in a hermaphrodite.

2.3.4 *Mouse*

The mouse is a promising animal model for spermatogenesis in man, because the testis and the development of male germ cells are highly similar in mouse and man. A substantial number of male sterile mutations in the mouse have been described (Handel 1987; Chubb 1989; Mouse Genome Database 1996). The phenotypes of new insertional mutations and mutations created by knockout technology and reverse genetics are revealing new genes that affect male fertility, sometimes rather surprisingly (TBASE, see Jacobson and Anagnostopoulos 1996; Elliott and Cooke 1997). Also, changes in the dosage of genes that are ubiquitously expressed can cause male sterility. Several highly conserved genes involved in cell cycle control, DNA mismatch repair and recombination have been identified that are essential for the completion of meiosis in male mice (Sassone-Corsi 1997). The significance of these findings for understanding meiotic arrest in sterile male patients seems promising and needs to be explored.

However, the systematic, genome-wide screening for male sterile mutations in the mouse suffers from practical and financial limitations. Most mutations have been identified fortuitously by pleiotropic effects on the phenotype, such as changes in morphology, coat pigmentation or behaviour. Hence, many male sterile mutations described in the mouse are pleiotropic (Handel 1987). Currently, some 300 male sterile strains of mice are commercially available (Bhasin et al. 1994), but most mutations have not been mapped and have not been sufficiently characterized at the phenotypic level, and the genes involved have not been identified.

2.3.5 *Zebra Fish*

Another vertebrate model organism, the zebra fish (*Danio rerio*), is becoming increasingly popular for the systematic identification of genes involved in complex developmental processes, such as dorsoventral patterning, gastrulation, somite formation, and the development of the brain and the notochord (Felsenfeld 1996). Embryonic lethal mutations are efficiently induced by chemical mutagens during different stages of male germ cell development (Solnica-Krezel et al. 1994), but the potential for systematically investigating the genetic basis of spermatogenesis has not yet been explored.

2.3.6 *Drosophila*

Drosophila stands out as a model organism with a history of almost 100 years of intensive genetic study (Ashburner 1989). Recently, it has become clear that the analysis of mutant phenotypes in *Drosophila* can help to identify genes implicated in human disease (reviewed by Banfi et al. 1996). In contrast to the mouse, systematic genetic screens for male sterile mutations are feasible in *Drosophila*, allowing the recovery of all types of male sterile mutations on the Y chromosome, the X chromosome and on all autosomes. Male sterile mutations are defined by a total lack of progeny if mutant males are allowed to mate with fertile, virgin females for an appropriate period. Most studies have focussed on *D. melanogaster* and *D. hydei*. Mutations can be induced by X-rays, chemical mutagens, or (in the case of *D. melanogaster*) by insertion of the *P*-element transposon. In both species, X chromosomal and autosomal recessive male sterile mutations are kept without difficulty in balanced strains – as are the few dominant mutations. Y chromosomal male sterile mutations can be conveniently maintained in strains carrying translocated Y chromosomal fragments, and Y chromosomes carrying male sterile mutations can be propagated in females, since the Y chromosome of *Drosophila* is not involved in sex determination (see Hackstein 1987 for review). Finally, it has become possible to identify enhancers that specifically function in the germ cells and the various somatic cells of the testis using a *P*-element transposon containing a *lacZ* reporter gene (Gönczy et al. 1992). In addition, mutational insertions of *P*-elements can be used to identify genes that are required for male fertility. The expression patterns of such genes in the various tissues of the sterile flies can be easily studied using the *lacZ* reporter gene (Castrillon et al. 1993).

The testis of *Drosophila* can be dissected easily. The coiled testis tube contains many cysts of developing germ cells (64 postmeiotic spermatids per cyst in *D. melanogaster*, and 32 in *D. hydei*); within a cyst the germ cells develop in synchrony. All stages of male germ cell differentiation are found in a single testis. After gentle squashing, the different stages can be examined under a light microscope using phase contrast optics, with well-defined cytogenetic landmarks for many stages (Lindsley and Tokuyasu 1980; Hennig 1985; Hackstein et al. 1990; Fuller 1993; Cenci et al. 1994). For instance, the normal completion of meiosis can be followed by counting the number of spermatid nuclei and mitochondrial derivatives per cyst. Equal diameters of nuclei are indicative of a correct chromosome segregation, and the formation of equally sized nebenkerne (which result from a fusion of all mitochrondria of a spermatid) is indicative of normal cytokinesis (cf. Hackstein et al. 1990).

2.3.6.1
In Flies, the Y Chromosome Carries Only a Few Genes Essential for Spermatogenesis

In contrast to mammals, the Y chromosome of *Drosophila* does not carry genes that have a role in the determination of the male sex. However, also in flies the Y chromosome harbours only a few male fertility genes, although it represents some 8 to 10% of the haploid genome. In *D. melanogaster*, male sterile mutations on this chromosome have been assigned to six complementation groups (named *ks-1*, *ks-2*, *kl-1*, *kl-2*, *kl-3*, *kl-5*), and in *Drosophila hydei* to at least seven (named *A*, *B*, *C*, *N*, *O*, *P*, *Q*) . Several of these fertility genes form giant lampbrush loops during the primary spermatocyte stage (for review see Hess and Meyer 1968; Hennig 1985; Hackstein 1987; Hackstein and Hochstenbach 1995). Characteristically, flies carrying a mutation in a Y chromosomal fertility gene produce immotile, but well differentiated sperm that appear normal under the light microscope. Electron microscopy reveals that the axonemes of flies mutated in the genes *kl-3* or *kl-5* (*D. melanogaster)* or genes *A* or *N* (*D. hydei*), respectively, lack the outer dynein arms of the peripheral flagellar microtubules. Comparable abnormalities are found in the axonemes of the spermatozoa of males suffering from the immotile cilia syndrome (Afzelius et al. 1975). Gene *kl-5* and gene *A* apparently encode sperm-specific variants of the dynein β heavy chain protein (Goldstein et al. 1982; Gepner and Hays 1993; Kurek et al. 1998). The absence of this component of the flagellar motor is the obvious cause for the immotility of the mutant spermatozoa.

There is no evidence that the sex chromosomes of *Drosophila* evolved from a pair of homologous chromosomes. Therefore, we have proposed that the *Drosophila* Y chromosome has evolved from a supernumerary chromosome, that during evolution acquired one or several male fertility genes (Hackstein et al. 1996). We have argued that the Y chromosomes of certain species of *Drosophila*, such as *D. affinis*, might represent early stages of Y chromosome evolution, since they are fertile both as XY and as X0 males. The Y chromosome of such species is dispensable for spermatogenesis and, consequently, all fertility genes that are located on the Y chromosomes of species as *D. melanogaster* and *D. hydei* must be located on the X chromosome or on the autosomes in species with fertile X0 males. In *D. affinis*, this is indeed the case because X0 males express a dynein β heavy chain protein that is orthologous to the Y chromosomal homolog encoded by the gene *kl-5* in *D. melanogaster* and by gene *A* in *D. hydei*, respectively (U. Lammermann, A.M. Reugels and H. Bünemann, unpubl. data).

2.3.6.2 Hundreds of Genes on the X Chromosome and Autosomes Are Involved in Spermatogenesis of *Drosophila*

Genetic screens in *D. melanogaster* and *D. hydei*, as they have been performed in various laboratories, have provided evidence for the existence of hundreds of male fertility genes on the X-chromosome and the autosomes (reviewed by Hackstein 1987, 1991). In *D. melanogaster*, approximately 600 X-linked male sterile mutations have been recovered (it remains unclear how many of these are allelic, see Lifschytz 1987 for review). In addition, 400 autosomal recessive male sterile mutations have been recovered by one of the authors in a screen of 9500 second and third chromosomes after treatment with EMS, a powerful chemical mutagen. About 100 male sterile mutations resulted from the screening of more than 3000 lines that were obtained after mutagenesis with the *P*-element transposon (Hackstein 1991; J. H. P. Hackstein, unpubl. results). In *D. hydei*, similar results have been obtained after chemical mutagenesis and X-ray irradiation. The screening of 16000 X chromosomes and autosomes 2, 3 and 4 yielded 365 male sterile mutants (Hackstein et al. 1990). All mutants were analyzed for testis morphology, cytology, sperm differentiation and their ability to transfer sperm to the receptaculum seminis of female flies (Hackstein 1991). Complementation tests between mutants that exhibit similar phenotypes revealed only one case of potential allelism among the mutants recovered in our screens (J.H.P. Hackstein, unpubl.), and we could only detect one case of allelism to a previously described mutation (i.e. *benign gonial cell neoplasm* of *D. melanogaster*, see Gateff 1982). The lack of recovery of multiple alleles indicates that the screens have been far from a saturation of the tested chromosomes with male sterile mutations, suggesting that many male sterile loci have remained undetected. Moreover, the complementation tests did not provide evidence in favour of polygenic male sterility. About 90% of male sterile mutations do not affect female fertility (Hackstein et al. 1990).

There are several ways to estimate how many genes in *Drosophila* can mutate to a male sterile phenotype. First, by direct extrapolation of the results of the different screens, we estimate that the number of X-linked and autosomal recessive genes that can mutate to male sterility is at least 600 (cf. Lindsley and Lifschytz 1972; Lindsley 1982). A second extrapolation is based on the observation that in 30–40% of the lines carrying an X-linked temperature sensitive lethal mutation, the mutant allele causes male sterility when male flies, that have been reared at the permissive temperature, are shifted to the nonpermissive temperature (Shellenbarger and Cross 1977; Lifschytz and Yakobovitz 1978; Lindsley 1982). On the basis of these observations, Lindsley (1982) estimates that there are 1800 to 2400 genes with male sterile alleles. A third, completely different extrapolation is based on the finding that after experimental induction of mutations, the ratio of male sterile to lethal mutations is always 15–20%, irrespective of the method of mutagenization (Lindsley and Lifschytz 1972; Watanabe and Lee 1977; Cooley et al. 1988). Because 75% of the estimated 15000 genes in the *Drosophila* genome are believed to be mutable to a lethal allele, this implies that some 1700–2250 genes can be regarded as putative male steriles (Lindsley and Tokuyasu 1980; Hackstein 1987). Thus, three different and independent estimates lead to the conclusion that mutations in 4 to 16% of the genes of *Drosophila* can cause male sterility!

2.3.6.3 Phenotypic Analysis of Sterile Male Flies Suggests That Most of These Genes Are Expressed in the Male Germ Cells

The involvement of so many genes in spermatogenesis raises the question of whether all these genes are expressed in the male germ cells. The answer is that most genes are. The phenotypic analysis of the male sterile mutants reveals that only a few male sterile mutants might lack progeny due to morphological modifications of the testis, the genital tract or the copulation organs (Lindsley and Tokuyasu 1980; Hackstein et al. 1990; Castrillon et al. 1993). There is also little evidence that behavioural abnormalities are a frequent cause of infertility, because, with a few exceptions, sterile male flies that are capable of producing motile spermatozoa, are also able to transfer sperm cells to females. Also, mutations causing a paternal lethal effect during early embryogenesis have been identified in both species. For example, the gene *ms(3)K81* of *D. melanogaster* (Yasuda et al. 1995) seems to encode a sperm-specific product that is required for the participation of the male pronucleus in the first embryonic nuclear divisions. However, mutations with a paternal lethal effect are extremely rare.

Table 1. Classification of monogenic male sterile phenotypes in *Drosophila* and comparison with disorders of male reproductive function in subfertile patients

Phenotypes in *D. melanogaster* and *D. hydei* [a]	Comparable disorders in man? [b]
Paternal-effect lethality	Yes
– Fully motile sperm, transfer to female storage organs, entry of sperm into the egg, but developmental arrest at early embryogenesis	
Fertilization defective	Yes
– Fully motile sperm, transfer to female storage organs, but no entry of sperm into the egg	
Abnormal behaviour	?
– Fully motile sperm, normal copulatory organs, but no transfer to female storage organs	
Copulatory organs defective [c]	Yes
– Fully motile sperm, no transfer to female storage organs, modified copulatory organs	
Paragonia (accessory glands) defective [c]	Yes
– Abnormal morphology and cytology	
Abnormal function of testis	Yes
– Testis too small	
– Testis of adult males retains larval shape	
– Degeneration of somatic cell types	
Germ cells defective	
– Gonial cells absent or present in low number	Yes
– Proliferating gonial cells ("gonial tumor") [c]	Yes
– Abnormal number of germ cells per cyst	?
– Cytology of germ cells abnormal	Yes
– Arrest at the various stages of meiotic prophase	Yes
– Postmeiotic development	
– – Topology of spermatid components abnormal	Yes
– – Heterochronic development of spermatid components [c]	?
– – Particular sperm component absent or defective	Yes
– – Developmental arrest	Yes
– – Sperm immotility	Yes

[a] cf. Kiefer (1973); Lifschytz (1987); Hackstein et al. (1990); Hackstein (1991); Castrillon et al. (1993); Fuller (1993); male sterility is defined by the lack of progeny when mutant males are allowed to mate with fertile, virgin females for an appropriate period.

[b] cf. Wong et al. (1973); Afzelius et al. (1975); Colgan et al. (1980); Baccetti et al. (1981); Zamboni (1987); Baccetti (1991); Skakkebaek et al. (1994); Simerly et al. (1995); de Kretser (1997).

[c] Only recovered in *D. melanogaster*, not in *D. hydei*.

The vast majority (about 80%) of the *Drosophila* mutants produce elongated, but immotile sperm that degenerate prior to individualization – regardless of the method of mutagenization (Hackstein et al. 1990; Hackstein 1991; Castrillon et al. 1993). Using an electron microscope, disorganized flagella and abnormal mitochondrial derivatives are seen (Fuller 1993). Because of the degeneration, it is difficult to discriminate between deviations from normal spermatogenesis and apoptotic phenomena (Kiefer 1973). In general, such mutants have a normal morphology and cytology of the testis, suggesting that the vast majority of the genes that can mutate to male sterility are expressed and required in the male germ cells (Lindsley and Tokuyasu 1980; Hackstein et al. 1990; Hackstein 1991).

Some 20% of the male sterile mutants have more or less distinct germ cell phenotypes that are similar to those described for mouse or human (Table 1). Some of these mutants possess low numbers of germ cells in a poorly developed testis, others interfere with gonial proliferation, meiosis, the differentiation of the mitochondrial derivatives, the shaping and condensation of the sperm nucleus, or spermatid differentiation. Compilations of selected mutants are given by Kiefer (1973), Lifschytz (1987), Hackstein (1991), Fuller (1993) and Castrillon et al. (1993) for *D. melanogaster*, and by Hackstein et al. (1990) for *D. hydei*. Two important conclusions can be drawn from the detailed phenotypical analysis of the mutants with distinct, conspicuous phenotypes. First, male sterile mutations can decouple the development of the different components of the germ cell, and second, there are regulatory genes that control fundamental choices in cellular fate during sperm development. In addition, a third conclusion has emerged from the analysis of the male sterile mutants: many male sterile mutations are pleiotropic.

2.2.6.4
Sperm Components Differentiate by Independent Programs in *Drosophila*

As illustrated by mutants in both *D. hydei* (Hackstein et al. 1990) and *D. melanogaster* (Hackstein 1991), spermatogenesis can be described as the coordinate execution of the individual developmental programs of the different components of the sperm cell. For instance, mutations that cause an early postmeiotic arrest of the shaping of the nucleus do not necessarily interfere with the differentiation of the sperm tail. Also, the spermatid nucleus and the flagellum can differentiate in absence of the mitochondrial derivative. These observations argue for the independent differentiation of the sperm cell components. A similar conclusion can be drawn from the study of male sterile alleles of the β2 tubulin gene in *D. melanogaster* that produce partially functional isoforms. β2 tubulin is expressed exclusively in the testis in both *D. melanogaster* and *D. hydei* (Michiels et al. 1989), and participates in several types of microtubule arrays during male germ cell development: the meiotic spindle, the axoneme, the cytoplasmic microtubules that surround the mitochondrial derivative and the perinuclear microtubule bundles that mediate the morphological changes of the sperm nucleus (Kemphues et al. 1982; Fuller 1993). A mosaic mode of sperm development is revealed, for example, in males homozygous for the *B2t*^{7} allele. In such males, nuclear shaping appears to be normal, but elongation of the sperm tail is severely affected (Fuller et al. 1988; Fuller 1993). In addition, replacing more than 20% of the testis-specific β2 tubulin by the mesoderm-specific β3 tubulin isoform leads to the disruption of axoneme assembly, but it does not affect the function of the other microtubule arrays during male germ cell development (Hoyle and Raff 1990). Thus, the differentiation of nucleus, mitochondrial derivative and axoneme is controlled by independent morphogenetic programs. This may explain why some 80% of male sterile mutants display a common male sterile phenotype in the light microscope: disorganized, elongating spermatids that degenerate before individualization. Apparently, most mutations affect the execution of only a certain program rather mildly, and the other programs continue.

2.3.6.5 Genetic Switches Operating During Male Germ Cell Development in *Drosophila*

A second conclusion is that the mutant screens allow the identification of genes that control crucial steps in the development of male germ cells. Mutations in such genes lead to an arrest of spermatogenesis at particular stages (Fuller 1993). The first switch occurs when stem cells devide to form either stem cells or gonial cells. The testes of sterile males carrying mutations of the genes *benign germ cell neoplasm* (Gateff 1982) and *bag-of-marbles* (McKearin and Spradling 1990) are filled with numerous small cells that resemble stem cells or gonial cells. Females homozygous for mutations in either gene are sterile as well, indicating that this switch, which might be coupled with the determination of the sex of the germ line cells, operates in both the male and female germ line (Pauli and Mahowald 1990). The second switch is identified by mutations that affect the progression from gonial divisions to the primary spermatocyte stage of cell growth and extensive gene expression. The third switch is identified by mutations in at least five genes that prevent the primary spermatocytes from entering meiotic divisions and postmeiotic spermatid differentiation. Mutations in the gene *degenerative spermatocyte* (Endo et al. 1996) affect the switch from primary spermatocyte development to meiosis, just before chromosome condensation has begun. Mutations in the genes *always early*, *cannonball*, *meiosis I arrest* and *ms(3)spermatocyte arrest* (Lin et al. 1996) cause an arrest of the meiotic cycle just after chromosome condensation has begun. Postmeiotic spermatid stages are absent in the testis of males homozygous for mutant alleles of these five genes.

2.3.6.6
Many Male Sterile Mutations in Flies Are Pleiotropic

Cytology and molecular genetic analysis reveal that pleiotropy is a hallmark for many genes required for spermatogenesis in *Drosophila* (Hackstein 1991). First, many visible mutations are associated with male sterility, as is evident from the catalogue of mutations of *Drosophila* compiled by Lindsley and Zimm (1992). For example, mutant alleles of the X-linked gene *downy* cause thin bristles in hemizygous males. In addition, spermatogenesis in mutant downy males is blocked early during the primary spermatocyte stage (Kiefer 1973). Second, Lindsley (1982) observed that 30% of temperature-sensitive lethal mutations are male sterile when flies are reared at the permissive temperature and then shifted to the restrictive temperature (also see Lifschytz and Yakobovitz 1978). The finding that many male sterile mutations identify genes with vital functions in other tissues demonstrates that these genes are not specific for spermatogenesis. This is consistent with the finding that male sterile mutations in *Drosophila* often have reduced viability, although they exhibit very distinct germline phenotypes (Hackstein 1991). Third, as expected from the pleiotropic nature of many male sterile mutations, it has been found by several methods that many genes that are expressed in the testis are also expressed in other tissues or organs. Gönczy et al. (1992) selected fertile and viable fly strains that carried a single *P* element enhancer detector insertion for *lacZ* reporter gene expression in the testis. In the majority of these enhancer detector strains, *lacZ* was also expressed in other tissues. Also, the analysis of tissue-specific cDNA libraries has demonstrated that many genes that are expressed in the testis or in the male germ cells are also expressed in other tissues (FlyBase 1994).

3
The Comparative Genetics of Male Germ Cell Differentiation in Flies and Man

The systematic genetic dissection of spermatogenesis in *Drosophila* has shown that an enormous number of genes can mutate to male sterile alleles. Estimates for this number range from 600 to 2400. Most genes are represented by autosomal recessive alleles, about 90% of which do not affect female fertility. Analysis of the male sterile mutations under the light microscope and by molecular and genetic methods has revealed that expression of most of these genes is required in the male germ cells, and that many genes are pleiotropic. The different components of the sperm cell differentiate by independent morphogenetic programs. Mutations have been identified that lead to a developmental arrest of spermatogenesis, for example at the entry into meiosis. We now have to discuss the implications of these findings for the analysis of the genetics of human male subfertility.

3.1 How Many Male Fertility Genes Exist in Man?

It seems very unlikely that the number of genes that are needed to make a spermatozoon in man is lower than in *Drosophila*. First, as argued above, human spermatozoa do not show a lower structural complexity (Fawcett 1975; Lindsley and Tokuyasu 1980). Second, the genes necessary for the control of gonial and meiotic divisions in man will not be less numerous than in flies since cell cycle control has been highly conserved throughout evolution (Page and Orr-Weaver 1997). For example, genes that are essential for meiosis are known to be conserved in *D. melanogaster* and humans (and in yeast as well): the *twine*/cdc25 phosphatase (Alphey et al. 1992; Courtot et al. 1992; White-Cooper et al. 1993), the cdc2 cell-cycle kinase (Sigrist et al. 1995), and the genes *pelota* (Eberhart and Wasserman 1995) and *boule* (Eberhart et al. 1996). In male flies carrying mutations in these genes the meiotic divisions do not take place (but mutant males can form abnormal spermatids with 4N nuclei that elongate and differentiate to a limited degree, see Hackstein 1991). Interestingly, the product of the *boule* gene, a member of the hnRNPG family of RNA-binding proteins, shares substantial amino acid identity with the *dazla* gene on chromosome 17 of the mouse and with the *AZF* candidate gene family *DAZ* at Yq11.23 on the human Y chromosome (Cooke and Elliott 1997). Deletions in this region are found in 13–16% of azoospermic male patients (Reijo et al. 1995; Najmabadi et al. 1996). Also, the genes that control entry into meiosis in male flies (Endo et al. 1996; Lin et al. 1996) may be conserved in humans, since maturation arrest at meiosis I is found in over 10% of azoospermic males (Wong et al. 1973; Colgan et al. 1980).

Third, the spectrum of male sterile phenotypes in man is very similar to that in flies (Table 1), and our description of spermatogenesis in *Drosophila* as the independent execution of the different developmental programs of the sperm components (Hackstein et al. 1990; Hackstein 1991) may also apply to humans. For example, Flörke-Gerloff et al. (1984) describe two infertile brothers with round-headed spermatozoa that lack the acrosome, but with normal, motile tails. In addition, there are several descriptions of sterile males with all sperm lacking heads, but again with highly motile tails (Zamboni 1987). As in *Drosophila*, sperm flagella can form in the absence of sperm nuclei. Apparently, there are substantial similarities in the genetic control of spermatogenesis in flies and man.

Given these considerations, we think that it is permitted to derive an estimate for the number of human male fertility genes from the minimum estimate for the fruit fly. Using a naive extrapolation, the human genome should harbour at least some 600–2400 genes that can mutate to male sterile alleles. However, it is very likely that this number is an underestimate, since humans possess about five to six times more genes than *Drosophila* (the much higher difference in genome size between fly and man, i.e. 1.7×10^8 base pairs vs.

3.0×10^9 base pairs, is largely due to an excess of non-coding DNA in the human genome). The results of the various genome sequencing projects suggest that the differences between the eukaryotic model organisms only rarely reflect the acquisition of a large number of totally "new" genes. The presumed higher genetic complexity of larger genomes is likely to be the result of gene duplications that generate redundancy and allow a subsequent functional divergence (Ohno 1970; Clayton et al. 1997; Cooke et al. 1997; Mewes et al. 1997). This evolutionary strategy has become evident, for example, for the homeotic genes that control morphogenesis and organogenesis throughout the animal kingdom (Carroll 1995).

There are several other reasons to believe that the number of genes that can mutate to male sterile alleles is higher in humans, since spermatogenesis in mammals is more complex than in the fly. First, human testes are much more complex than the testes of *Drosophila*. Quite a number of somatic cell types can be found in the mammalian testis, i.e. Sertoli cells, Leydig cells, myoid cells, fibroblasts, blood vessel cell types (Guraya 1987; Griswold 1995). In addition, functional interactions exist between germ cells and Sertoli cells, which provide nutrients and encodrine factors to the germ cells (Kierszenbaum 1994; Griswold 1995). Male mice deficient for certain genes that are expressed almost exclusively in Sertoli cells, such as the *RXRβ* retinoid receptor gene, are sterile (Elliott and Cooke 1997; Sassone-Corsi 1997). In *Drosophila*, all somatic cells in the testis belong to the cyst cell lineage, and there are no cells comparable to Sertoli cells (Lindsley and Tokuyasu 1980; Gönczy and DiNardo 1996). Male sterile mutations in genes that are expressed exclusively in the somatic cyst cells have not yet been described in *Drosophila*.

Second, there are some fundamental differences between meiosis in *Drosophila* males and in man. There is no recombination in *Drosophila* males (reviewed by McKee and Handel 1993). Pairing of the X and Y chromosomes (which do not become inactivated during meiotic prophase) is mediated by the 240 base pair intergenic spacer repeats within the ribosomal RNA gene arrays located on both chromosomes. Pairing of the autosomes is mediated by ill-defined pairing sites that are distributed along the euchromatin (McKee 1996). Meiosis in male mammals might require the activity of more genes than in *Drosophila*. In mammals, synaptonemal complexes, which consist of at least eight proteins, are formed during meiosis (Heyting 1996). In addition, mammalian meiocytes possess recombination nodules, which also contain several proteins (Plug et al. 1998). For example, one protein is encoded by the *mlh1* gene that exhibits a substantial homology to the *Escherichia coli mutL* DNA mismatch repair gene. Mutations in this and other conserved genes that cause failure of chromosome pairing, chiasma formation and meiotic arrest in male mice (Sassone-Corsi 1997) might be unique for mammals. Such genes may not have their counterparts in *Drosophila*.

Third, in mammals, maternally and paternally inherited methylation patterns are erased in primordial germ cells, and distinctive, sex-specific patterns of differential cytosine methylation are imprinted during oogenesis and spermatogenesis (Chaillet et al. 1991; Ariel et al. 1994). In contrast, there is no evidence for methylation of cytosines in the DNA of *Drosophila* (Urieli-Shoval et al. 1982), and imprinting based on the parental-specific changes in methylation patterns of certain genes is not known to occur during spermatogenesis in *Drosophila*. Thus, the genes that are required in mammals for the sex-specific imprinting of parental alleles may be absent in *Drosophila*.

In conclusion, it is very likely that the human genome contains thousands of genes that can mutate to male sterile alleles. As in the fly, the vast majority of these genes are likely to be represented by mutant alleles on the X chromosome and the autosomes. Only a few genes that are essential for spermatogenesis are located on the Y chromosome. About ten distinct gene families have been mapped within the male-specific, nonrecombining, euchromatic part of the human Y chromosome, all of which are expressed exclusively in the testis (Affara et al. 1996; Lahn and Page 1997). The significance of these genes for male fertility has not been fully explored, but may be substantial, as small deletions in the long arm of the Y chromosome (most of them less than one megabase) have been detected by molecular methods in up to 18% of the azoospermic and severely oligospermic male patients selected for study (Reijo et al. 1995; Najmabadi et al. 1996; Vogt et al. 1996; Pryor et al. 1997). Nevertheless, due to the repetitive nature of these genes, the establishment of genotype-phenotype relationships is not without difficulty, as illustrated by the hunt for *AZF* candidate genes (Cooke and Elliott 1997). Moreover, the homologies with candidate genes on the Y chromosome of the mouse are very limited (Elliott and Cooke 1997).

3.2 Why Does Such a Large Fraction of All Genes Participate in Spermatogenesis?

The function of many genes that can be mutated to male sterility in *Drosophila* is not restricted to spermatogenesis, and this may be true in humans as well. For about half of the entries from the OMIM database of inherited disorders in man that are associated with male subfertility, the effect on male reproduction is but one aspect of a more complex, pleiotropic phenotype. It remains unclear why so many genes are expressed in the testis and why so many mutants – the majority is likely to represent hypomorphic alleles – cause dysfunction of the male germ cells without causing lethality in most of the other cell types.

A speculative explanation is based on links between transcription and DNA repair. There is high premeiotic transcriptional activity in the testis as illustrated by the accumulation of high levels of RNA polymerase II and TATA-binding protein in male germ cells of rodents (Schmidt and Schibler 1995). The RNA pol II complex seems to scan transcribed regions for DNA damage

(Buratowski 1993). Strong support for a link between transcription and DNA repair comes from the finding that the 89-kilodalton subunit of the human transcription initiation factor TFII H is a DNA helicase encoded by the *ERCC3* gene (Schaeffer et al. 1993). The *ERCC3* gene is mutated in patients suffering from xeroderma pigmentosum (group B), a DNA excision repair disorder. Remarkably, *ERCC3* is the human homologue of the *haywire* gene of *Drosophila*, sharing 66% amino acid identity (Mounkes et al. 1992). Male sterile alleles of *haywire* affect meiosis and other microtubule-based events during spermatogenesis (Regan and Fuller 1988, 1990), possibly by interfering with the transcription of genes that are highly expressed during spermatogenesis, such as the β2-tubulin gene. The transcription of a large fraction of the genome during male germ cell development would allow an improvement of the DNA repair efficiency throughout the genome – if transcription is not part of a repair mechanism itself. Both aspects may require RNA binding proteins such as the products of the *RB97D* gene of *D. melanogaster* (Karsch-Mizrachi and Haynes 1993; Heatwole and Haynes 1996) and the *RBM* and *DAZ* genes on the human Y chromosome (Cooke and Elliott 1997). These genes are only expressed in male germ cells. Mutations in the *RB97D* gene cause sterility in male flies, and *DAZ* or *RBM* are deleted in about 13–16% of azoospermic males (Reijo et al. 1995; Najmabadi et al. 1996; Pryor et al. 1997).

In addition, the spontaneous mutation rate might be intrinsically higher in the male germ line than in the female germ line. This might be caused by the higher number of cell divisions that are required for male germ cells to become functional spermatozoa (Vogel and Motulsky 1997). Evidence for a mutational bias in the male germ line has come from a study of the chromo-helicase-DNA-binding protein gene located on homologous regions of the Z and W chromosomes of birds. In birds, females are the heterogametic sex, with W being the female-specific chromosome. Ellegren and Fridolfsson (1997) determined the nucleotide substitutions in 615 bp of coding sequence and 230–275 bp of intronic sequence of this gene by comparing the homologous copies on Z and W chromosomes in three species of birds. This comparison suggests that the Z chromosome, which is transmitted twothirds of the time in the male germ line and onethird of the time in the female germ line, has a fourfold higher mutation rate for synonymous substitutions in coding DNA sequences, and 6.5-fold higher in intronic sequences compared with the W chromosome, which is transmitted only in the egg. Therefore, a more efficient DNA repair mechanism would be required in the male germ line in order to avoid a mutational bias, not only in birds, but also in *Drosophila* and mammals (cf. Crow 1997).

Finally, studies concerning the basis of reproductive isolation and Haldane's rule in the *Drosophila simulans* clade revealed that hybrid male sterility plays a much more eminent role compared with hybrid female sterility or hybrid viability of both sexes. Haldane's rule states that "When in the F1 offspring of two different animal races one sex is absent, rare or sterile, that sex is the heterozygous sex" (reviewed by Wu et al. 1996). A recent explanation of Haldane's rule is based on the introduction of marked autosomal segments from *D. mauritiana* into the sibling species *D. simulans* (Hollocher and Wu 1996; True et al. 1996). From such introgressions, it appeared that hybrid male sterility evolves ten times faster than either hybrid female sterility or hybrid viability. The number of genes causing hybrid male sterility between *D. simulans* and *D. mauritiana* is estimated at 120 (Wu et al. 1996). One has to conclude that more genes are needed for spermatogenesis than oogenesis, and, in addition, that genes that function in male reproduction mutate more often than genes involved in female reproduction.

It is remarkable that studies of mRNA complexity in plants have shown that about twothirds of the estimated 30000 different mRNAs expressed in vegetative tissues are also expressed in pollen, as shown for *Tradescantia paludosa* (Willing and Mascarenhas 1984) and *Zea mays* (Willing et al. 1988). Thus, in both animals and plants, a large fraction of the genes are active in the male reproductive cell, and the expression of most of these genes is not restricted to the male germ cell. Whatever the explanation may be for the high number of genes that are expressed in the male germ line and that can eventually mutate to male sterile alleles, this seems to represent an ancient and widespread phenomenon among sexually reproducing eukaryotes.

4 Population Studies and Mutations Affecting Male Fertility in Flies and Man

Genetic and molecular genetic analysis of spermatogenesis in *Drosophila* has revealed that hundreds, if not thousands, of genes are needed to make functional sperm, and that pleiotropy is a hallmark of most of the male sterile mutations. All available data suggest that spermatogenesis in humans is likely to require the activity of several hundreds to thousands of genes as well. Is there any evidence from systematic population studies in favour of such an enormous number of male sterile mutations in flies and humans?

Only a few data are available about the number of male sterile mutations in natural populations of *Drosophila*. Individual flies from natural populations can be outcrossed in the laboratory to flies with balancer chromosomes. In the subsequent generations the "wild-type" chromosomes can be made homozygous and tested for effects on male fertility (for example, see Loukas et al. 1980). However, estimates for the genetic load by male sterile mutations are greatly influenced by founder effects, genetic drift and the effective population size.

In human populations such an approach is impossible. Because of founder effects and random genetic drift, a highly divergent genetic load and different kinds of male sterile mutations can be expected. The data available to test this expectation are rather limited, but there are some studies that reveal different spectra of clinical manifestations of male subfertility in the various populations (Thomas and Jamal 1995). Studies on the effects of consanguinity can provide some clues about the genetic load by mutations that cause male subfertility, but unfortunately, most such studies have payed little attention to the fertility of the offspring of consanguineous unions. In a few populations with a high proportion of consanguineous marriages it could be shown that "infertility of the couples" is significantly different from control marriages (Vogel and Motulsky 1997). However, this parameter is not only determined by male subfertility, but also by female subfertility and foetal waste. Recently, Zlotogora (1997) observed an increased attendance of infertility clinics by the male offspring from consanguineous unions in Palestinian Arabs, possibly indicating an increased incidence of male subfertility. It may not be a coincidence that consanguinty of the parents is mentioned in many of the described cases of male subfertility (see, for example, Rott 1979; Chaganti et al. 1980).

5 Concluding Remarks

Genetic and molecular genetic investigations of spermatogenesis in the fruit fly *Drosophila* strongly suggest that several thousands of genes in the human genome can mutate to cause male subfertility. Almost all of these genes await identification. Deletions of the *AZF* gene on the human Y chromosome originate de novo at frequencies of about 10^{-4}, probably by unequal sister chromatid exchange between repetitive DNA sequences (Reijo et al. 1995; Vogt et al. 1996). *AZF* deletions may explain the subfertility in up to 16% of azoospermic patients, but they are not transmitted to the next generation (except by the use of intracytoplasmic sperm injection, ICSI, technology). In contrast, the genetic load imposed by the thousands of X-linked and autosomal recessive mutations affecting male fertility has a much greater impact: such mutant alleles remain in the population because they are transmitted by heterozygous carriers.

Even intensive studies of human populations for genetically controlled idiopathic male subfertility cannot provide information about the plethora of loci involved in male germ cell development. The unprecedented genetic heterogeneity of male subfertility and the pleiotropic nature of most mutations, in combination with small family sizes, will seriously hamper the application of conventional strategies for the mapping and cloning of disease genes. The pleiotropy of mutations affecting male fertility makes it impossible to predict a priori functions in spermatogenesis from DNA sequence homologies (e.g. Mounkes et al. 1992). Only the combination of systematic genetic and molecular genetic studies in model animals with flagellated sperm, such as *Drosophila* and mouse, will allow the identification of candiate genes. In a next step, by DNA sequence comparisons, human homologs can be identified (cf. Banfi et al. 1996), and their potential impact on fertility in the affected offspring of consanguineous unions can be studied. For example, germ cell-derived testicular tumors in man may have a hereditary basis (Leahy et al. 1995). Male sterile mutations that cause similar phenotypes are known in *Drosophila* (Gateff 1982; McKearin and Spradling 1990). Genetic defects leading to spermatogenic arrest phenotypes in man, which are observed in more than 10% of azoospermic male patients, may be similar to mutations identified in *Drosophila* (Endo et al. 1996; Lin et al. 1996). Human homologs of the transmembrane GTPase encoded by the *fuzzy onions* gene, that is required for mitochondrial fusion in *Drosophila* sperm (Hales and Fuller 1997), may also be required for facilitating contacts between mitochondria in human sperm. The unravelling of the genetics of human male subfertility is a great challenge that will require coordinated multidisciplinary efforts of scientists studying spermatogenesis in the most divergent model organisms.

Acknowledgements. We are greatly indebted to Hans Bünemann for permission to cite unpublished results from his laboratory. We thank Mary Ann Handel, Elisabeth Hauschteck-Jungen, Steven L'Hernault, Christiane Kirchhoff, Gerard te Meerman, Peter Pearson, Pino Poddighe, Jacques Scheres, Richard Sinke and Helmut Zacharias for helpful comments and discussions.

References

Affara N, Bishop C, Brown W, Cooke H, Davey P, Ellis N, Graves JM, Jones M, Mitchell M, Rappold G, Tyler-Smith C, Yen P, Lau Y-FC (1996) Report of the second international workshop on Y chromosome mapping 1995. Cytogenet Cell Genet 73:33–76

Afzelius BA, Eliasson R (1979) Flagellar mutants in man: on the heterogeneity of the immotile-cilia syndrome. J Ultrastruct Res 69:43–52

Afzelius BA, Eliasson R, Johnson Ø, Lindholmer C (1975) Lack of dynein arms in immotile human spermatozoa. J Cell Biol 66:225–232

Aitken RJ, Baker HWG, Irvine DS (1995) On the nature of semen quality and infertility. Hum Reprod 10:248–249

Akhmanova A, Bindels PCT, Xu J, Miedema K, Kremer H, Hennig W (1995) Structure and expression of histone H3.3 genes in *Drosophila melanogaster* and *Drosophila hydei*. Genome 38:586–600

Akhmanova A, Miedema K, Hennig W (1996) Identification and characterization of the *Drosophila* histone H4 replacement gene. FEBS Lett 388:219–222

Akhmanova A, Miedema K, Wang Y, van Bruggen M, Berden JH, Moudrianakis EN, Hennig W (1997) The localization of histone H3.3 in germ line chromatin of *Drosophila* males as established with a histone H3.3-specific antiserum. Chromosoma 106:335–347

Alphey L, Jimenez J, White-Cooper H, Dawson I, Nurse P, Glover DM (1992) *twine*, a cdc25 homolog that functions in the male and female germline of *Drosophila*. Cell 69:977–988

Ariel M, Cedar H, McCarrey J (1994) Developmental changes in methylation of spermatogenesis-specific genes include reprogramming in the epididymis. Nat Genet 7:59–63

Ashburner M (1989) *Drosophila*. A laboratory handbook. Cold Spring Harbor Lab Press, Cold Spring Harbor, New York

Baccetti B (ed) (1991) Comparative spermatology 20 years after. Raven Press, New York
Baccetti B, Afzelius BA (1976) The biology of the sperm cell. S Karger, Basel
Baccetti B, Burrini AG, Pallini V, Renieri T (1981) Human dynein and sperm pathology. J Cell Biol 88:102–107
Baltz JM, Williams PO, Cone RA (1990) Dense fibers protect mammalian sperm against damage. Biol Reprod 43:485–491
Banfi S, Borsani G, Rossi E, Bernard L, Guffanti A, Rubboli F, Marchitiello A, Giglio S, Coluccia E, Zollo M, Zuffardi O, Ballabio A (1996) Identification and mapping of human cDNAs homologous to *Drosophila* mutant genes through EST database searching. Nat Genet 13: 167–174
Bhasin S, de Kretser DM, Baker HWG (1994) Pathophysiology and natural history of male infertility. J Clin Endocrinol Metab 79:1525–1529
Braun RE, Behringer RR, Peschon JJ, Brinster RL, Palmiter RD (1989) Genetically haploid spermatids are phenotypically diploid. Nature 337:373–376
Buratowski S (1993) DNA repair and transcription: the helicase connection. Science 260:37–38
Burfeind P, Hoyer-Fender S (1991) Sequence and developmental expression of a mRNA encoding a putative protein of rat sperm outer dense fibers. Dev Biol 148:195–204
Caldwell KA, Handel MA (1991) Protamine transcript sharing among postmeiotic spermatids. Proc Natl Acad Sci USA 88:2407–2411
Carroll SB (1995) Homeotic genes and the evolution of arthropods and chordates. Nature 376: 479–485
Castrillon DH, Gönczy P, Alexander S, Rawson R, Eberhart CG, Viswanathan S, DiNardo S, Wasserman SA (1993) Toward a molecular genetic analysis of spermatogenesis in *Drosophila melanogaster*: characterization of male-sterile mutants generated by single *P* element mutagenesis. Genetics 135:489–505
Cenci G, Bonaccorsi S, Pisano C, Verni F, Gatti M (1994) Chromatin and microtubule organization during premeiotic, meiotic and early postmeiotic stages of *Drosophila melanogaster* spermatogenesis. J Cell Biol 107:3521–3534
Chaganti RSK, Jhanwar SC, Ehrenbard LT, Kourides IA, Williams JJ (1980) Genetically determined asynapsis, spermatogenic degeneration, and infertility in men. Am J Hum Genet 32: 833–848
Chaillet JR, Vogt TF, Beier DR, Leder P (1991) Parental-specific methylation of an imprinted transgene is established during gametogenesis and progressively changes during embryogenesis. Cell 66:77–83
Chubb C (1989) Genetically defined mouse models of male infertility. J Androl 10:77–88
Clayton RA, White O, Ketchum KA, Venter JC (1997) The first genome from the third domain of life. Nature 387:459–462
Colgan TJ, Bedard YC, Strawbridge HTG, Buckspan MB, Klotz PG (1980) Reappraisal of the value of testicular biopsy in the investigation of infertility. Fertil Steril 33:56–60
Cooke HJ, Elliott DJ (1997) RNA-binding proteins and human male infertility. Trends Genet 13:87–89
Cooke J, Nowak MA, Boerlijst M, Maynard-Smith J (1997) Evolutionary origins and maintenance of redundant gene expression during metazoan development. Trends Genet 13:360–364
Cooley L, Berg C, Spradling A (1988) Controlling *P* element insertional mutagenesis. Trends Genet 4:254–258
Courtot C, Frankhauser C, Simanis V, Lehner CF (1992) The *Drosophila* cdc25 homologue *twine* is required for meiosis. Development 116:405–416
Crow JW (1997) Molecular evolution – who is in the driver's seat? Nat Genet 17:129–130
Czyglik F, Mayaux M-J, Guihard-Moscato M-L, David G, Schwartz D (1986) Lower sperm characteristics in 36 brothers of infertile men, compared with 545 controls. Fertil Steril 45: 255–258
De Braekeleer M, Dao T-N (1991) Cytogenetic studies in male infertility: a review. Hum Reprod 6:245–250

de Kretser DM (1997) Male infertility. Lancet 349:787–790
Denis H, Lacroix J-C (1993) The dichotomy between germ line and somatic line, and the origin of cell mortality. Trends Genet 9:7–11
Dutcher SK (1995) Flagellar assembly in two hundred and fifty easy-to-follow steps. Trends Genet 11:398–404
Eberhart CG, Wasserman SA (1995) The *pelota* locus encodes a protein required for meiotic cell division: an analysis of G2/M arrest in *Drosophila* spermatogenesis. Development 121: 3477–3486
Eberhart CG, Maines JZ, Wasserman SA (1996) Meiotic cell cycle requirement for a fly homologue of human *Deleted in Azoospermia*. Nature 381:783–785
Ellegren H, Fridolfsson A-K (1997) Male-driven evolution of DNA sequences in birds. Nat Genet 17:182–184
Elliott DJ, Cooke HJ (1997) The molecular genetics of male infertility. BioEssays 19:801–809
Elliott DJ, Millar MR, Oghene K, Ross A, Kiesewetter F, Pryor J, McIntyre M, Hargreave TB, Saunders PTK, Vogt PH, Chandley AC, Cooke H (1997) Expression of *RBM* in the nuclei of human germ cells is dependent on a critical region of the Y chromosome long arm. Proc Natl Acad Sci USA 94:3848–3853
Endo K, Akiyama T, Kobayashi S, Okada M (1996) *Degenerative spermatocyte*, a novel gene encoding a transmembrane protein required for the initiation of meiosis in *Drosophila* spermatogenesis. Mol Gen Genet 253:157–165
Erickson RP (1990) Post-meiotic gene expression. Trends Genet 6:264–269
Escalier D, David G (1984) Pathology of the cytoskeleton of the human sperm flagellum: axonemal and peri-axonemal anomalies. Biol Cell 50:37–52
Fawcett DW (1975) The mammalian spermatozoon. Dev Biol 44:394–436
Felsenfeld AL (1996) Defining the boundaries of zebrafish developmental genetics. Nat Genet 14:258–263
Flörke-Gerloff S, Töpfer-Petersen E, Müller-Esterl W, Mansouri A, Schatz R, Schirren C, Schill W, Engel W (1984) Biochemical and genetic investigation of round-headed spermatozoa in infertile men including two brothers and their father. Andrologia 16:187–202
FlyBase (1994) FlyBase – the *Drosophila* database. Available from the ftp.bio.indiana.edu network server and Gopher site. Nucleic Acids Res 22:3456–3458
Foe VE, Odell GM, Edgar BA (1993) Mitosis and morphogenesis in the *Drosophila* embryo: point and counterpoint. In: Martinez-Arias A, Bate M (eds) The development of *Drosophila melanogaster*. Cold Spring Harbor Lab Press, Cold Spring Harbor, New York, pp 149–300
Fuller MT (1993) Spermatogenesis. In: Martinez-Arias A, Bate M (eds) The development of *Drosophila melanogaster*. Cold Spring Harbor Lab Press, Cold Spring Harbor, New York, pp 71–147
Fuller MT, Caulton JH, Hutchens JA, Kaufman TC, Raff EC (1988) Mutations that encode partially functional β2 tubulin subunits have different effects on structurally different microtubule arrays. J Cell Biol 107:141–152
Gateff E (1982) Gonial cell neoplasm of genetic origin affecting both sexes of *Drosophila melanogaster*. In: Burger MM, Weber R (eds) Progress in clinical and biological research, vol 85. Embryonic development, part B: Cellular aspects. Alan R Liss, New York, pp 621–632
Gepner J, Hays T (1993) A fertility region on the Y chromosome of *Drosophila melanogaster* encodes a dynein microtubule motor. Proc Natl Acad Sci USA 90:11 132–11 136
Goldstein LSB, Hardy RW, Lindsley DL (1982) Structural genes on the Y chromosome of *Drosophila melanogaster*. Proc Natl Acad Sci USA 79:7405–7409
Gönczy P, DiNardo S (1996) The germ line regulates somatic cyst cell proliferation and fate during *Drosophila* spermatogenesis. Development 122:2437–2447
Gönczy P, Viswanathan S, DiNardo S (1992) Probing spermatogenesis in *Drosophila* with *P*-element enhancer detectors. Development 114:89–98
Griswold MD (1995) Interactions between germ cells and Sertoli cells in the testis. Biol Reprod 52:211–216

Guraya SS (1987) Biology of spermatogenesis and spermatozoa in mammals. Springer, Berlin Heidelberg New York

Hackstein JHP (1987) Spermatogenesis in *Drosophila*. In: Hennig W (ed) Spermatogenesis. Genetic aspects. Results and problems in cell differentiation, vol 15. Springer, Berlin Heidelberg New York, pp 63–116

Hackstein JHP (1991) Spermatogenesis in *Drosophila*. A genetic approach to cellular and subcellular differentiation. Eur J Cell Biol 56:151–169

Hackstein JHP, Hochstenbach R (1995) The elusive fertility genes of *Drosophila*: the ultimate haven for selfish genetic elements. Trends Genet 11:195–200

Hackstein JHP, Beck H, Hochstenbach R, Kremer H, Zacharias H (1990) Spermatogenesis in *Drosophila hydei*: a genetic survey. Roux's Arch Dev Biol 199:251–280

Hackstein JHP, Hochstenbach R, Hauschteck-Jungen E, Beukeboom LW (1996) Is the Y chromosome of *Drosophila* an evolved supernumerary chromosome? BioEssays 18:317–323

Hales KG, Fuller MT (1997) Developmentally regulated mitochondrial fusion mediated by a conserved, novel predicted GTPase. Cell 90:121–129

Handel MA (1987) Genetic control of spermatogenesis in mice. In: Hennig W (ed) Spermatogenesis. Genetic aspects. Results and problems in cell differentiation, vol 15. Springer, Berlin Heidelberg New York, pp 1–62

Hauschteck-Jungen E, Hartl DL (1982) Defective histone transition during spermiogenesis in heterozygous segregation distorter males of *Drosophila melanogaster*. Genetics 101:57–69

Heatwole VM, Haynes SR (1996) Association of RB97D, an RRM protein required for male fertility, with a Y chromosome lampbrush loop in *Drosophila melanogaster*. Chromosoma 105: 285–292

Hecht NB (1993) Gene expression during male germ cell development. In: Desjardins C, Ewing LL (eds) Cell and molecular biology of the testis. Oxford Univ Press, New York , pp 400–432

Hennig W (1985) Y chromosome function and spermatogenesis in *Drosophila hydei*. Adv Genet 23:179–234

Hess O, Meyer GF (1968) Genetic activities of the Y chromosome in *Drosophila* during spermatogenesis. Adv Genet 14:171–223

Heyting C (1996) Synaptonemal complexes: structure and function. Curr Opin Cell Biol 8: 389–396

Hollocher H, Wu C-I (1996) The genetics of reproductive isolation in the *Drosophila simulans* clade: X vs. autosomal effects and male vs. female effects. Genetics 143:1243–1255

Hoyle HD, Raff EC (1990) Two *Drosophila* beta tubulin isoforms are not functionally equivalent. J Cell Biol 111:1009–1026

Jacobson D, Anagnostopoulos A (1996) Internet resources for transgenic or targeted mutation research. Trends Genet 12:117–118 (http://www.gdb.org/Dan/tbase.html)

Karsch-Mizrachi I, Haynes SR (1993) The *Rb97D* gene encodes a potential RNA-binding protein required for spermatogenesis in *Drosophila*. Nucleic Acids Res 21:2229–2235

Kemphues KJ, Kaufman TC, Raff, RA, Raff EC (1982) The testis-specific β-tubulin subunit in *Drosophila melanogaster* has multiple functions in spermatogenesis. Cell 31:655–670

Kiefer BI (1973) Genetics of sperm development in *Drosophila*. In: Ruddle FH (ed) Genetic mechanisms of development. Academic Press, New York, pp 47–102

Kierszenbaum AL (1994) Mammalian spermatogenesis in vivo and in vitro: a partnership of spermatogenic and somatic cell lineages. Endocr Rev 15:116–134

Koulischer L, Schoysman R, Gillerot Y (1982) Chromosomes méiotiques et infertilité masculine: évaluation des resultats. J Génét Hum 30:81–99

Kremer H, Hennig W, Dijkhof R (1986) Chromatin organization in the male germ line of *Drosophila hydei*. Chromosoma 94:147–161

Kurek R, Reugels AM, Glatzer KH, Bunemann H (1998) The Y chromosomal fertility factor *Threads* in *Drosophila hydei* harbors a functional gene encoding an axonemal dynein β heavy chain protein. Genetics 149:1363–1376

Lahn BT, Page DC (1997) Functional coherence of the human Y chromosome. Science 278: 675–680

Lange R, Krause W, Engel W (1997) Analyses of meiotic chromosomes in testicular biopsies of infertile patients. Hum Reprod 12:2154–2158

Leahy MG, Tonks S, Moses JH, Brett AR, Huddart R, Forman D, Oliver RTD, Bishop DT, Bodmer JG (1995) Candidate regions for a testicular cancer susceptibility gene. Hum Mol Genet 4: 1551–1555

L'Hernault SW (1997) Spermatogenesis. In: Riddle DL, Blumenthal T, Meyer BJ, Priess JR (eds) *C. elegans* II, Cold Spring Harbor Lab Press, Cold Spring Harbor, New York, pp 271–294

Lifschytz E (1987) The developmental program of spermiogenesis in *Drosophila*: a genetic analysis. Int Rev Cytol 109:211–258

Lifschytz E, Yakobovitz N (1978) The role of X-linked lethal and viable male-sterile mutations in male gametogenesis of *Drosophila melanogaster*: genetic analysis. Mol Gen Genet 161: 275–284

Lilford R, Jones AM, Bishop TD, Thornton J, Mueller R (1994) Case-control study of whether subfertility in men is familial. Br Med J 309:570–573

Lin T-Y, Viswanathan S, Wood C, Wilson PG, Wolf N, Fuller M (1996) Coordinate developmental control of the meiotic cell cycle and spermatid differentiation in *Drosophila* males. Development 122:1331–1341

Lindsley DL (1982) The genetics of male fertility. Drosophila Inf Serv 58:2

Lindsley DL, Grell EH (1969) Spermiogenesis without chromosomes in *Drosophila melanogaster*. Genetics 61 Suppl (1):69–78

Lindsley DL, Lifschytz E (1972) The genetic control of spermatogenesis in *Drosophila*. In: Beatty RA, Gluecksohn-Waelsch S (eds) The genetics of the spermatozoon, Proc Int Symp Edinburgh. Bogtrykkeriet Forum, Copenhagen, pp 203–222

Lindsley DL, Tokuyasu KT (1980) Spermatogenesis. In: Ashburner M, Wright TRF (eds) The genetics and biology of *Drosophila*, vol 2b. Academic Press, London, pp 225–294

Lindsley DL, Zimm GG (1992) The genome of *Drosophila melanogaster*. Academic Press, New York

Loukas M, Krimbas CG, Sourdis J (1980) The genetics of *Drosophila subobscura* populations XIII. A study of lethal allelism. Genetica 54:197–207

Luck DJL (1984) Genetic and biochemical dissection of the eukaryotic flagellum. J Cell Biol 98: 789–794

McCarrey JR (1993) Development of the germ cell. In: Desjardins C, Ewing LL (eds) Cell and molecular biology of the testis. Oxford Univ Press, New York, pp 58–89

McKearin D, Spradling AC (1990) *Bag-of-marbles*: a *Drosophila* gene required to initiate both male and female gametogenesis. Genes Dev 4:2242–2251

McKee BD (1996) The license to pair: identification of meiotic pairing sites in *Drosophila*. Chromosoma 105:135–141

McKee BD, Handel MA (1993) Sex chromatin, recombination and chromatin conformation. Chromosoma 102:71–80

Meistrich ML, Bucci LR, Trostle-Weige PK, Brock WA (1985) Histone variants in rat spermatogonia and primary spermatocytes. Dev Biol 112:230–240

Mewes HW, Albermann K, Bähr M, Frishman D, Gleissner A, Hani J, Heumann K, Kleine K, Maierl A, Oliver SG, Pfeiffer F, Zollner A (1997) Overview of the yeast genome. Nature 387 Suppl: 7–65

Michiels F, Gasch A, Kaltschmidt B, Renkawitz-Pohl R (1989) A 14 bp promoter element directs the testis-specificity of the *Drosophila* β2 tutulin gene. EMBO J 8:1559–1565

Mounkes LC, Jones RS, Liang B-C, Gelbart W, Fuller MT (1992) A *Drosophila* model for xeroderma pigmentosum and Cockayne's syndrome: *haywire* encodes the fly homolog of *ERCC3*, a human excision repair gene. Cell 71:925–937

Mouse genome database (MGD), mouse genome informatics project (1996) (Bar Harbor, Me.: The Jackson Laboratory). World Wide Web URL: http://www.informatics.jax.org/

Najmabadi H, Huang V, Yen P, Subbarao MN, Bhasin D, Banaag L, Naseeruddin S, de Kretser DM, Baker HWG, McLachlan RI, Loveland KA, Bhasin S (1996) Substantial prevalence of microdeletions of the Y chromosome in infertile men with idiopathic azoospermia and oligospermia detected using a sequence-tagged site-based mapping strategy. J Clin Endocrinol Metab 81:1347–1352

Ohno S (1970) Evolution by gene duplication. Springer, Berlin Heidelberg New York

Online mendelian inheritance in man (OMIM), Center for Medical Genetics, Johns Hopkins University (Baltimore, MD) and Natural Center for Biotechnology Information, National Library of Medicine (Bethesda, MD) (1997). World Wide Web URL: http://www.ncbi.nlm.nih.gov/omim/

Page AW, Orr-Weaver TL (1997) Stopping and starting the meiotic cell cycle. Curr Opin Genet Dev 7:23–31

Pauli D, Mahowald AP (1990) Germ-line sex determination in *Drosophila melanogaster*. Trends Genet 6:259–264

Piperno G, Huang B, Ramanis Z, Luck DJL (1981) Radial spokes of *Chlamydomonas* flagella: polypeptide composition and phosphorylation of stalk components. J Cell Biol 88:73–79

Plug AW, Peters AHFM, Keegan KS, Hoekstra MF, de Boer P, Ashley T (1998) Changes in protein composition of meiotic nodules during mammalian meiosis. J Cell Sci 111:413–423

Poccia D (1986) Remodeling of nucleoproteins during gametogenesis, fertilization, and early development. Int Rev Cytol 105:1–65

Pryor J, Kent-First M, Muallem A, Van Bergen AH, Nolten WE, Meisner L, Roberts KP (1997) Microdeletions in the Y chromosome of infertile men. N Engl J Med 336:534–539

Regan CL, Fuller MT (1988) Interacting genes that affect microtubule function: the *nc2* allele of the *haywire* locus fails to complement mutations in the testis-specific β-tubulin gene of *Drosophila*. Genes Dev 2:82–92

Regan CL, Fuller MT (1990) Interacting genes that affect microtubule function in *Drosophila melanogaster*: two classes of mutation revert the failure to complement between hay^{nc2} and mutations in tubulin genes. Genetics 125:77–90

Reijo R, Lee T-Y, Salo P, Alagappan R, Brown LG, Rosenberg M, Rozen S, Jaffe T, Straus D, Hovatta O, de la Chapelle A, Silber S, Page DC (1995) Diverse spermatogenic defects in humans caused by Y chromosome deletions encompassing a novel RNA-binding protein gene. Nat Genet 10:383–393

Roeder GS (1995) Sex and the single cell: meiosis in yeast. Proc Natl Acad Sci USA 92:10450–10456

Rott H-D (1979) Kartagener's syndrome and the syndrome of immotile cilia. Hum Genet 46:249–261

Sassone-Corsi P (1997) Transcriptional checkpoints determining the fate of male germ cells. Cell 88:163–166

Schaeffer L, Roy R, Humbert S, Moncollin V, Vermeulen W, Hoeijmakers JHJ, Chambon P, Egly J-M (1993) DNA repair helicase: a component of BTF2 (TFIIH) basic transcription factor. Science 260:58–63

Schäfer M, Börsch D, Hülster A, Schäfer U (1993) Expression of a gene duplication encoding conserved sperm tail proteins is translationally regulated in *Drosophila melanogaster*. Mol Cell Biol 13:1708–1718

Schäfer M, Nayernia K, Engel W, Schäfer U (1995) Translational control in spermatogenesis. Dev Biol 172:344–352

Schmidt EE, Schibler U (1995) High accumulation of components of the RNA polymerase II transcription machinery in rodent spermatids. Development 121:2373–2383

Shellenbarger DL, Cross DP (1977) A new class of male-sterile mutations with combined temperature-sensitive lethal effects in *Drosophila melanogaster*. Genetics 86:s358

Sigrist S, Ried G, Lehner CF (1995) Dmcdc2 kinase is required for both meiotic divisions during *Drosophila* spermatogenesis and is activated by the *Twine*/cdc25 phosphatase. Mech Dev 53:247–260

Simerly C, Wu G-J, Zoran S, Ord T, Rawlins R, Jones J, Navara C, Gerrity M, Rinehart J, Binor Z, Asch R, Schatten G (1995) The paternal inheritance of the centrosome, the cell's microtubule-organizing center, in humans, and the implications for fertility. Nat Med 1:47–52

Skakkebaek NE, Giwercman A, de Kretser D (1994) Pathogenesis and management of male infertility. Lancet 343:1473–1479

Solnica-Krezel L, Schier AF, Driever W (1994) Efficient recovery of ENU-induced mutations form the zebrafish germline. Genetics 136:1401–1420

Theriot JA (1996) Worm sperm and advances in cell locomotion. Cell 84:1–4

Thomas JO, Jamal A (1995) Primary testicular causes of infertility. Do environmental and sociocultural factors have a role? Trop Geogr Med 47:203–205

True JR, Weir BS, Laurie CC (1996) A genome-wide survey of hybrid incompatibility factors by the introgression of marked segments of *Drosophila mauritiana* chromosomes into *Drosophila simulans*. Genetics 142:819–837

Urieli-Shoval S, Gruenbaum Y, Sedat J, Razin A (1982) The absence of detectable methylated bases in *Drosophila melanogaster* DNA. FEBS Lett 146:148–152

Van Assche E, Bonduelle M, Tournaye H, Joris H, Verheyen G, Devroey P, Van Steirteghem A, Liebaers I (1996) Cytogenetics of infertile men. Hum Reprod 11 Suppl(4):1–26

Vogel F, Motulsky AG (1997) Human genetics: problems and approaches. Springer, Berlin Heidelberg New York

Vogt PH, Edelmann E, Kirsch S, Henegariu O, Hirschmann P, Kiesewetter F, Köhn FM, Schill WB, Farah S, Ramos C, Hartmann M, Hartschuh W, Meschede D, Behre HM, Castel A, Nieschlag E, Weidner W, Gröne H-J, Jung A, Engel W, Haidl G (1996) Human Y chromosome azoospermia factors (*AZF*) mapped to different subregions in Yq11. Hum Mol Genet 5: 933–943

Watanabe TK, Lee WH (1977) Sterile mutation in *Drosophila melanogaster*. Genet Res 30: 107–113

White-Cooper H, Alphey L, Glover DM (1993) The *cdc25* homologue *twine* is required for only some aspects of the entry into meiosis in *Drosophila*. J Cell Sci 106:1035–1044

Willing RP, Mascarenhas JP (1984) Analysis of the complexity and diversity of mRNAs from pollen and shoots of *Tradescantia*. Plant Physiol 75:865–868

Willing RP, Basche D, Mascarenhas JP (1988) An analysis of the quantity and diversity of messenger RNAs from pollen and shoots of *Zea mays*. Theor Appl Genet 75:751–753

Wong T-W, Straus FH, Warner NE (1973) Testicular biopsy in the study of male infertility. Arch Pathol 95:151–159

Wu C-I, Johnson NA, Palopoli MF (1996) Haldane's rule and its legacy: why are there so many sterile males? Trends Ecol Evol 11:281–284

Yasuda GK, Schubiger G, Wakimoto BT (1995) Genetic characterization of *ms(3)K81*, a paternal effect gene of *Drosophila melanogaster*. Genetics 140:219–229

Zamboni L (1987) The ultrastructural pathology of the spermatozoon as a cause of infertility: the role of electron microscopy in the evaluation of semen quality. Fertil Steril 48:711–734

Zlotogora J (1997) Genetic disorders among Palestinian Arabs: 1. Effects of consanguinity. Am J Med Genet 68:472–475

Subject Index

Results and Problems in Cell Differentiation, Vol. 28
McElreavey (Ed.): The Genetic Basis of Male Infertility

Printing (Computer to Film): Saladruck, Berlin
Binding: Lüderitz & Bauer, Berlin